INTRODUCTION TO LINEAR PROGRAMMING

MONOGRAPHS AND TEXTBOOKS IN PURE AND APPLIED MATHEMATICS

Other Volumes in Preparation

INTRODUCTION TO LINEAR PROGRAMMING

Applications and Extensions

RICHARD B. DARST

Colorado State University
Fort Collins, Colorado

CRC Press
Taylor & Francis Group
Boca Raton London New York

CRC Press is an imprint of the
Taylor & Francis Group, an informa business

Library of Congress Cataloging-in-Publication Data

Darst, Richard B.
  Introduction to linear programming: applications and extensions / Richard
B. Darst.
    p.   cm. – (Monographs and textbooks in pure and applied mathematics;
141)
  Includes bibliographical references and index.
  ISBN 0-8247-8383-2
  1. Linear programming. I. Title. II. Series.
T57.74.D37 1990
519.7'2–dc20   90-14012
                                                                         CIP

MARCEL DEKKER, INC.
270 Madison Avenue, New York, New York 10016

Current printing (last digit):
10 9 8 7 6 5

To Jean
My friend and wife

Preface

Linear programming (LP) is an exceptionally useful modeling tool. The tremendous variety of situations to which LP applies is remarkable, and George Dantzig's discovery of the simplex method for solving linear programs is a beautiful and useful blend of geometry and linear algebra. Although a model for a "real world" situation is much more useful if you can solve it, most LP models are tedious or impossible to solve by hand. However, software packages based on variations of the simplex method are now readily available. Good LP software is a very powerful analytical tool, and an understanding of the simplex method can help you use the software effectively. This book can be used without LP software, but it is to your advantage to use an LP software package and actually solve the problems.

The two basic goals of this book are (1) to help you learn to use LP and (2) to give you a clear, elementary introduction to the simplex method. When you finish reading Chapters 2–5 and you have a problem to solve, you should be able to recognize whether LP applies and formulate an appropriate model if it does. A third goal is to introduce, briefly, networks, dynamic programming, quadratic programming, and quadratic functions. Some problems can be modeled several different ways and some models can be solved several different ways; the best formulation and method of solution will depend on the software available, and software availability is currently increasing dramatically. Some of the basic ideas related to these additional topics are discussed in Chapters 5, 7, 8, and 9. Much about LP and these other topics is merely mentioned or not discussed at all, but the Reading List at the end of the book provides an entrée to the vast literature.

Chapter 1 contains topics in linear algebra that are used subsequently. Skim Chapter 1 to see what it contains, then move on to Chapter 2, and go back to parts of Chapter 1 as needed.

Three examples are used in Chapter 2 to introduce several important aspects of LP. Chapters 2 and 5 contain extensive lists of exercises. These exercises illustrate the utility of LP and provide opportunities to practice formulating and solving LP models. Doing these exercises is an intrinsic part of learning from this book; they will help you learn how to model many types of situations. Chapters 2, 3, 4, and 5 conclude with a discussion of some exercises, and Appendixes 2 and 3 contain additional solutions for exercises in Chapters 2 and 5, respectively.

Chapters 3 and 4 contain an introductory theoretical discussion of LP. Elementary geometric and algebraic properties of an LP model are presented in Chapter 3. These properties underlie the simplex method. Two and three dimensional pictures can be drawn that expose the geometry of the feasible set for an LP and that can relate its geometric and algebraic properties. The algebraic properties carry over to higher dimensions, where it is difficult to visualize objects geometrically. Algebraic properties are used in Chapter 4 to develop the simplex method.

The modeling theme returns in Chapter 5, which contains further discussion of types of problems that fit into LP format, including transportation problems in Sections 5.6 and 5.7 and some network problems in Section 5.7. The issue of integer values for variables comes up frequently in the problems. In Sections 5.4 and 5.5, the software used permits some variables to be valued as zero-one integers to solve the problems. Dynamic programming is introduced in Section 5.8. Section 5.9 considers stability of the set of solutions of an LP and sensitivity of the value of the objective function of a stable LP with respect to changes in its parameters.

The discussion of duality which began in Chapter 2 continues in Chapter 6, providing a bridge to Chapter 7, where quadratic programming is discussed. In Chapter 8 three methods to minimize a quadratic function are presented; many nonlinear programming algorithms use various combinations of these methods. Chapter 8 gives you a foundation for dealing with nonlinear programming.

Algorithms for several network problems are discussed in Chapter 9.

This book evolved while I was teaching a variety of engineering mathematics courses at Colorado State University. I gladly take this opportunity to thank students in these courses for many helpful contributions and to express my appreciation to Maria Allegra and Robin J. Wishnie of Marcel Dekker, Inc., for their considerate and efficient support.

Richard B. Darst

Contents

Contents

INTRODUCTION TO LINEAR PROGRAMMING

1

Introduction to Systems of Linear Equations (Linear Systems) and Related Properties of Matrices

The topics presented in this introductory chapter are used to formulate and analyze models for a wide variety of practical problems in subsequent chapters. If you have been exposed to systems of linear equations and matrix operations, you may wish merely to skim this chapter to pick up the notation and then proceed to Chapter 2, where linear programming is introduced. In this chapter we discuss solving linear systems by row reduction, matrix operations, rank, and linear independence. Two examples are used to motivate definitions and illustrate concepts.

1.1 LINEAR SYSTEMS

The discussion of linear systems begins with Example 1.1.

Example 1.1. Solve the linear system

(1) $2x_1 + 3x_2 = 3$
(2) $-x_1 + 2x_2 = 4.$

Solve means to find all two-dimensional vectors $x = \begin{bmatrix} x_1 \\ x_2 \end{bmatrix}$ that satisfy both equations (1) and (2); here x is a pair x_1, x_2 of numbers written in column form. Except for the names x_1 and x_2 of the variables (unknowns), all the

1

information that we need to solve the system is given by the parameters of the system; the parameters of the system are the coefficients of the variables and the numbers on the right-hand side of the equations. For Example 1.1, all of this information is contained in the augmented matrix $[\begin{smallmatrix} 2 & 3 \\ -1 & 2 \end{smallmatrix} | \begin{smallmatrix} 3 \\ 4 \end{smallmatrix}]$, which is composed of the coefficient matrix $[\begin{smallmatrix} 2 & 3 \\ -1 & 2 \end{smallmatrix}]$ followed by the right-hand-side vector $[\begin{smallmatrix} 3 \\ 4 \end{smallmatrix}]$. Examples 1.1 and 1.2 below illustrate how to solve the *general m × n linear system*:

$\mathcal{L}$:

(1) $a_{11}x_1 + a_{12}x_2 + \cdots + a_{1j}x_j + \cdots + a_{1n}x_n = b_1$
(2) $a_{21}x_1 + a_{22}x_2 + \cdots + a_{2j}x_j + \cdots + a_{2n}x_n = b_2$
         $\vdots$
(i) $a_{i1}x_1 + a_{i2}x_2 + \cdots + a_{ij}x_j + \cdots + a_{in}x_n = b_i$
         $\vdots$
(m) $a_{m1}x_1 + a_{m2}x_2 + \cdots + a_{mj}x_j + \cdots + a_{mn}x_n = b_m.$

Solving this linear system means finding all n-dimensional column vectors

$$x = \begin{bmatrix} x_1 \\ \vdots \\ x_j \\ \vdots \\ x_n \end{bmatrix}$$

that satisfy the equations (1), (2), . . . , (m). The $a_{i,j}$'s and b_i's are (specific) numbers called *parameters* of the system. Again, except for the names $x_1, x_2, \ldots, x_n$ of the variables, all the relevant information about the system is contained in the *augmented matrix*

$$m \left\{ \begin{bmatrix} a_{11} & \cdots & a_{1j} & \cdots & a_{1n} & b_1 \\ \vdots & & \vdots & & \vdots & \vdots \\ a_{i1} & \cdots & a_{ij} & \cdots & a_{in} & b_i \\ \vdots & & \vdots & & \vdots & \vdots \\ a_{m1} & \cdots & a_{mj} & \cdots & a_{mn} & b_m \end{bmatrix} \right. = [A \,|\, b]$$

composed of the $m \times n$ *coefficient matrix A* followed by the m-dimensional column vector b. In Example 1.1, $m = n = 2$, $A = [\begin{smallmatrix} 2 & 3 \\ -1 & 2 \end{smallmatrix}]$, and $b = [\begin{smallmatrix} 3 \\ 4 \end{smallmatrix}]$.

Before an algorithm that solves linear systems is presented, a few pertinent observations are made, and these observations are applied to Example 1.1 to illustrate how the algorithm works. If all the coefficients of one of the variables in the general linear system were zero, that variable would have no effect on the system; it would have been irrelevant for the problem and it would not have been included in the list of variables. The labeling (naming) of the variables is for our convenience, so you can interchange columns in the coefficient matrix with impunity; however, when you interchange columns, you must keep track of the changes in location of the corresponding variables. Similarly, you can interchange rows in the augmented matrix. Also, both multiplying a row in the augmented matrix by a nonzero number and adding a row to another row in the matrix yield an augmented matrix for an equivalent system; the original system and the associated modified system have the same solutions. Example 1.1 is used below to illustrate.

The matrix modifications

$$
\begin{bmatrix} 2 & 3 & | & 3 \\ -1 & 2 & | & 4 \end{bmatrix} \xrightarrow[1/2(1)]{} \begin{bmatrix} 1 & \frac{3}{2} & | & \frac{3}{2} \\ -1 & 2 & | & 4 \end{bmatrix} \xrightarrow[+(1)\,\text{to}\,(2)]{} \begin{bmatrix} 1 & \frac{3}{2} & | & \frac{3}{2} \\ 0 & \frac{7}{2} & | & \frac{11}{2} \end{bmatrix}
$$

$$
\xrightarrow[2/7(2)]{} \begin{bmatrix} 1 & \frac{3}{2} & | & \frac{3}{2} \\ 0 & 1 & | & \frac{11}{7} \end{bmatrix} \xrightarrow[-3/2(2)\,\text{to}\,(1)]{} \begin{bmatrix} 1 & 0 & | & -\frac{6}{7} \\ 0 & 1 & | & \frac{11}{7} \end{bmatrix},
$$

correspond to the following list of operations applied to the two equations

(1) $2x_1 + 3x_2 = 3$
(2) $-x_1 + 2x_2 = 4.$

1. Multiply (1) by $\frac{1}{2}$. [Now (1) is $x_1 + \frac{3}{2}x_2 = \frac{3}{2}$.]
2. Add (1) to (2). [Now (2) is $\frac{7}{2}x_2 = \frac{11}{2}$.]
3. Multiply (2) by $\frac{2}{7}$. [Now (2) is $x_2 = \frac{11}{7}$.]
4. Add $-\frac{3}{2}(2)$ to (1). [Now (1) is $x_1 = -\frac{6}{7}$.]

The system is now in the equivalent form

$$x_1 = -\frac{6}{7}$$

$$x_2 = \frac{11}{7},$$

from which we can read the solution

$$x = \begin{bmatrix} -\frac{6}{7} \\ \frac{11}{7} \end{bmatrix}.$$

Interchanging rows, multiplying a row by a nonzero constant, and adding a multiple of one row to another row are called *elementary row operations*. They correspond to interchanging two equations, multiplying an equation by a nonzero constant, and adding a multiple of one equation to another equation. To verify that operations do not change the set of solutions to a linear system, consider the general linear system $\mathscr{L}$ that is displayed on page 2. Notice that if c is a nonzero number, the equation

$$a_{i1}x_1 + \cdots + a_{ij}x_j + \cdots + a_{in}x_n = b_i \tag{i}$$

and the equation

$$c(a_{i1}x_1 + \cdots + \alpha_{ij}x_j + \cdots + a_{in}x_n) = cb_i, \tag{c(i)}$$

or equivalently, $(ca_{i1})x_1 + \cdots + (ca_{ij})x_j + \cdots + (ca_{in})x_n = cb_i$, have the same solutions. If we add c(i) to (j) to get

$$c(a_{i1}x_1 + \cdots + a_{in}x_n) + (a_{j1}x_1 + \cdots + a_{jn}x_n) = cb_i + b_j, \tag{c(i) + (j)}$$

or equivalently, $(ca_{i1} + a_{j1})x_1 + \cdots + (ca_{in} + a_{jn})x_n = cb_i + b_j$, we see that solutions to (i) and (j) are solutions to (i) and c(i) + (j). Conversely, solutions to (i) and c(i) + (j) are solutions to (i) and (j). For instance, in Example 1.1 the linear systems

(1) $2x_1 + 3x_2 = 3$
(2) $-x_1 + 2x_2 = 4$

and the linear system

$\frac{1}{2}$(1) $x_1 + \frac{3}{2}x_2 = \frac{3}{2}$
(2) $-x_1 + 2x_2 = 4$

and the linear system

$\frac{1}{2}$(1) $x_1 + \frac{3}{2}x_2 = \frac{3}{2}$
$\frac{1}{2}$(1) + (2) $\frac{7}{2}x_2 = \frac{11}{2}$

and so on, have the same sets of solutions.

Row reduction is a procedure for solving linear systems which involves systematic application of elementary row operations. Row reduction is composed of two algorithms. We begin in the next section by listing the steps in the first algorithm.

1.2 ROW ECHELON ALGORITHM

1. Put $i = 1$, $j = 1$.
2. Find the first equation (k) with $k \geq i$ and $a_{kj} \neq 0$.
3. If $k > i$, interchange (k) and (i). (Now $a_{ij} \neq 0$.)
4. Multiply (i) by $1/a_{ij}$. (Now $a_{ij} = 1$.)
5. For $i < k \leq m$, subtract a_{kj}(i) from (k). (Now $a_{kj} = 0$, $k > i$.)
6. Put $j_i = j$.
7. If $i = m$, go to 12.
8. Put $i = i + 1$, $j = j + 1$.

9. If $j = n + 1$, go to 10; else, if $a_{kj} = 0$ for $k \geq i$, put $j = j + 1$ and to to 9; else, go to 2.
10. (The equations are dependent and we check to see if there is a solution.)
11. If $b_k = 0$ for $k \geq i$, then (there are solutions) go to 12; else, (there are no solutions) proclaim "There are no solutions." and stop.
12. Proclaim "There are solutions; system is ready for the reduce echelon algorithm."

Except for the latter proclamation, the row echelon algorithm applied to Example 1.1 corresponds to the first three items on its solution list; it puts the system in the augmented matrix form

$$\begin{bmatrix} 1 & \frac{3}{2} & \Big| & \frac{3}{2} \\ 0 & 1 & \Big| & \frac{11}{7} \end{bmatrix}.$$

Let us apply row echelon to another example.

Example 1.2. Solve the linear system

$$x_1 + x_2 + 2x_3 + x_4 - x_5 = 1$$
$$x_1 + x_2 + 2x_3 - x_4 + x_5 = 2$$
$$x_1 + x_2 + x_3 + 2x_4 + 3x_5 = 1.$$

Row echelon for Example 1.2 will be given as a list of row operations and then by a corresponding sequence of augmented matrices.

Subtract (1) from (2) and subtract (1) from (3).
Interchange (3) and (2).
Multiply (2) by -1.
Multiply (3) by $-\frac{1}{2}$.

$$\begin{bmatrix} 1 & 1 & 2 & 1 & -1 & \Big| & 1 \\ 1 & 1 & 2 & -1 & 1 & \Big| & 2 \\ 1 & 1 & 1 & 2 & 3 & \Big| & 1 \end{bmatrix} \xrightarrow[-(1)\text{ to }(3)]{-(1)\text{ to }(2)} \begin{bmatrix} 1 & 1 & 2 & 1 & -1 & \Big| & 1 \\ 0 & 0 & 0 & -2 & 2 & \Big| & 1 \\ 0 & 0 & -1 & 1 & 4 & \Big| & 0 \end{bmatrix}$$

$$\xrightarrow{(2)\leftrightarrow(3)} \begin{bmatrix} 1 & 1 & 2 & 1 & -1 & \Big| & 1 \\ 0 & 0 & -1 & 1 & 4 & \Big| & 0 \\ 0 & 0 & 0 & -2 & 2 & \Big| & 1 \end{bmatrix}$$

$$\xrightarrow{-1(2)} \begin{bmatrix} 1 & 1 & 2 & 1 & -1 & 1 \\ 0 & 0 & 1 & -1 & -4 & 0 \\ 0 & 0 & 0 & -2 & 2 & 1 \end{bmatrix}$$

$$\xrightarrow{-1/2(3)} \begin{bmatrix} 1 & 1 & 2 & 1 & -1 & 1 \\ 0 & 0 & 1 & -1 & -4 & 0 \\ 0 & 0 & 0 & 1 & -1 & -\frac{1}{2} \end{bmatrix}$$

The positions $\{(1,j_1), (2,j_2), (3,j_3)\} = \{(1,1), (2,3), (3,4)\}$ are called *corner points* in the augmented matrix. Corner-point entries are equal to 1. When (i,j_i) is a corner point, all entries in positions to its left in row i are equal to zero. Also, $j_1 < j_2 < \cdots$.

1.3 ROW REDUCTION

After applying row echelon and finding that a solution exists, a second algorithm, called the *reduce echelon algorithm*, uses elementary row operations to put zeros in the entries that are above the corner points in the augmented matrix. Row echelon followed by reduce echelon is called reduced row echelon or Gauss–Jordan reduction or any of several other names; we'll call it *row reduction* and denote it by "$\to \cdots \to$".

The following sequence of matrices displays the reduce echelon algorithm applied to Example 1.2:

$$\xrightarrow[\substack{(3) \text{ to } (2) \\ -(3) \text{ to } (1)}]{} \begin{bmatrix} 1 & 1 & 2 & 0 & 0 & \frac{3}{2} \\ 0 & 0 & 1 & 0 & -5 & -\frac{1}{2} \\ 0 & 0 & 0 & 1 & -1 & -\frac{1}{2} \end{bmatrix} \xrightarrow{-2(2) \text{ to } (1)} \begin{bmatrix} 1 & 1 & 0 & 0 & 10 & \frac{5}{2} \\ 0 & 0 & 1 & 0 & -5 & -\frac{1}{2} \\ 0 & 0 & 0 & 1 & -1 & -\frac{1}{2} \end{bmatrix}.$$

The corresponding list is

Add (3) to (2) and add $-(3)$ to 1 [to put zeros above corner point (3,4)]. Add $-2(2)$ to (1) [to put zeros above corner point (2,3)].

Variables that do not correspond to corner points [e.g., variables whose indices do not appear in the list $\{j_1,j_2,j_3\} = \{1,3,4\}$ in Example 1.2] are called *free variables*. When the free variables are given (specific) values, the other variables are (uniquely) determined, for example,

$$\begin{aligned} x_1 &= -x_2 - 10x_5 + \tfrac{5}{2} \\ x_3 &= 5x_5 - \tfrac{1}{2} \\ x_4 &= x_5 - \tfrac{1}{2}. \end{aligned}$$

Thus the general solution x to Example 1.2 can be written

$$
x = \begin{bmatrix} x_1 \\ x_2 \\ x_3 \\ x_4 \\ x_5 \end{bmatrix} = \begin{bmatrix} \frac{5}{2} & -x_2 - 10x_5 \\ & x_2 \\ -\frac{1}{2} & + 5x_5 \\ -\frac{1}{2} & + x_5 \\ & x_5 \end{bmatrix} = \begin{bmatrix} \frac{5}{2} \\ 0 \\ -\frac{1}{2} \\ -\frac{1}{2} \\ 0 \end{bmatrix} + x_2 \begin{bmatrix} -1 \\ 1 \\ 0 \\ 0 \\ 0 \end{bmatrix} + x_5 \begin{bmatrix} -10 \\ 0 \\ 5 \\ 1 \\ 1 \end{bmatrix} ;
$$

the latter form x will be explained in subsequent sections.
Applying row reduction to a matrix transforms the matrix to a form called *reduced row echelon form*.

Exercises

1. Which of the following matrices are in reduced row echelon form?

$$
A = \begin{bmatrix} 1 & 0 & 0 & -3 \\ 0 & 1 & 0 & 4 \\ 0 & 0 & 1 & 2 \end{bmatrix}, \quad
B = \begin{bmatrix} 1 & 0 & 0 & 5 \\ 0 & 1 & 0 & -4 \\ 0 & 0 & -1 & 3 \end{bmatrix}, \quad
C = \begin{bmatrix} 1 & 1 & 0 & 5 \\ 0 & 1 & 0 & 4 \\ 0 & 0 & 2 & 3 \end{bmatrix},
$$

$$
D = \begin{bmatrix} 1 & 0 & 0 & 2 \\ 0 & 1 & 0 & -1 \\ 0 & 0 & 0 & 0 \\ 0 & 0 & 0 & 1 \end{bmatrix}, \quad
E = \begin{bmatrix} 1 & 0 & 0 \\ 0 & 1 & 3 \\ 0 & 0 & 0 \end{bmatrix}, \quad
F = \begin{bmatrix} 1 & 2 & -3 \\ 0 & 1 & 0 \\ 0 & 0 & 0 \end{bmatrix}.
$$

2. Let

$$
A = \begin{bmatrix} 1 & 0 & 3 \\ -3 & 1 & 4 \\ 4 & 2 & 2 \\ 5 & -1 & 5 \end{bmatrix}.
$$

Find the matrices obtained by performing the following elementary row operations on A.
   (a) Interchanging the second and fourth rows.
   (b) Multiplying the third row by 3.
   (c) Adding -3 times the first row to the fourth row.

3. Let

$$
A = \begin{bmatrix} 2 & 0 & 4 & 2 \\ 3 & -2 & 5 & 6 \\ -1 & 3 & 1 & 1 \end{bmatrix}.
$$

Find the matrices obtained by performing the following elementary row operations on A.
(a) Interchanging the second and third rows.
(b) Multiplying the second row by -4.
(c) Adding 2 times the third row to the first row.

4. Given

$$A = \begin{bmatrix} 1 & -2 & 0 & 2 \\ 2 & -3 & -1 & 5 \\ 1 & 3 & 2 & 5 \\ 1 & 1 & 0 & 2 \end{bmatrix}.$$

(a) Find the matrix in row echelon form that is row equivalent to A.
(b) Find the matrix in reduced row echelon form that is row equivalent to A.

5. Use row reduction to find all solutions to the given linear systems.

(a) $x_1 + x_2 + 2x_3 = -1$
    $x_1 - 2x_2 + x_3 = -5$
    $3x_1 + x_2 + x_3 = 3.$

(b) $x_1 + x_2 + 3x_3 + 2x_4 = 7$
    $2x_1 - x_2 \quad\quad + 4x_4 = 8$
    $\quad\quad 3x_2 + 6x_3 \quad\quad = k.$

(c) $x_1 + x_2 + 2x_3 + 3x_4 = 13$
    $x_1 + 2x_2 + x_3 + x_4 = 8$
    $3x_1 + x_2 + x_3 - x_4 = 1.$

(d) $x_1 + x_2 + x_3 + x_4 = 6$
    $2x_1 + x_2 - x_3 \quad\quad = 3$
    $3x_1 + x_2 + 2x_4 = 6.$

(e) $x_1 + \quad\quad x_2 = 3$
    $x_1 + (a^2 - 8)x_2 = a.$

6. Solve the linear system with the given augmented matrix by row reduction.

(a) $\begin{bmatrix} 1 & 1 & 1 & | & 0 \\ 1 & 1 & 0 & | & 3 \\ 0 & 1 & 1 & | & 1 \end{bmatrix}.$ (b) $\begin{bmatrix} 1 & 2 & 3 & | & 0 \\ 1 & 1 & 1 & | & 0 \\ 5 & 7 & 9 & | & 0 \end{bmatrix}.$ (c) $\begin{bmatrix} 1 & 1 & 3 & -3 & | & 0 \\ 0 & 2 & 1 & -3 & | & 3 \\ 1 & 0 & 2 & -1 & | & -1 \end{bmatrix}.$

(d) $\begin{bmatrix} 1 & 1 & 2 & 1 & -1 & | & 1 \\ 1 & 1 & 2 & 1 & -1 & | & 2 \\ 1 & 1 & 1 & 2 & 3 & | & 1 \end{bmatrix}.$ (e) $\begin{bmatrix} 1 & 1 & 2 & 1 & -1 & | & 1 \\ 2 & 2 & 3 & 3 & 2 & | & 2 \\ 1 & 1 & 1 & 2 & 3 & | & 1 \end{bmatrix}.$

7. Given

$$A = \begin{bmatrix} 1 & 1 & 0 & -1 \\ 1 & 0 & -1 & 0 \\ 1 & -3 & 1 & 0 \end{bmatrix}, \quad x = \begin{bmatrix} x_1 \\ x_2 \\ x_3 \\ x_4 \end{bmatrix}, \quad b = \begin{bmatrix} 0 \\ 0 \\ 6 \end{bmatrix}.$$

(a) Write the linear system corresponding to the matrix equation $Ax = b$.

(b) Solve the linear system $Ax = b$ by row reduction.

8. Suppose that x is the general solution to the linear system $Ax = b$. Write the general solution to the associated homogeneous system $Ax = 0$ in the following cases.

(a) $x = \begin{bmatrix} 1 - x_3 \\ \frac{2}{3} + x_4 \\ x_3 \\ x_4 \\ 2 \end{bmatrix}$. (b) $x = \begin{bmatrix} x_2 \\ x_2 \\ 1 + x_4 \\ x_4 \\ 3 \end{bmatrix}$. (c) $x = \begin{bmatrix} 1 - x_4 \\ x_4 \\ x_3 \\ x_4 \\ 1 \end{bmatrix}$.

9. Write a list of steps for the reduce echelon algorithm.

1.4 MATRIX OPERATIONS

The definitions and matrix operations introduced below are very handy.

An n-dimensional column vector x is a list of n numbers arranged in column format:

$$x = \begin{bmatrix} x_1 \\ \vdots \\ x_j \\ \vdots \\ x_n \end{bmatrix}.$$

Similarly, an n-dimensional row vector y is a list of n numbers arranged in row format:

$$y = [y_1, \ldots, y_j, \ldots, y_n].$$

It will be very convenient to have both column and row format available.

If s is a number, w and x are n-dimensional column vectors, and y and z are n-dimensional row vectors, the following definitions apply:

$$sx = s \begin{bmatrix} x_1 \\ \vdots \\ x_n \end{bmatrix} = \begin{bmatrix} sx_1 \\ \vdots \\ sx_n \end{bmatrix} \quad \text{and}$$

$$w + x = \begin{bmatrix} w_1 \\ \vdots \\ w_n \end{bmatrix} + \begin{bmatrix} x_1 \\ \vdots \\ x_n \end{bmatrix} = \begin{bmatrix} w_1 + x_1 \\ \vdots \\ w_n + x_n \end{bmatrix}.$$

Similarly,

$$sy = s[y_1, \ldots, y_n] = [sy_1, \ldots, sy_n] \quad \text{and}$$

$$y + z = [y_1, \ldots, y_n] + [z_1, \ldots, z_n] = [y_1 + z_1, \ldots, y_n + z_n];$$

$$x^t = [x_1, \ldots, x_n] \quad (t \text{ denotes transpose}),$$

$$y^t = \begin{bmatrix} y_1 \\ \vdots \\ y_n \end{bmatrix}, \quad \text{and}$$

$$yx = [y_1, \ldots, y_n] \begin{bmatrix} x_1 \\ \vdots \\ x_n \end{bmatrix} = y_1 x_1 + \cdots + y_n x_n = \sum_{j=1}^{n} y_j x_j.$$

Notice that the transpose operation interchanges column and row format.

Let R^n denote the set of n-dimensional column vectors and R_n denote the set of n-dimensional row vectors. An $m \times n$ matrix A is a list of $(m)(n)$ numbers in the following format:

$$A = \begin{bmatrix} a_{11} & \cdots & a_{1j} & \cdots & a_{1n} \\ \vdots & & \vdots & & \vdots \\ a_{i1} & \cdots & a_{ij} & \cdots & a_{in} \\ \vdots & & \vdots & & \vdots \\ a_{m1} & \cdots & a_{mj} & \cdots & a_{mn} \end{bmatrix} = [a_{ij}]$$

If s is a number, A and P are $\mathbf{m} \times \mathbf{n}$ matrices, and Q is an $n \times p$ matrix, the following definitions apply.

$$sA = s[a_{ij}] = [sa_{ij}], \qquad A + P = [a_{ij}] + [p_{ij}] = [a_{ij} + p_{ij}], \qquad \text{and}$$

$$AQ = \begin{bmatrix} a_{11} & \cdots & a_{1n} \\ a_{i1} & \cdots & a_{in} \\ a_{m1} & \cdots & a_{mn} \end{bmatrix} \begin{bmatrix} q_{11} & \cdots & q_{1j} & \cdots & q_{1p} \\ \vdots & & \vdots & & \vdots \\ q_{n1} & \cdots & q_{nj} & \cdots & q_{np} \end{bmatrix} = R = [r_{ij}],$$

where R is a $m \times p$ matrix with i,j entry

$$r_{ij} = \sum_{k=1}^{n} a_{ik}q_{kj}.$$

For example,

$$xy = \begin{bmatrix} x_1 \\ \vdots \\ x_n \end{bmatrix} [y_1, \ldots, y_n] = \begin{bmatrix} x_1y_1, & x_1y_2, & \cdots, & x_1y_n \\ x_2y_1, & x_2y_2, & \cdots, & x_2y_n \\ \vdots & & & \vdots \\ x_ny_1, & x_ny_2, & \cdots, & x_ny_n \end{bmatrix}.$$

Let's apply these definitions to the general $m \times n$ linear system. Since

$$Ax = \begin{bmatrix} a_{11}x_1 + \cdots + a_{1n}x_n \\ a_{i1}x_1 + \cdots + a_{in}x_n \\ \vdots & \vdots \\ a_{m1}x_1 + \cdots + a_{mn}x_n \end{bmatrix},$$

we may write the $m \times n$ linear system in the compact form $Ax = b$. Another useful way to write Ax is

$$Ax = x_1 \begin{bmatrix} a_{11} \\ \vdots \\ a_{m1} \end{bmatrix} + x_2 \begin{bmatrix} a_{12} \\ \vdots \\ a_{m2} \end{bmatrix} + \cdots + x_n \begin{bmatrix} a_{1n} \\ \vdots \\ a_{mn} \end{bmatrix}.$$

If we denote the columns of A by $a_1, \ldots, a_n$, then

$$Ax = x_1a_1 + x_2a_2 + \cdots + x_na_n = [a_1, \ldots, a_n] \begin{bmatrix} x_2 \\ \vdots \\ x_n \end{bmatrix}.$$

Similarly, if y is an m-dimensional row vector and we denote the rows of A by $r_1, \ldots, r_m$, then

$$yA = [y_1, \ldots, y_m] \begin{bmatrix} r_1 \\ \vdots \\ r_m \end{bmatrix} = y_1r_1 + \cdots + y_mr_m = \sum_{i=1}^{m} y_ir_i.$$

To conserve space, a n-dimensional column vector x will often be written in the (row) form

$$x = \begin{bmatrix} x_1 \\ \vdots \\ x_n \end{bmatrix} = (x_1, \ldots, x_n).$$

Order of Vectors

Given $x = (x_1, \ldots, x_n)$ in R^n and $y = (y_1, \ldots, y_n)$ in R^n,

$x = y$ means $x_i = y_i$, $i = 1, \ldots, n$,

$x \geq y$ means $x_i \geq y_i$, $i = 1, \ldots, n$,   and

$x > y$ means $x_i > y_i$, $i = 1, \ldots, n$.

For example, (1,0) is not $=$ (0,0), (1,0) is $\geq$(0,0), and (1,0) is not $>$ (0,0).

Transpose

Finally, the *transpose*, denoted by A^t, of a $m \times n$ matrix A is the $n \times m$ matrix whose rows are the columns of A, in the same order: $A^t = [a_{ij}^t]$, where $a_{ij}^t = a_{ji}$.

Exercise

1. Given

$$A = \begin{bmatrix} 1 & 2 & 3 \\ 2 & 1 & 4 \end{bmatrix}, \quad B = \begin{bmatrix} 1 & 0 \\ 2 & 1 \\ 3 & 2 \end{bmatrix}, \quad C = \begin{bmatrix} 3 & -1 & 3 \\ 4 & 1 & 5 \\ 2 & 1 & 3 \end{bmatrix},$$

$$D = \begin{bmatrix} 3 & -2 \\ 2 & 4 \end{bmatrix}, \quad E = \begin{bmatrix} 2 & -4 & 5 \\ 0 & 1 & 4 \\ 3 & 2 & 1 \end{bmatrix}, \quad \text{and} \quad F = \begin{bmatrix} -4 & 5 \\ 2 & 3 \end{bmatrix}.$$

If possible, compute:

(a) $C + B$. (b) BA. (c) $(A)^t D$. (d) $D(A^t)$. (e) AB.
(f) $A + C$. (g) $(B^t)F$. (h) $F(A^t)$. (i) $A + E$. (j) BD.
(k) DB. (l) $A + B^t$. (m) AF. (n) $E + A$. (o) FA.
(p) $A^t + B$.

1.5 RANK

The *row rank* of a linear system is the number of nonzero rows that remain after row echeloning $[A \mid b]$; the row rank of A is the number of nonzero rows that remain in A. Thus the system has solutions if and only if the row rank of $[A \mid b]$ equals the row rank of A.

Example 1.3. Consider the linear system

$$x_1 + 2x_2 + 3x_3 = 1$$
$$x_1 + x_2 - x_3 = 2$$
$$2x_1 + 3x_2 + 2x_3 = k,$$

where k will be specified later. We apply row echelon to the augmented matrix:

$$\begin{bmatrix} 1 & 2 & 3 & 1 \\ 1 & 1 & -1 & 2 \\ 2 & 3 & 2 & k \end{bmatrix} \xrightarrow[\substack{-(1)\,\text{to}\,(2) \\ -2(1)\,\text{to}\,(3)}]{} \begin{bmatrix} 1 & 2 & 3 & 1 \\ 0 & -1 & -4 & 1 \\ 0 & -1 & -4 & k-2 \end{bmatrix}$$

$$\xrightarrow[-(2)\,\text{to}\,(3)]{} \begin{bmatrix} 1 & 2 & 3 & 1 \\ 0 & -1 & -4 & 1 \\ 0 & 0 & 0 & k-3 \end{bmatrix}$$

and find that this system has a solution if and only if $k = 3$ because the row rank of $A = 2$ and the row rank of $[A \mid b]$ is 3 unless $k - 3 = 0$, in which case the row rank of $[A \mid b] = 2$.

1.6 IDENTITY AND INVERSE MATRICES

Suppose that the row rank of A is m. Let B denote the $m \times m$ matrix whose columns are the columns of A that contain corner points after row echelon is carried out; the same row operations that row reduce A transform B to an *identity matrix* I: a matrix with 1's on the diagonal and 0's elsewhere. In Example 1.1,

$$B = A \rightarrow \begin{bmatrix} 1 & 0 \\ 0 & 1 \end{bmatrix} = I$$

and in Example 1.2, columns 1, 3, and 4 of A compose

$$B = \begin{bmatrix} 1 & 2 & 1 \\ 1 & 2 & -1 \\ 1 & 1 & 2 \end{bmatrix} \rightarrow \cdots \rightarrow \begin{bmatrix} 1 & 0 & 0 \\ 0 & 1 & 0 \\ 0 & 0 & 1 \end{bmatrix} = I.$$

In Example 1.1, the row operations that took A to I took

$$b = \begin{bmatrix} 3 \\ 4 \end{bmatrix} \quad \text{to} \quad \begin{bmatrix} -\frac{6}{7} \\ \frac{11}{7} \end{bmatrix} \quad \text{and} \quad A\begin{bmatrix} -\frac{6}{7} \\ \frac{11}{7} \end{bmatrix} = \begin{bmatrix} 3 \\ 4 \end{bmatrix}.$$

In Example 1.2, the row operations that took B to I took

$$\begin{bmatrix} 1 \\ 2 \\ 1 \end{bmatrix} \quad \text{to} \quad \begin{bmatrix} \frac{5}{2} \\ -\frac{1}{2} \\ -\frac{1}{2} \end{bmatrix} \quad \text{and} \quad B\begin{bmatrix} \frac{5}{2} \\ -\frac{1}{2} \\ -\frac{1}{2} \end{bmatrix} = \begin{bmatrix} 1 \\ 2 \\ 1 \end{bmatrix}.$$

In general, *if B is an $m \times m$ matrix of rank m, v is an m-dimensional column vector, and $[B \mid v] \rightarrow \cdots \rightarrow [I \mid w]$, then $Bw = v$.* For example, let's take B as in Example 1.2, take

$$v_1 = \begin{bmatrix} 1 \\ 0 \\ 0 \end{bmatrix}, \quad v_2 = \begin{bmatrix} 0 \\ 1 \\ 0 \end{bmatrix}, \quad \text{and} \quad v_3 = \begin{bmatrix} 0 \\ 0 \\ 1 \end{bmatrix},$$

and apply row reduction (simultaneously) to $[B \mid I] = [B \mid v_1, v_2, v_3]$;

$$\begin{bmatrix} 1 & 2 & 1 & | & 1 & 0 & 0 \\ 1 & 2 & -1 & | & 0 & 1 & 0 \\ 1 & 1 & 2 & | & 0 & 0 & 1 \end{bmatrix} \xrightarrow[\substack{-(1) \text{ to } (2) \\ -(1) \text{ to } (3)}]{} \begin{bmatrix} 1 & 2 & 1 & | & 1 & 0 & 0 \\ 0 & 0 & -2 & | & -1 & 1 & 0 \\ 0 & -1 & 1 & | & -1 & 0 & 1 \end{bmatrix}$$

$$\xrightarrow[(2) \leftrightarrow (3)]{} \begin{bmatrix} 1 & 2 & 1 & | & 1 & 0 & 0 \\ 0 & -1 & 1 & | & -1 & 0 & 1 \\ 0 & 0 & -2 & | & -1 & 1 & 0 \end{bmatrix}$$

$$\xrightarrow[\substack{-(2) \\ -1/2(3)}]{} \begin{bmatrix} 1 & 2 & 1 & | & 1 & 0 & 0 \\ 0 & 1 & -1 & | & 1 & 0 & -1 \\ 0 & 0 & 1 & | & \frac{1}{2} & -\frac{1}{2} & 0 \end{bmatrix}$$

$$\xrightarrow[\substack{(3) \text{ to } (2) \\ -(3) \text{ to } (1)}]{} \begin{bmatrix} 1 & 2 & 0 & | & \frac{1}{2} & \frac{1}{2} & 0 \\ 0 & 1 & 0 & | & \frac{3}{2} & -\frac{1}{2} & -1 \\ 0 & 0 & 1 & | & \frac{1}{2} & -\frac{1}{2} & 0 \end{bmatrix}$$

$$\xrightarrow[-2(2) \text{ to } (1)]{} \begin{bmatrix} 1 & 0 & 0 \\ 0 & 1 & 0 \\ 0 & 0 & 1 \end{bmatrix}\begin{array}{|ccc} -\frac{5}{2} & \frac{3}{2} & 2 \\ \frac{3}{2} & -\frac{1}{2} & -1 \\ \frac{1}{2} & -\frac{1}{2} & 0 \end{array}$$

$$= [I \mid w_1, w_2, w_3],$$

where

$$w_1 = \begin{bmatrix} -\frac{5}{2} \\ \frac{3}{2} \\ \frac{1}{2} \end{bmatrix}, \qquad w_2 = \begin{bmatrix} \frac{3}{2} \\ -\frac{1}{2} \\ -\frac{1}{2} \end{bmatrix}, \qquad w_3 = \begin{bmatrix} 2 \\ -1 \\ 0 \end{bmatrix};$$

$$Bw_1 = v_1, \qquad Bw_2 = v_2, \qquad \text{and} \qquad Bw_3 = v_3,$$

or, more compactly, $B[w_1, w_2, w_3] = [Bw_1, Bw_2, Bw_3] = [v_1, v_2, v_3] = I$. Put

$$B^{-1} = [w_1, w_2, w_3] = \begin{bmatrix} -\frac{5}{2} & \frac{3}{2} & 2 \\ \frac{3}{2} & -\frac{1}{2} & -1 \\ \frac{1}{2} & -\frac{1}{2} & 0 \end{bmatrix}.$$

Then $[B \mid I] \to \cdots \to [I \mid B^{-1}]$ and $BB^{-1} = I$; B^{-1} is called B *inverse*; B^{-1} is quite useful. For example, given any three-dimensional column vector v, the vector $w = B^{-1}v$ is a solution to the linear system $Bw = v$ because $B(B^{-1}v) = (BB^{-1})v = Iv = v$. To show that $B^{-1}v$ is the only solution to the system $Bw = v$, we will use the following fact, which we do not prove.

Fact

Given a $m \times m$ matrix B of rank m, there is exactly one $m \times m$ matrix W such that $BW = I$: $W = B^{-1}$. Moreover, $B^{-1}B = I$.

Exercises

1. (a) Check that $B^{-1}B = I$ in Example 1.2.
   (b) compute B^{-1} and check that $B^{-1}B = BB^{-1} = I$ in Example 1.1.
2. If possible, compute the inverses of the matrices given in Exercise 1 of Section 1.4.

The fact that $B^{-1}B = I$ can be used to show that $w = B^{-1}v$ is the only solution to the system $Bw = v$ as follows: Suppose that y is also a solution to the system, and put $z = w - y$. Then $Bz = B(w - y) = Bw - By = v - v = 0$. Thus $0 = B^{-1}0 = B^{-1}(Bz) = (B^{-1}B)z = Iz = z$.

The preceding argument relates to the important concept of linear independence defined below.

1.7 LINEAR INDEPENDENCE

Given a positive integer k and a positive integer j; k-dimensional vectors $z_1, \ldots, z_j$ are said to be *linearly independent* when a linear combination $y_1 z_1 + \cdots + y_j z_j$ of them (the y_i's are numbers) is equal to the zero vector (if and) only if $y_1 = y_2 = \cdots = y_j = 0$ (i.e., only when all the y_i's are zero). Alternatively, the set $\{z_1, \ldots, z_j\}$ is said to be a linearly independent set of (k-dimensional) vectors when $\{\Sigma_{i=1}^{j} y_i z_i = 0\} \Leftrightarrow \{y = (y_1, \ldots, y_j) = 0\}$.

Looking at Example 1.2 again, let c_1, c_2, and c_3 denote the columns of B. Then

$$By = [c_1, c_2, c_3] \begin{bmatrix} y_1 \\ y_2 \\ y_3 \end{bmatrix} = y_1 c_1 + y_2 c_2 + y_3 c_3,$$

so the columns of B are linearly independent because $By = 0$ if and only if $y = 0$.

In example 1.2,

$$y = \begin{bmatrix} y_1 \\ y_2 \\ y_3 \end{bmatrix} = \begin{bmatrix} x_1 \\ x_3 \\ x_4 \end{bmatrix} \quad \text{and} \quad B = \begin{bmatrix} 1 & 2 & 1 \\ 1 & 2 & -1 \\ 1 & 1 & 2 \end{bmatrix}.$$

Thus

$$c_1 = \begin{bmatrix} 1 \\ 1 \\ 1 \end{bmatrix}, \quad c_2 = \begin{bmatrix} 2 \\ 2 \\ 1 \end{bmatrix}, \quad \text{and} \quad c_3 = \begin{bmatrix} 1 \\ -1 \\ 2 \end{bmatrix};$$

consequently,

$$By = y_1 \begin{bmatrix} 1 \\ 1 \\ 1 \end{bmatrix} + y_2 \begin{bmatrix} 2 \\ 2 \\ 1 \end{bmatrix} + y_3 \begin{bmatrix} 1 \\ -1 \\ 2 \end{bmatrix}.$$

Looking at the equation $By = v$, we know that $y = B^{-1}v$ is the only solution and we know that row reduction, which we denote by "$\rightarrow \cdots \rightarrow$," takes the augmented matrix $[B \,|\, v]$ to $[I \,|\, B^{-1}v]$. For example, if

$$v = \begin{bmatrix} 1 \\ 2 \\ 1 \end{bmatrix}, \quad \text{then} \quad B^{-1}v = \begin{bmatrix} \frac{5}{2} \\ -\frac{1}{2} \\ -\frac{1}{2} \end{bmatrix}$$

because

$$\left[B \left| \begin{array}{c} 1 \\ 2 \\ 1 \end{array} \right. \right] \rightarrow \cdots \rightarrow \left[I \left| \begin{array}{c} \frac{5}{2} \\ -\frac{1}{2} \\ -\frac{1}{2} \end{array} \right. \right].$$

Thus

$$\frac{5}{2}\begin{bmatrix} 1 \\ 1 \\ 1 \end{bmatrix} - \frac{1}{2}\begin{bmatrix} 2 \\ 2 \\ 1 \end{bmatrix} - \frac{1}{2}\begin{bmatrix} 1 \\ -1 \\ 2 \end{bmatrix} = \begin{bmatrix} 1 \\ 2 \\ 1 \end{bmatrix}.$$

Similarly, putting $v = $ column 2 and then column 5 in the original coefficient matrix for Example 1.2, we have

$$\left[B \left| \begin{array}{c} 1 \\ 1 \\ 1 \end{array} \right. \right] \rightarrow \cdots \rightarrow \left[I \left| \begin{array}{c} 1 \\ 0 \\ 0 \end{array} \right. \right]$$

and

$$\left[B \left| \begin{array}{c} -1 \\ 1 \\ 3 \end{array} \right. \right] \rightarrow \cdots \rightarrow \left[I \left| \begin{array}{c} 10 \\ -5 \\ -1 \end{array} \right. \right],$$

so

$$1\begin{bmatrix} 1 \\ 1 \\ 1 \end{bmatrix} + 0\begin{bmatrix} 2 \\ 2 \\ 1 \end{bmatrix} + 0\begin{bmatrix} 1 \\ -1 \\ 2 \end{bmatrix} = \begin{bmatrix} 1 \\ 1 \\ 1 \end{bmatrix} \qquad \text{or} \qquad a_1 = a_2$$

and

$$10\begin{bmatrix} 1 \\ 1 \\ 1 \end{bmatrix} - 5\begin{bmatrix} 2 \\ 2 \\ 1 \end{bmatrix} - 1\begin{bmatrix} 1 \\ -1 \\ 2 \end{bmatrix} = \begin{bmatrix} -1 \\ 1 \\ 3 \end{bmatrix} \qquad \text{or} \qquad 10a_1 - 5a_3 - a_4 = a_5.$$

Rewriting, we have

$$-a_1 + a_2 = A \begin{bmatrix} -1 \\ 1 \\ 0 \\ 0 \\ 0 \end{bmatrix} = 0$$

and

$$-10a_1 + 5a_3 + a_4 + a_5 = A \begin{bmatrix} -10 \\ 0 \\ 5 \\ 1 \\ 1 \end{bmatrix} = 0.$$

Consequently, $z_1 = (-1,1,0,0,0)$ and $z_2 = (-10,0,5,1,1)$ are solutions to the *associated homogeneous equation* $Ax = 0$; these solutions are linearly independent because if

$$x_2z_1 + x_5z_2 = x_2(-1,1,0,0,0) + x_5(-10,0,5,1,1)$$
$$= (-x_2 - 10x_5, x_2, 5x_5, x_5, x_5) = 0 = (0,0,0,0,0),$$

then $x_2 = 0$ and $x_5 = 0$. Going back to Section 1.3, we recall that the general solution to

$$Ax = \begin{bmatrix} 1 \\ 2 \\ 1 \end{bmatrix}$$

is $x = (\frac{5}{2},0,-\frac{1}{2},-\frac{1}{2},0) + x_2z_1 + x_5z_2$; $(\frac{5}{2},0,-\frac{1}{2},-\frac{1}{2},0)$ is a particular solution to

$$Ax = \begin{bmatrix} 1 \\ 2 \\ 1 \end{bmatrix}$$

and $x_2z_1 + x_5z_2$ is the general solution to the associated homogeneous equation $Ax = 0$. We will look at Example 1.2 again in Section 1.9.

We leave the interested reader with the project of showing (1) that given any three-dimensional row vector z, $u = zB^{-1}$ is the only solution to the linear system $uB = z$; (2) if we denote the rows of B by r_1, r_2, and r_3, then $uB = u_1 r_1 + u_2 r_2 + u_3 r_3$; and (3) the rows of B are linearly independent.

These same ideas apply when B is an $m \times m$ matrix of rank m; if

$$[B \mid I] = \begin{bmatrix} & & 1 & 0 & & & \cdots & 0 \\ & & 0 & 1 & 0 & \cdots & & 0 \\ B & & \vdots & 0 & & & & \vdots \\ & & & & & & & 0 \\ & & 0 & & & \cdots & 0 & 1 \end{bmatrix}$$

$$\rightarrow \cdots \rightarrow [I \mid B^{-1}],$$

then $BB^{-1} = I$ and $w = B^{-1}v$ satisfies the equation $Bw = v$ for

$$v = \begin{bmatrix} v_1 \\ \vdots \\ v_m \end{bmatrix}.$$

We have asserted that $B^{-1}B = I$, so the same arguments that we used when $m = 3$ apply and the following statements are valid:

1. Given $v = (v_1, \ldots, v_m)$, $w = B^{-1}v$ is the only solution to $Bw = v$.
2. The columns of B are linearly independent.
3. Given $z = [z_1, \ldots, z_m]$, $u = zB^{-1}$ is the only solution to $uB = z$.
4. The rows of B are linearly independent.

1.8 REARRANGEMENT

Rearrangement is a type of reorganization that will be used in the simplex method in Chapter 4. Example 1.2 is used again below to illustrate; referring to Section 1.6, B is composed of the first, third, and fourth columns of A. Rearrange the columns a_1, a_2, a_3, a_4, a_5 of $A = [a_1, a_2, a_3, a_4, a_5]$ so that the columns of B appear first; the matrix defined by this rearrangement is denoted below by $\tilde{A}$:

$$\tilde{A} = [a_1, a_3, a_4, a_2, a_5] = [B \mid V],$$

where

$$V = [a_2, a_5] = \begin{bmatrix} 1 & -1 \\ 1 & 1 \\ 1 & 3 \end{bmatrix}.$$

Check that

$$Ax = \tilde{A}(x_1,x_3,x_4,x_2,x_5) = B(x_1,x_3,x_4) + V(x_2,x_5)$$

and that

$$\tilde{A} = [B\,|\,V] \to \cdots \to [I\,|\,B^{-1}a_2,B^{-1}a_5] = [I\,|\,B^{-1}V],$$

where

$$B^{-1}V = \begin{bmatrix} 1 & 10 \\ 0 & -5 \\ 0 & -1 \end{bmatrix}.$$

Now take another look at the second augmented matrix in Section 1.3; notice that $B^{-1}V$ is composed of the second and fifth columns of that matrix.

1.9 SOLUTIONS TO LINEAR SYSTEMS

In Section 1.3 we found that for Example 1.2, x_2 and x_5 were free variables in the solutions $x = (x_1,x_2,x_3,x_4,x_5)$ to $Ax = b$, where

$$A = \begin{bmatrix} 1 & 1 & 2 & 1 & -1 \\ 1 & 1 & 2 & -1 & 1 \\ 1 & 1 & 1 & 2 & 3 \end{bmatrix} \quad \text{and} \quad b = \begin{bmatrix} 1 \\ 2 \\ 1 \end{bmatrix}.$$

The basic variables x_1, x_3, and x_4 were determined from the values of the free variables by the equations

$$x_1 = \tfrac{5}{2} - x_2 - 10x_5$$
$$x_3 = -\tfrac{1}{2} \qquad + 5x_5$$
$$x_4 = -\tfrac{1}{2} \qquad + x_5.$$

Thus

$$x = \begin{bmatrix} x_1 \\ x_2 \\ x_3 \\ x_4 \\ x_5 \end{bmatrix} = \begin{bmatrix} \tfrac{5}{2} - x_2 - 10x_5 \\ x_2 \\ -\tfrac{1}{2} \qquad + 5x_5 \\ -\tfrac{1}{2} \qquad + x_5 \\ x_5 \end{bmatrix}$$

$$= \begin{bmatrix} \frac{5}{2} \\ 0 \\ -\frac{1}{2} \\ -\frac{1}{2} \\ 0 \end{bmatrix} + x_2 \begin{bmatrix} -1 \\ 1 \\ 0 \\ 0 \\ 0 \end{bmatrix} + x_5 \begin{bmatrix} -10 \\ 0 \\ 5 \\ 1 \\ 1 \end{bmatrix}$$

is the general solution to $Ax = b$. Notice that $(\frac{5}{2},0,-\frac{1}{2},-\frac{1}{2},0)$ is a (particular) solution to $Ax = b$, while $(-1,1,0,0,0)$ and $(-10,0,5,1,1)$ are linearly independent solutions to the *associated homogeneous system* $Ax = 0$.

Example 1.2 is typical of the general linear system. The general solution x to a linear system $Ax = b$ can be written in the form $x = x^p + x^h$, where x^p is the particular solution obtained by putting all the free variables equal to zero and x^h is the general solution to the associated homogeneous system $Ax = 0$. In Example 1.2,

$$x^p = (\tfrac{5}{2},0,-\tfrac{1}{2},-\tfrac{1}{2},0)$$

and

$$x^h = x_2(-1,1,0,0,0) + x_5(-10,0,5,1,1).$$

EXERCISES

1.1 Given a homogeneous linear system $Ax = 0$:
  (a) Show that if vectors y and z are solutions and t and u are numbers, then $x = ty + uz$ is a solution.
  (b) Explain why the vectors corresponding to the free variables in Example 1.2 are linearly independent.

1.2 Apply row reduction to:

  (a) $\begin{bmatrix} 1 & \frac{2}{3} & 1 & | & 1 \\ \frac{1}{2} & 1 & 0 & | & 1 \end{bmatrix}$. (b) $\begin{bmatrix} 1 & \frac{2}{3} & | & 1 \\ \frac{1}{2} & 1 & | & 1 \end{bmatrix}$. (c) $\begin{bmatrix} 1 & 1 & | & 1 \\ \frac{1}{2} & 0 & | & 1 \end{bmatrix}$.

  (d) $\begin{bmatrix} 1 & \frac{2}{3} & 1 & 1 & 0 & | & 1 \\ 0 & 1 & \frac{1}{2} & 0 & 1 & | & 1 \end{bmatrix}$.

1.3 (a) Apply row reduction to $\begin{bmatrix} 1 & -1 & | & -1 \\ 0 & -2 & | & 1 \end{bmatrix}$.

  (b) Rearrange the columns of the augmented matrix

  $$[A \mid b] = [a_1,a_2,a_3,a_4,a_5 \mid b] = \begin{bmatrix} -1 & -1 & 2 & 1 & 0 & | & 1 \\ 1 & -2 & 1 & 0 & 1 & | & -1 \end{bmatrix}$$

  in the order $[a_4,a_2,a_1,a_3,a_5 \mid b]$ and apply row reduction.

(c) Verify that $-\frac{3}{2}\begin{bmatrix} 1 \\ 0 \end{bmatrix} - \frac{1}{2}\begin{bmatrix} -1 \\ -2 \end{bmatrix} = \begin{bmatrix} -1 \\ 1 \end{bmatrix}$.

(d) Verify that $\begin{bmatrix} -1 \\ 1 \end{bmatrix} + \frac{1}{2}\begin{bmatrix} -1 \\ -2 \end{bmatrix} + \frac{3}{2}\begin{bmatrix} 1 \\ 0 \end{bmatrix} = 0$.

(e) Verify that $A(1,\frac{1}{2},0,\frac{3}{2},0) = 0$.

1.4 Let

$$B = \begin{bmatrix} 1 & 3 & 3 & 6 \\ 0 & 1 & 4 & 0 \\ 1 & 2 & 3 & 6 \\ 6 & 2 & 2 & 0 \end{bmatrix} = [a_1,a_2,a_3,a_4] \quad \text{and} \quad b = \begin{bmatrix} 16 \\ 8 \\ 14 \\ 13 \end{bmatrix}.$$

(a) Compute B^{-1}.

(b) Solve the linear system $Bx = b$.

(c) Verify that

$$\begin{bmatrix} 1 \\ 0 \\ 1 \\ 6 \end{bmatrix} + 2\begin{bmatrix} 3 \\ 1 \\ 2 \\ 2 \end{bmatrix} + \frac{3}{2}\begin{bmatrix} 3 \\ 4 \\ 3 \\ 2 \end{bmatrix} + \frac{3}{4}\begin{bmatrix} 6 \\ 0 \\ 6 \\ 0 \end{bmatrix} = \begin{bmatrix} 16 \\ 8 \\ 14 \\ 13 \end{bmatrix}.$$

1.5 (a) Find all solutions to the linear system

$$\begin{aligned} x_1 + x_2 \qquad\qquad &= 3 \\ x_2 + x_3 \qquad &= 1 \\ x_3 + x_4 &= 1. \end{aligned}$$

(b) Show that each solution of this linear system satisfies the equation $x_1 + x_2 + x_3 + x_4 = 4$.

1.6 Put $k = 6$ in Exercise 5b of Section 1.3 and show that vectors corresponding to the free variables x_3 and x_4 are linearly independent solutions to the asociated homogeneous system.

1.7 Consider the general $m \times n$ linear system $Ax = b$, where A is an $m \times n$ matrix and b is an m-dimensional column vector. Suppose that both the matrix A and the augmented matrix $[A \mid b]$ have same rank, say k, so that the system has solutions. Suppose that after row reduction, the pivot points are $(1,j_1), (2,j_2), \ldots ,(k,j_k)$, where $j_1 < j_2 < \cdots < j_k$.

Consider the corresponding column vectors $a_{j_1},a_{j_2}, \ldots ,a_{j_k}$ in A.

(a) Show that these vectors are linearly independent.

(b) Show that every column vector in A is a linear combination of these vectors.

(c) Show that the system $Ax = v$ has a solution if, and only if, v is a linear combination of these vectors.

(d) Show (1) if $k = m$, then the rows of A are linearly independent, and (2) if the rows of A are not linearly independent, then $k < n$.

2
Introduction to Linear Programming

Linear programming (LP) encompasses several activities. LP includes re-cognizing whether or not a problem that you wish to solve can be modeled effectively as a linear problem. If a linear model is appropriate for the problem, LP includes formulating an accurate LP model (also called sim-ply an LP), solving the LP, and analyzing the implications of the solution of the LP model of the problem. Programming has artistic and technical aspects, and like learning to paint or play a game, one learns to program by programming. Reading can help you get started, but you need to participate to become good at it.

Comparison with Tennis

Rod Laver and Roy Emerson, with Barry Tarsis, wrote a book called *Tennis for the Bloody Fun of It*. The essence of the book is given in three of their remarks:

(a) Enjoy it.
(b) It's an individual game. You can get all the instruction you want, but in the final accounting, you have to do what works best for you (p. 46).
(c) There are three basic principles that apply pretty much to all the strokes in tennis (p. 47): (1) footwork (get ready), (2) watch the ball, and (3) hit "through" the ball.

In my opinion, programming boils down similarly to (a), (b), and the following modification of (c):

(1) *Understand the situation* (get ready).
(2) *Formulate an effective model.*
(3) *Solve the model and interpret the implications of the model for the situation.*

Introduction

In this chapter we use three examples and several exercises to give you experience answering the following two questions intelligently when you have a problem to solve.

1. Can I use LP to help solve the problem?
2. If LP is appropriate, how do I formulate an effective LP model?

Please notice the two little words *use* and *do*. In this chapter we emphasize learning by using LP to do exercises. The three examples help you get started; they introduce some vocabulary and show you how to formulate a LP model.

Homework

Effective learning here requires your active participation in setting up and solving models for the problems, not just passively reading my solutions! For best results you should do most of the exercises in Chapters 2 and 5.

Discussion and Answers

To provide feedback, Chapter 2 and some of the other chapters conclude with a discussion and answers for some of the exercises in that chapter; you will also find more solutions for the Chapter 2 and Chapter 5 problems in Appendixes 2 and 3, respectively. However, it is to your advantage not to look at my comments about a problem before you have tried diligently to solve the problem. After you have solved the problem, look at my solution; some ideas are introduced in the solutions because I believe that you learn them better when you see them used in a problem on which you have worked.

Multiple Models and/or Solutions

There are several valid LP formulations for many situations, so your model may look very different from mine, yet both of us may have valid models. In some cases we might get different solutions to a problem even if we use the same model because we might be using different software or entering the model in different orders. Multiple solutions, and how to look for them, are discussed later.

LP Software

The simplex method will be introduced in Chapter 4; variations of it are used in several readily available LP software packages. A good LP package is a powerful tool for analyzing problems, and an understanding of the simplex method can help you utilize the potential of LP software. I strongly recommend that you use an LP package to actually solve the problems; some LP packages, including the one that I used to solve the LP models in this book, cost about as much as a textbook.

Examples 2.1 to 2.3 introduce the basic ingredients of linear programming. During the discussion of these three examples, you will learn the three basic steps in formulating an LP model and you will see how the notation and terminology of linear algebra can be employed to state the model succinctly. After LP models for the three examples are formulated in Sections 2.1 to 2.3, the situations addressed in these examples are examined from a second (dual) perspective in Section 2.4. In each case the original problem is called the *primal* problem and the LP that arises from the second perspective is called the *dual* problem. The three pairs of LP models for these three examples are of two types. These two types of pairs of LP models are called the symmetric primal–dual pair (of LP models) and the standard primal–dual pair. More is said about these two primal–dual pairs in Section 2.5; a result from Section 2.5 is used very incisively in Section 4.5.

2.1 EXAMPLE 2.1: A PRODUCTION PROBLEM
Information Provided

An organization has an abundance of two types of crude oil, called light crude and dark crude. It also has a refinery in which it can process light crude for $25 per barrel and dark crude for $17 per barrel. Processing yields fuel oil, gasoline, and jet fuel as indicated in the following table.

| | | Input | |
		Light crude	Dark crude
	Fuel oil	0.21	0.55
Output	Gasoline	0.5	0.3
	Jet fuel	0.25	0.1

The organization requires 3,000,000 barrels of fuel oil, 7,000,000 barrels of gasoline, and 5,000,000 barrels of jet fuel.

Understanding the Situation

Part of programming is interpreting what people are trying to tell you and feeding your impressions back to those people until you believe you understand what they are trying to tell you. For example, the entry 0.21 in this table seems to imply that each unit of light crude oil that you process produces 0.21 unit of fuel oil. Let's suppose that their objective is to process enought oil to meet the listed demands at the minimum processing cost. Then we wish to determine how much light and dark crude they should process to meet the requirements with minimal processing cost.

Formulating an LP Model

Formulating a linear program to model a situation involves three basic steps:

1. *Specify the decision variables.*
2. *Specify the constraints.*
3. *Specify the objective function.*

Decision Variables

For Example 2.1 we must decide how much light crude oil to process and how much dark crude oil to process. Thus we can specify a set of decision variables for Example 2.1 as follows. Let

L = number of millions of barrels of light crude to process
D = number of millions of barrels of dark crude to process.

Notice that *decision variables represent numbers.* I cannot overemphasize the importance of step 1. Specifying the *decision variables* forces you to *focus on exactly what decisions you need to make.* You are required to interpret the information you are given. This is not a "plug the numbers in a formula" activity; it is creative and fun, not mechanical.

The next two steps are to specify a set of *constraints* that *model the requirements* and to specify an appropriate *objective function to model the cost*, which we wish to minimize. The form and order of these specifications depend on our choice of decision variables and our thought processes at the time when we are making the model. Let's consider the constraints first.

Constraints

I interpreted the table to imply that each million barrels of light crude processed produces 0.21 million barrels of fuel oil, and so on. Thus processing L million barrels of light crude and D million barrels of dark crude produces $0.21L + 0.55D$ barrels of fuel oil. Since 3 million barrels are required, we have the constraint $0.21L + 0.55D = 3$; this constraint is a linear equation in L and D. Similar equations apply for gasoline and jet fuel. Putting these equations together gives us the linear system

$$0.21L + 0.55D = 3 \quad \text{(fuel oil)}$$
$$0.50L + 0.30D = 7 \quad \text{(gasoline)}$$
$$0.25L + 0.10D = 5 \quad \text{(jet fuel).}$$

Necessity of Inequality Constraints

Unfortunately, because processing either type of crude oil produces at least twice as much gasoline as jet fuel, there is no way to produce 5,000,000 barrels of jet fuel without producing at least 10,000,000 barrels of gasoline. Thus there is no feasible solution to this linear system and we are forced to formulate our constraints as a system of linear inequalities:

$$0.21L + 0.55D \geq 3$$
$$0.50L + 0.30D \geq 7$$
$$0.25L + 0.10D \geq 5.$$

Standard Form for Constraints

I would prefer to have the constraints in the form of a system of linear equations because I know how to solve systems of linear equations, so let's put them in that form by introducing *surplus variables* F, G, and J to denote the number of millions of barrels of surplus fuel oil, gasoline, and jet fuel produced. The preceding system of linear inequalities is replaced by the following system of linear equations:

$$0.21L + 0.55D = 3 + F$$
$$0.50L + 0.30D = 7 + G$$
$$0.25L + 0.10D = 5 + J,$$

which can be rewritten in *standard form*:

$$0.21L + 0.55D - F = 3$$
$$0.50L + 0.30D - G = 7$$
$$0.25L + 0.10D - J = 5.$$

Objective Function

The cost of processing L million barrels of light crude and D million barrels of dark crude is $25L + 17D$ millions of dollars. The objective function for this problem is the linear function of L and D given by the formula

$$25L + 17D.$$

We put the pieces together below.

Standard Form LP

$$\text{minimize} \quad 25L + 17D$$
$$\text{subject to} \quad 0.21L + 0.55D - F = 3$$
$$0.50L + 0.30D - G = 7$$
$$0.25L + 0.10D - J = 5.$$

The latter formulation of Example 2.1 is called a *standard-form* LP. A standard-form LP has the form: Minimize a linear function of the decision variables subject to a system of linear equations being satisfied by the decision variables, plus *implicit constraints* requiring that the decision variables be nonnegative.

Implicit Constraints

The following implicit constraints are always present in an LP unless it is explicitly state otherwise:

All decision variables are required to be greater than or equal to 0.

Canonical Form

Some authors use the name *canonical LP*, or *canonical form LP*, to denote an LP that we have called a standard-form LP. Sometimes "standard form" or "canonical form" refers to an LP with equality constraints and a "maximize" objective, so be aware of these variations in terminology when you read about or discuss LP.

Before presenting a vector-matrix formulation of Example 2.1, I am going to make an informal comment about vectors and matrices.

Comment About Vectors and Matrices

A *vector* is simply a list of numbers. This simple concept is extremely useful both conceptually and computationally. Imagine a case register drawer with eight compartments, labeled 1 to 8, for pennies, nickels, dimes, quarters, $1 bills, $5 bills, $10 bills and $20 bills, respectively. Suppose that I count the number of pieces of money in each compartment and write those numbers in a list. If that list is (3,5,2,4,11,4,1,6), how much money is in the drawer? If you trust my counting and your answer is $162.48, you understand how vectors work. We can either write the list as a row of numbers: a row vector, or as a column of numbers: a column vector. Column vectors take more space on a page, so we use "(. . .)" to denote column vectors written in row format; for example, the preceding vector, representing a list of the numbers of the various items in the drawer, really looks like this:

$$\begin{bmatrix} 3 \\ 5 \\ 2 \\ 4 \\ 11 \\ 4 \\ 1 \\ 6 \end{bmatrix}.$$

A *matrix* is a rectangular array of numbers. We can think of a matrix as a row of column vectors or as a column of row vectors when it is convenient to do so.

Example 2.1 is discussed below from a vector-matrix perspective to illustrate how vectors and matrices can be used to write an LP in a very compact form.

Vector-Matrix Formulation of Example 2.1

Continuing with Example 2.1, the unit output vectors

$$\begin{bmatrix} 0.21 \\ 0.5 \\ 0.25 \end{bmatrix} \quad \text{and} \quad \begin{bmatrix} 0.55 \\ 0.3 \\ 0.1 \end{bmatrix}$$

provide us with a list of the number of units of fuel oil, gasoline, and jet fuel produced by processing 1 unit of light and dark crude oil. Suppose that we let

$$A = \begin{bmatrix} 0.21 & 0.55 \\ 0.5 & 0.3 \\ 0.25 & 0.1 \end{bmatrix}$$

denote the 3×2 matrix whose columns are the unit output vectors of light and dark crude oil, and we let

$$x = \begin{bmatrix} L \\ D \end{bmatrix}$$

denote the column vector with components L and D. Then

$$L \begin{bmatrix} 0.21 \\ 0.5 \\ 0.25 \end{bmatrix} + D \begin{bmatrix} 0.55 \\ 0.3 \\ 0.1 \end{bmatrix} = \begin{bmatrix} 0.21 & 0.55 \\ 0.5 & 0.3 \\ 0.25 & 0.1 \end{bmatrix} x = Ax$$

lists the numerical output (in millions of barrels) of input x. The requirements that at least 3, 7, and 5 million barrels of fuel oil, gasoline, and jet fuel be produced can be expressed algebraically in vector format as follows:

$$Ax \geq b, \qquad x \geq 0,$$

where

$$b = \begin{bmatrix} 3 \\ 7 \\ 5 \end{bmatrix}$$

lists the production requirements. To formulate the objective function, put $c = [25,17]$ and get the numerical cost

$$cx = [25,17] \begin{bmatrix} L \\ D \end{bmatrix} = 25L + 17D,$$

in millions of dollars, of processing x; cx is the number of millions of dollars required to process x. Now we have a linear program (LP) that models the situation:

$$\min \quad cx$$
$$\text{s.t.} \quad Ax \geq b, \qquad x \geq 0.$$

Recall that $(x_1, \ldots, x_n)$ denotes an n-dimensional column vector and $[x_1, \ldots, x_n]$ denotes an n-dimensional row vector.

Standard Form

To rewrite this problem using equality constraints, let F, G, and J denote the millions of barrels of surplus fuel oil, gasoline, and jet fuel produced and rewrite the problem as follows:

$$\min \quad [25,17,0,0,0] \begin{bmatrix} L \\ D \\ F \\ G \\ J \end{bmatrix} = \bar{c}\bar{x}$$

$$\text{s.t.} \quad \begin{bmatrix} 0.21 & 0.55 & -1 & 0 & 0 \\ 0.5 & 0.3 & 0 & -1 & 0 \\ 0.25 & 0.1 & 0 & 0 & -1 \end{bmatrix} \bar{x} = \bar{A}\bar{x} = b, \quad \bar{x} \geq 0,$$

where $\bar{x} = (L,D,F,G,J)$, $\bar{c} = [25,17,0,0,0]$, and $\bar{A} = [A \,|\, -I]$; F, G, and J are called *surplus variables*. They tell us how much surplus fuel oil, gasoline, and jet fuel will be produced and they permit us to replace "≥" constraints by "=" constraints. The latter formulation of Example 2.1,

$$\min \quad \bar{c}\bar{x}$$

$$\text{s.t.} \quad \bar{A}\bar{x} = b, \quad x \geq 0,$$

is called a *standard-form* LP.

Graphical Solution of Two-Variable Problems

When a small problem has only two decision variables, it can often be solved graphically. A graphical solution to the original LP model for Example 2.1 begins by graphing the jet fuel constraint (Fig. 1). The shaded region represents the set of points $x = (L,D)$ which satisfy the inequalities $0.25L + 0.10D \geq 5$, $L \geq 0$, and $D \geq 0$. This region has two corner points, (20,0) and (0,50). The value of the objective function, $25L + 17D$, at these corner points is $25(20) + 17(0) = 500$ and $25(0) + 17(50) = 850$, respectively. The *feasible region*, also called the *feasible set*, for an LP is the set of points that satisfy all the constraints simultaneously. If you draw in the other two constraints and look at the corresponding regions for them, you will see that the set which has been described above is the feasible region for the original LP model for Exercise 2.1. In Chapter 3 you will see that solutions always occur at corner points. Thus the solution we seek can be found simply by comparing the values, 500 and 850, of the objective function at the two corner points (0,50) and (20,0). Thus 20 million barrels of light crude and no dark crude should be processed, at a cost of $500 million, in order to meet the demands at a minimum processing cost of $500 million.

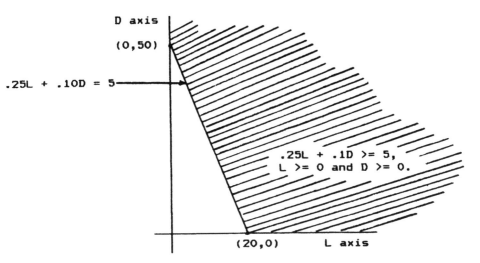

Figure 1   Graph of the jet fuel constraint for Example 2.1.

The shape of the feasible set for an LP will be examined in Chapter 3. When only two decision variables are present, the feasible set can be visualized as a convex (a term defined in Chapter 3) set in the plane (two dimensions). When there are three decision variables, the feasible set can still be visualized in three dimensions; but when an LP model has four or more decision variables, most people cannot "see" the feasible set in higher dimensions. Fortunately, linear algebra is available as a tool and corner points can be described algebraically; according to Exercise 3.9, corner points are the basic feasible solutions defined in Section 3.2 and used in Chapter 4.

Summation notation, defined below, is used in Examples 2.2 and 2.3, which follow. Example 2.2 has n (a generic name for an arbitrary but fixed positive integer) decision variables; $n = 7$ in the illustration in Section 2.2.

Summation Notation

Given n quantities $Q_1, Q_2, \ldots, Q_i, \ldots, Q_n$ that can be added, the sum

$Q_1 + Q_2 + \cdots + Q_i + \cdots + Q_n$ is denoted by $\displaystyle\sum_{i=1}^{n} Q_i = \sum_{j=1}^{n} Q_j = \cdots$:

The index i is a symbol that represents a generic integer between 1 and n. Instead of i, we can use j or any other symbol except n or Q; n

represents the number of terms to be added and Q has also been assigned a meaning in the expression. Example 2.2 is a classical example; it is usually modeled by an LP with the same form as Example 2.1.

2.2 EXAMPLE 2.2: A DIET PROBLEM

A large institution wishes to formulate a diet to meet a given set of nutritional requirements at minimal cost. Suppose that n foods are available and m nutritional components of the foods are to be considered. Label the foods $f_1, \ldots, f_n$ and label the nutrients $N_1, \ldots, N_m$. The costs and nutritional characteristics of the foods, and the nutritional requirements of the diet, are provided in the following format. A unit of f_j costs c_j money units and contains a_{ij} units of N_i; at least b_i units of N_i are required for the diet. An LP model for the diet problem is developed below.

Recall that there are three basic steps in formulating an LP model:

1. *Specify the decision variables.*
2. *Specify the constraints.*
3. *Specify the objective function.*

In this problem we wish to specify decision variables to answer the question: How many units of each food should we put into the diet? Thus we let

$$x_j = \text{number of units of } f_j \text{ to put in the diet}, \quad 1 \le j \le n,$$

and step 1 is completed; $x = (x_1, \ldots, x_n)$ is a list of how many units of each food to put in the diet.

In step 2 we must specify constraints so that the diet meets the given nutritional requirements. To this end we observe that the m-dimensional column vector

$$a_j = \begin{bmatrix} a_{1j} \\ a_{2j} \\ \vdots \\ a_{ij} \\ \vdots \\ a_{mj} \end{bmatrix}$$

provides a list of the numbers of units of the relevant nutrients in 1 unit of f_j. The $m \times n$ matrix

$$A = \begin{bmatrix} a_{11} & a_{12} & \cdots & a_{1j} & \cdots & a_{1n} \\ a_{21} & a_{22} & \cdots & a_{2j} & \cdots & a_{2n} \\ \vdots & & & \vdots & & \vdots \\ a_{i1} & a_{i2} & \cdots & a_{ij} & \cdots & a_{in} \\ \vdots & & & \vdots & & \vdots \\ a_{m1} & a_{m2} & \cdots & a_{mj} & \cdots & a_{mn} \end{bmatrix} \overset{\text{def}}{=} [a_{ij}]$$

provides a nutritional table for the problem.

To specify constraints so that the diet meets the given nutritional requirements, we observe that using x_j units of food f_j, (for) $1 \le j \le n$, will provide

$a_{i1}x_1 + a_{i2}x_2 + \cdots + a_{ij}x_j + \cdots + a_{in}x_n$ units of nutrient N_i, $1 \le i \le m$.

Consequently, the constraints are modeled by requiring that

$$Ax \ge b, \quad \text{where } x = \begin{bmatrix} x_1 \\ x_2 \\ \vdots \\ x_n \end{bmatrix} \quad \text{and} \quad b = \begin{bmatrix} b_1 \\ b_2 \\ \vdots \\ b_m \end{bmatrix},$$

because Ax is a list of the nutritional output of the diet x and b is a list of the nutritional requirements to be met. Another useful interpretation of the constraints follows: The m-dimensional column vector a_j is a (relevant) nutritional list of 1 unit of food f_j, the m-dimensional column vector $x_j a_j$ is a nutritional list of x_j units of f_j, and the m-dimensional column vector

$$\sum_{j=1}^{n} x_j a_j = x_1 a_1 + \cdots + x_n a_n = Ax$$

is, again, a nutritional list of the diet x.

It remains to complete step 3 by specifying the objective function to represent the cost of the diet, which we wish to minimize. But the cost is simply $cx = [c_1, c_2, \ldots, c_n]x = c_1 x_1 + \cdots + c_n x_n$. So we have constructed an LP model for this diet problem:

$$\min \quad cx$$
$$\text{s.t.} \quad Ax \ge b, \quad x \ge 0.$$

Let's look at an example.

Illustration. Suppose that seven foods—green beans, soybeans, cottage cheese, Twinkies, vegetable juice, spaghetti, and veggie supreme cheese pizza—are available. Suppose that units of these foods cost 0.35, 0.2, 0.28, 0.2, 0.15, 0.99, and 2.49 dollars, respectively. Suppose that four properties of each food—calories, protein, carbohydrates, and vitamins—are required to be considered, and that the following properties matrix A and requirements vector b are provided:

$$A = \begin{bmatrix} 1 & 3 & 3 & 6 & 1 & 14 & 16 \\ 0 & 1 & 4 & 0 & 0 & 8 & 8 \\ 1 & 2 & 3 & 6 & 0 & 14 & 14 \\ 6 & 2 & 2 & 0 & 3 & 7 & 13 \end{bmatrix}, \quad b = \begin{bmatrix} 16 \\ 8 \\ 14 \\ 13 \end{bmatrix}.$$

The four numbers listed in a column of the matrix represent the numbers of units of the four properties in 1 unit of the food corresponding to that column; for instance, the third column corresponds to cottage cheese, so each unit of cottage cheese contains 3 units of calories, 4 units of protein, 3 units of carbohydrates, and 2 units of vitamins.

A unit of pizza has been designed to meet the dietary requirements exactly. A meal composed of 1 unit of green beans, 2 units of soybeans, $\frac{3}{2}$ units of cottage cheese, and $\frac{3}{4}$ unit of Twinkie also fits the minimal requirements of the diet exactly; this meal costs \$1.32, while pizza costs \$2.49. Notice that a meal composed of 1 unit of spaghetti and 2 units of vegetable juice will also meet the requirements exactly, at a cost of \$1.29. Which of these three meals would you choose?

The transportation problem is another classical example; a standard-form LP model for it is developed in Example 2.3, and it is discussed further in Sections 5.6, 5.7, and 9.9.

2.3 EXAMPLE 2.3: A TRANSPORTATION PROBLEM

The goal of a transportation problem is to ship quantities of a material from a set of supply points to a set of demand points at minimal cost. A model for a transportation problem consists of supplying five lists—a list of *supply points*, a list of *demand points*, a list of the *quantities* of the material *available at* the *supply points*, a list of the *quantities* of the material *required at* the *demand points*, and a list of the *costs* of transporting a unit of material from a supply point to a demand point—and an a priori constraint—that the *total supply is required to equal the total demand*.

When we have the five lists with the a priori constraint satisfied, Example 2.3 tells us how to put the model into a standard primal LP format. Consequently, writing a tableau that contains the lists and satisfies the constraint provides a model for a transportation problem.

Transportation problems have a special structure that can be used to design algorithms for their solution; these algorithms generally solve transportation problems faster than algorithms that are used in LP software. So if you need to solve many and/or large transportation problems, you might wish to consider special software, which will be called a TRANS package, designed specifically for transportation problems. The algorithm presented in Section 9.10 will also solve transportation problems, but it does not take some of their special structure into account. Just as some LP software packages take LP models without requiring that they be in standard form, some TRANS packages do not require that the a priori constraint be satisfied. However, for purposes of discussion, it is required to be satisfied by a transportation model; this requirements is relaxed in Section 5.6, where variations of the transportation problem are discussed. Section 5.6 focuses on a tableau formulation of the five lists that compose a transportation problem. Example 2.3 is used to introduce a standard-form LP model for a transportation problem.

Example 2.3. A well-known fast-food chain has 2300 franchise restaurants and four potato farms. Label the restaurants R_j, $1 \le j \le 2300$, and the farms F_i, $1 \le i \le 4$. Suppose that F_i can supply s_i tons of potatoes per year and R_j can use d_j tons of potatoes per year, where $\Sigma_{i=1}^4 s_i = \Sigma_{j=1}^{2300} d_j$. Suppose that it costs c_{ij} dollars to ship a ton of potatoes from F_i to R_j. We can set up an LP to model this problem as follows. To specify the decision variables, let x_{ij} denote the number of tons of potatoes to be shipped from F_i to R_j. Then the 2304 constraints are given by

$$\sum_{j=1}^{2300} x_{ij} = s_i, \quad i = 1,2,3,4, \quad \text{and}$$

$$\sum_{i=1}^{4} x_{ij} = d_j, \quad j = 1,2,\ldots,2300.$$

The first four constraints say that the available supply, s_i, is shipped from F_i, $i = 1,2,3,4$. The remaining 2300 constraints say that the desired number, d_j, of tons of potatoes is shipped to R_j, $j \le 2300$. The objective function to be minimized is $\Sigma_{i=1}^4 \Sigma_{j=1}^{2300} c_{ij} x_{ij}$. Put

$$
x = \begin{bmatrix} x_{1,1} \\ x_{1,2} \\ x_{1,3} \\ \vdots \\ x_{1,2300} \\ x_{2,1} \\ \vdots \\ x_{2,2300} \\ x_{3,1} \\ \vdots \\ x_{4,2300} \end{bmatrix}, \quad b = \begin{bmatrix} s_1 \\ s_2 \\ s_3 \\ s_4 \\ d_1 \\ d_2 \\ \vdots \\ d_{2300} \end{bmatrix}, \quad c = [c_{11}, c_{12}, \ldots, c_{1,2300}, c_{2,1}, \ldots, c_{4,2300}],
$$

and

$$
A = \begin{bmatrix}
\overbrace{11 \cdots 1}^{2300} & \overbrace{00 \cdots 0}^{2300} & \overbrace{00 \cdots 0}^{2300} & \overbrace{00 \cdots 0}^{2300} \\
00 \cdots 0 & 11 \cdots 1 & 00 \cdots 0 & 00 \cdots 0 \\
00 \cdots 0 & 00 \cdots 0 & 11 \cdots 1 & 00 \cdots 0 \\
00 \cdots 0 & 00 \cdots 0 & 00 \cdots 0 & 11 \cdots 1 \\
10 \cdots 0 & 10 \cdots 0 & 10 \cdots 0 & 10 \cdots 0 \\
01 \cdots 0 & 01 \cdots 0 & 01 \cdots 0 & 01 \cdots 0 \\
\vdots & \vdots & \vdots & \vdots \\
0 \cdots 01 & 0 \cdots 01 & 0 \cdots 01 & 0 \cdots 01
\end{bmatrix}
\begin{matrix} \left.\vphantom{\begin{matrix}1\\1\\1\\1\end{matrix}}\right\}4 \\ \\ \left.\vphantom{\begin{matrix}1\\1\\1\\1\end{matrix}}\right\}2300 \end{matrix}.
$$

Thus we have the standard-form LP

$$
\begin{array}{ll}
\min & cx \\
\text{s.t.} & Ax = b, \quad x \geq 0,
\end{array}
$$

to model this transportation problem.
Please *do* the following exercise.

Exercise

1. Write the objective function and the seven constraint equations for the numerical example involving blueberries that is discussed in Section 5.6.

2.4 DUALITY

An LP has a specific algebraic form. There is a corresponding algebraic form called the *dual* LP. The original problem is called the *primal* problem, and the primal together with the dual is called a *primal–dual pair*;

the primal is the first LP in the pair and the dual LP is the second LP in the pair. When an LP is a model of a "real world" situation, very often there is a different (dual) perspective of the situation that is modeled by the dual LP. Knowing the existence and form of the dual LP provides a vantage point from which to look for a dual interpretation of the situation; examples are provided below by exhibiting dual linear programs for Examples 2.1 to 2.3.

Dual LP for Example 2.1

A related (dual) problem for the original (primal) problem will be introduced by considering Example 2.1 from a different perspective. Suppose that a group of neighbors has surpluses of fuel oil, gasoline, and jet fuel which they want to sell for as many dollars as possible. They wish to decide on prices at which to offer to sell these commodities to the organization. From this perspective, the decision variables can be written as follows. Let PF, PG, and PJ equal the number of million dollars to charge (price) for 1 million barrels of fuel oil, gasoline, and jet fuel. The group's objective is to maximize the income from selling 3, 7, and 5 million barrels of fuel oil, gasoline, and jet fuel to the organization:

$$\max 3PF + 7PG + 5PJ = [PF,PG,PJ] \begin{bmatrix} 3 \\ 7 \\ 5 \end{bmatrix} = Pb,$$

where $P = [PF,PG,PJ]$. Constraints are imposed by the necessity to be competitive with the organization's cost of processing. For example, $25 million will process 1 million barrels of light crude and produce 0.21 million barrels of fuel oil, 0.5 million barrels of gasoline, and 0.25 million barrels of jet fuel. Thus the prices must be set so that the cost of buying the production of a unit of light crude oil is no greater than the cost of processing a unit of light crude oil:

$$0.21PF + 0.5PG + 0.25PJ = [PF,PG,PJ] \begin{bmatrix} 0.21 \\ 0.5 \\ 0.25 \end{bmatrix} \leq 25.$$

Similarly, considering dark crude, the prices also need to satisfy the inequality

$$[PF,PG,PJ] \begin{bmatrix} 0.55 \\ 0.3 \\ 0.1 \end{bmatrix} \leq 17.$$

These constraints can be summarized by the inequality $PA \leq c$. We have formulated a related or dual LP:

$$\max \quad Pb$$
$$\text{s.t.} \quad PA \leq c, \quad P = [PF, PG, PJ] \geq 0.$$

We reformulate the dual constraints as a system of linear equalities by introducing two nonnegative *slack variables* SL and SD and two equality constraints,

$$0.21PF + 0.50PG + 0.25PJ + SL \quad = 25$$

and

$$0.55PF + 0.30PG + 0.10PJ \quad + SD = 17,$$

which can be written in vector form as follows:

$$[PF,PG,PJ,SL,SD] \begin{bmatrix} 0.21 & 0.55 \\ 0.5 & 0.3 \\ 0.25 & 0.1 \\ 1 & 0 \\ 0 & 1 \end{bmatrix} = [25,17].$$

Slack variables change "≤" constraints to "=" constraints.

Dual LP for Example 2.2

A large corporation has lots of each of the nutrients in a warehouse. The corporation marketing group decides to try to get the institution to buy nutrients and feed nutrients instead of food. They want to price nutrients so that they maximize income and stay competitive; the price for the nutritional equivalent of a unit of each food must cost no more than a unit of that food.

To specify a set of decision variables for the dual problem, let $\lambda = [\lambda_1, \ldots, \lambda_i, \ldots, \lambda_m]$, where $\lambda_i =$ number of money units to charge for each unit of nutrient N_i. The objective is to maximize $\lambda b = \lambda_1 b_1 + \cdots + \lambda_m b_m$. The constraints are

$$[\lambda_1, \ldots, \lambda_i, \ldots, \lambda_m] \begin{bmatrix} a_{1j} \\ \vdots \\ a_{ij} \\ \vdots \\ a_{mj} \end{bmatrix} = \lambda_1 a_{1j} + \cdots + \lambda_i a_{ij} + \cdots + \lambda_m a_{mj} \leq c_j,$$

$$\text{(for)} \quad j = 1, 2, \ldots, n,$$

or, more compactly,

$$\lambda A \leq c, \quad \lambda \geq 0.$$

This dual LP model takes the form

$$\max \quad \lambda b$$
$$\text{s.t.} \quad \lambda A \leq c, \quad \lambda \geq 0.$$

Notice that our choice $\lambda = [\lambda_1, \ldots, \lambda_m]$ for the names of the dual decision variables was arbitrary. We could have used $y = [y_1, \ldots, y_m]$ instead of λ. Then the dual model would be stated

$$\max \quad yb$$
$$\text{s.t.} \quad yA \leq c, \quad y \geq 0.$$

Dual LP for Example 2.3

Suppose that an agribusiness wholesaler decides to approach the fast-food chain with a proposition. He will offer to buy the potatoes from the farms and sell potatoes to the restaurants. The wholesaler's decisions are which prices to offer in order to be competitive and maximize income. So let n_i denote the buying price he will offer to pay per ton to buy at F_i, and let p_j denote the selling price he will offer per ton at R_j. He wishes to

$$\max \quad \sum_{j=1}^{2300} d_j p_j - \sum_{i=1}^{4} s_i n_i = [-n_1, -n_2, -n_3, -n_4, p_1, \ldots, p_{2300}]b$$

$$\text{s.t.} \quad [-n_1, -n_2, -n_3, -n_4, p_1, p_2, \ldots, p_{2300}]A \leq c.$$

Put $y = [-n_1, -n_2, -n_3, -n_4, p_1, \ldots, p_{2300}]$. Then we have an LP model for the wholesaler:

$$\max \quad yb$$
$$\text{s.t.} \quad yA \leq c.$$

The constraints $yA \leq c$ can be rewritten

$$-n_i + p_j \leq c_{ij}, \qquad i \leq 4, \quad j \leq 2300.$$

These 9200 constraints say that the wholesaler will provide the "shipping" service from F_i to R_j for $\leq c_{ij}$ dollars per ton. Notice that we do not have $y \geq 0$ in this dual LP because buying at F_i generates a negative rate $-n_i$ of income per ton for the wholesaler.

Summary

LP models for three classical problems were developed in Sections 2.1 to 2.3. In this section a dual problem has been introduced for each of these problems, and LPs have been formulated to model the duals. Each of the three primal–dual pairs has one of the two forms listed below.

Primal	Pairs Dual	
min cx s.t. $Ax \geq b$, $x \geq 0$	max yb s.t. $yA \leq c$, $y \geq 0$	Symmetric pair
min cx s.t. $Ax = b$, $x \geq 0$	max yb s.t. $yA \leq c$	Standard pair

The specific format in which we choose to write an LP model is arbitrary. Because we have considered systems $Ax = b$ of linear equations in elementary linear algebra, we arbitrarily call the second pair in our list the standard primal–dual pair. The primal in this pair is called the *standard-form* LP. The primal in the first pair is called a symmetric primal. The dual in the first pair is called a symmetric dual, or *the* dual of the symmetric primal. Similarly, the LP

$$\max \quad yb$$
$$\text{s.t.} \quad yA \leq c$$

is arbitrarily called *the* dual of the standard-form LP

$$\min \quad cx$$
$$\text{s.t.} \quad Ax = b, \quad x \geq 0.$$

By introducing slack variables, surplus variables, and other appropriate modifications, any LP can be put in standard form. For instance, Example

2.1 showed how to write a symmetric primal LP in standard form by introducing surplus variables. A symmetric dual can be written in standard form by introducing slack variables and minimizing the negative of the objective, which is equivalent to maximizing the objective.
The LP

$$\text{max} \quad cx \tag{1}$$
$$\text{s.t.} \quad x \text{ is in } K$$

and the LP

$$\text{min} \quad -cx \tag{2}$$
$$\text{s.t.} \quad x \text{ is in } K$$

have the same solution sets. The LPs (1) and (2) have the same feasible set and optimal points, but the value of the objective function in (1) at an optimal point is the negative of the value in (2). Multiplying a constraint by -1 interchanges "$\geq$" constraints and "$\leq$" constraints. For example, the LP

$$\text{min} \quad cx$$
$$\text{s.t.} \quad Ax \leq b, \quad x \geq 0$$

can be written in symmetric primal form

$$\text{min} \quad cx$$
$$\text{s.t.} \quad \{-A\}x \geq \{-b\}, \quad x \geq 0$$

or symmetric dual form

$$\begin{Bmatrix} \text{max} & \{-c\}x \\ \text{s.t.} & Ax \leq b, \quad x \geq 0 \end{Bmatrix} \Leftrightarrow \begin{Bmatrix} \text{max} & x^t\{-c\}^t \\ \text{s.t.} & x^t A^t \leq b^t, \quad x^t \geq 0 \end{Bmatrix}.$$

Notice also that given a in R_n, x in R^n, and a number b_i,

$$ax = b_i \Leftrightarrow (-a)x = -b_i$$

and

$$ax \geq b_i \Leftrightarrow (-a)x \leq -b_i.$$

Thus we can always write *equality constraints* with the right side $b_i \geq 0$. This topic is discussed further in Appendix 1.

The next section contains two fundamental facts about standard and symmetric primal–dual pairs. These facts will be very useful.

2.5 TWO FUNDAMENTAL FACTS ABOUT STANDARD AND SYMMETRIC PRIMAL–DUAL PAIRS

1. If x is feasible for the primal and y is feasible for the dual, then

$$yb \leq y(Ax) = (yA)x \leq cx.$$

2. If x is feasible for the primal and y is feasible for the dual and $yb = cx$, then x and y are optimal solutions to the primal and dual (i.e., both are optimal solutions).

A point is *feasible* for an LP if it satisfies the constraints for that LP. For example, x is feasible for the symmetric primal LP

$$\min \quad cx$$
$$\text{s.t.} \quad Ax \geq b, \quad x \geq 0$$

if $x \geq 0$ and $Ax \geq b$. Fundamental fact 1 says that the maximum of the objective values of feasible points for the dual is $\leq$ the minimum of objective values of feasible points for the primal.

EXERCISES

The exercises in Chapter 2 give you opportunities to practice formulating LP models. Sometimes several valid LP models exist for a problem, so your solutions might look quite different from my solutions. These exercises are designed to introduce you to a variety of types of problems that can be modeled using LP; I recommend that you *do* many of them. *Do* some now, then come back and *do* some more as you read the next two chapters. A warning precedes the exercises.

Beware of Reading Assumptions into Problems: A Case History

Louie and Marie supervised a French bakery. Two of the products produced by the bakery were bread and cake. The bakery supplied nutritional needs for the community in which they lived. Cakes produced more profit, and market research indicated that they could be backlogged without serious dissatisfaction. When a shortage of supplies occurred, Marie assumed that bread could similarly be backlogged. Unfortunately, her assumption was not valid and, in fact, it led to a nonlinear situation with serious consequences: To respond, "Let them eat cake!" to the statement "The people have no bread" can lead to a loss of head.

2.1 In Example 2.1, show that any $x = (L,D)$ which produces 5 million barrels of jet fuel will produce at least 4.2 million barrels of fuel oil and will cost at least $500 million to process.

2.2 Suppose that a pipeline could be built to make a medium crude available for processing in Example 2.1. This medium crude can be processed for $22 per barrel and it had

$$\begin{bmatrix} 0.25 \\ 0.4 \\ 0.3 \end{bmatrix}$$

as unit output vector. How much could we reduce processing costs by building the pipeline?

2.3 A plastic manufacturer receives an order for four types of parts: $10,000P_1$, $8000P_2$, $6000P_3$, and $6000P_4$. These parts can be made on any one of three molding machines, which we label A, B, and C. There are up to 50 hours of machine time available on each machine to process the order. Machines A, B, and C have hourly operating costs of $30, $50, and $80, and they can produce parts according to the following table.

Part	Machine A	B	C
P_1	300	600	800
P_2	250	400	700
P_3	200	350	600
P_4	100	200	300

Formulate an appropriate LP to model this situation.

2.4 An agency has control of 1000 units of land, 2000 units of water, 20,000 units of money, and 8000 units of labor. The agency can use its resources to grow wheat, corn, and soybeans. Relevant information is tabulated below.

	Corn	Wheat	Soybeans
Metric tons produced per land unit	10	4	6
Units of water required per land unit	3	1	2
Units of money required per land unit	15	10	12
Units of labor required per land unit	8	6	10
Profit per metric ton	10	6	8

(a) Formulate an LP to model this situation.

(b) State a reasonable dual problem and formulate an appropriate LP to model it.

2.5 The Volkswagen Company produces three products: the bug, the superbug, and the van. The profit from each bug, superbug, and van is $1000, $1500, and $2000, respectively. It takes 15, 18, and 20 labor-hours to produce an engine; 15, 19, and 30 labor-hours to produce a body; and 10, 20, and 25 minutes to assemble a bug, superbug, and van. The engine works has 10,000 labor-hours available each week; the body works has 15,000 labor-hours available each week, and the assembly line has 168 hours available each week.

(a) Formulate an appropriate LP.

(b) Demographics change with time and BMW Corp. decides to increase production. BMW decides to approach Volkswagen about leasing engine works, body works, and assembly time. Formulate an LP to determine a reasonable offer.

2.6 The Douglas Company produces three products: lumber, plywood, and sawdust from two types of wood: fir and pine. A truckload of fir will produce 1000 board feet of lumber or 140 sheets of plywood and 15 bags of sawdust. A truckload of pine will produce 800 board feet of lumber or 160 sheets of plywood and 25 bags of sawdust. Fir lumber sells for 50 cents a board foot; pine lumber sells for 80 cents a board foot. Plywood sells for $5 a sheet. At least 1000 sheets of plywood must be produced and at least 5000 board feet of lumber must be produced. Ten truckloads of each of fir and pine are available.

(a) Formulate an appropriate LP.

(b) Winter comes and the company has 32 truckloads of pine logs and 72 truckloads of fir logs in stock. Douglas Co. can manufacture truckloads of lumber or cabin kits. A truckload of lumber requires one truckload of pine logs and four truckloads of fir logs; lumber sells for $4500 per truckload. Cabin kits require two truckloads of pine logs and four truckloads of fir logs; they sell for $6000. Five truckloads of lumber and three cabin kits have been ordered. Assume that all production can be sold. Formulate an appropriate LP.

2.7 The shipping department of an appliance manufacturer has 8 hours of preparation time and 8 hours of crating time available each day. It is processing dishwashers, refrigerators, and stoves. Dishwashers take 0.75 minute to prepare and 1.2 minutes to crate, refrigerators take 1.4 minutes to prepare and 1 minute to crate, and stoves take 1.1 minutes to prepare and 1.3 minutes to crate.

(a) Formulate an LP to maximize the number of units processed per day by the department.

(b) Formulate an LP that incorporates the additional requirement that the department process at least 45 dishwashers, 60 refrigerators, and 32 stoves per day.

2.8 A farmer requires that each of his cows receives between 16,000 and 18,000 calories, at least 2 kilograms of protein, and at least 3 grams of vitamins per day. Three kinds of feed are available. The following table lists their relevant characteristics per kilogram.

Feed	Cost	Calories	Protein (kg)	Vitamins (grams)
1	$0.8	3600	0.25	0.7
2	0.6	2000	0.35	0.4
3	0.2	1600	0.15	0.25

(a) Formulate an appropriate LP.

(b) Formulate an LP which includes the additional requirements that the mix of feeds contains at least 20% (by weight) of feed 1 and at most 50% (by weight) of feed 3.

2.9 A paper recycling machine can produce toilet paper, writing pads, and paper towels, which sell for 18, 29, and 25 cents and consume 0.5, 0.22, and 0.85 kilogram of newspaper and 0.2, 0.4, and 0.22 minute. Each day 10 hours and 1500 kilograms of newspaper are available, and at least 1000 rolls of toilet paper, 200 writing pads, and 400 rolls of paper towels are required. Formulate an appropriate LP.

2.10 The Delux Nut Company produces a deluxe mix composed of almonds, cashews, peanuts, and walnuts. The deluxe mix must contain at least 10% of each kind of nut; at least half of the mix must be composed of almonds and cashews; and it can contain at most 20% peanuts. The company also produces a companion mix which must contain at least 10% cashews, at least 30% walnuts, and between 20 and 40% raisins. Raisins are available in unlimited supply at a cost of $1.80 per kilogram. The quantities of nuts available and their costs in dollars per kilogram follow:

Nut	Price	Kilograms available
Almonds	$6.00	400
Cashews	5.30	200
Peanuts	1.40	600
Walnuts	4.20	300

A 100-gram package of deluxe mix sells for 1 dollar; a 100-gram package of companion mix sells for 59 cents. The company receives an order for 4000 packages of deluxe mix.

(a) Formulate an appropriate LP.

(b) After agreeing to fill the first order, the company receives an order for up to 5000 packages of companion mix. Formulate an LP to decide how to fill both orders. Compare your solutions to parts (a) and (b).

(c) Suppose that the company learns that the price of almonds has increased to $40 per kilogram. What effect does this new information have on their decisions?

2.11 An oil company has three storage tanks. Each tank is full of a type of oil, which can be sold for $15, $17, and $20 per barrel. The tanks hold 4000, 3000, and 5000 barrels. The performance numbers of the oils are 6.8, 7.4, and 8.1. The company has decided to empty the tanks by filling the following four orders and selling the rest of the oil straight out of the tanks. Assume that an oil produced by blending certain amounts of several ingredient oils will have a performance number that is the weighted average of the performance numbers of the component oils. This assumption is one of the critical assumptions of mixing problems; it permits us to write each performance number requirement as a linear inequality. It has been found to be an acceptable approximation for many of the blending activities of a refinery. Formulate an appropriate LP.

Order	Barrels required	Performance required
1	2000	At least 7
2	1500	At most 7.8
3	2500	Between 7.2 and 7.6
4	3000	7.4

2.12 A small steel company produces pipe in four sizes. It can produce pipe by either of two processes, each of which can run up to 40 hours per week. It receives an order. Relevant information is tabulated below.

Pipe size	Meters required	Process A		Process B	
		Meters/hour	Cost/hour	Meters/hour	Cost/hour
1	7000	25	$16	42	$25
2	6500	20	14	37	26
3	7500	15	10	34	27
4	8500	10	8	31	28

(a) How many whole weeks are needed to process the order?

(b) What will it cost to produce the order in the minimum number of weeks?

(c) Suppose that each process could be run a sixth 8-hour day each week at a cost per hour $4 higher than those listed in the table. How much would production costs increase if the company were to use some overtime to complete the order in 12 weeks?

2.13 Consider the following nutritional table.

Food	Cost per unit	Units of nutrient per unit of food				
		Protein	Carbohydrate	Fat	Vitamins	Calcium
1	$0.15	0	7	1	1	0
2	0.23	1	0	3	1	4
3	0.79	5	0	4	0	1
4	0.47	2	2	1	3	0
5	0.52	0	3	0	2	1

Formulate an LP to provide a diet containing at least 3 units of protein, at least 10 units of carbohydrate, at most 2 units of fat, at least 3 units of vitamins, and at least 2 units of calcium.

2.14 Fruit can be dried according to the following table in a dryer that can hold 1 cubic meter of fruit.

Hours	Relative volume		
	Grapes	Apricots	Plums
1	0.30	0.46	0.53
2	0.23	0.44	0.51
3	0.16	0.34	0.47
4	0.13	0.31	0.42

Formulate an LP to approximate the minimum time in which 20 meters of grapes, 10 meters of apricots, and 5 meters of plums can be dried to a volume of 10 meters.

2.15 The Brite-Lite Company receives an order for 78 floor lamps, 198 dresser lamps, and 214 table lamps from Condoski Corp. Brite-Lite ships orders in two types of containers. The first cost $15 and can hold 2 floor lamps and 4 table lamps or 2 floor lamps and 2 table lamps and 4 dresser lamps. The second type costs $25 and holds 3 floor lamps and 8 table lamps or 8 table lamps and 12 dresser lamps.

Formulate an LP to minimize the cost of the containers to hold the order.

2.16 The Model-Kit Company makes two types of kits. They have 215 engine assemblies, 525 axle assemblies, 440 balsa blocks, and 560 color packets in stock. Their earthmover kit contains 2 engine assemblies, 3 axle assemblies, 4 balsa blocks, and 2 color packets; its profit is $15. Their racing kit contains 1 engine assembly, 2 axle assemblies, 2 balsa blocks, and 3 color kits; its profit is $9.50.

(a) Formulate an LP to maximize Model-Kit's profit potential from its stock on hand.

(b) Suppose that sales have been slow and Sears offers to buy all that Model-Kit can supply if Model-Kit will sell all earthmover kits over 60 at a $5 discount and all racing kits over 50 at a $3 discount.

2.17 Let t be a (positive or negative) number. Show that $|t|$ is the value of the objective function at an optimal solution of the LP

$$\min \quad u + v$$
$$\text{s.t.} \quad t = u - v$$
$$u \geq 0$$
$$v \geq 0.$$

2.18 Formulate each of the following problems as a symmetric primal LP.

(a)

$$\min \quad 12q + |5r|$$
$$\text{s.t.} \quad q + 2r \geq 4$$
$$5q + 6r \leq 7$$
$$8q - 9r = 11.$$

(*Note*: Neither q nor r is required to be positive.)

(b)

$$\min \quad |a - 1| + 3|b|$$
$$\text{s.t.} \quad a + 2b \geq 2$$
$$a + b = 2.$$

(*Note*: Neither a nor b is required to be positive.)

(c) Repeat part (b) with $a + b = 2$ replaced by $|a| + |b| \leq 2$.

2.19 Referring to Exercise 2.7(a), sales projections indicate that the processing ratios $D/45 = R/60 = S/32$ are optimal. Formulate an LP that maximizes the objective

$$D + R + S - 20\left|\frac{D}{45} - \frac{R}{60}\right| - 20\left|\frac{D}{45} - \frac{S}{32}\right| - 20\left|\frac{R}{60} - \frac{S}{32}\right|.$$

2.20 Referring again to Exercise 2.7(a), formulate an LP that maximizes the objective

$$f(D,R,S) = \begin{cases} D + R + S & \text{if } D \le 50 \\ D + R + S - 0.5(D - 50) & \text{if } D > 50. \end{cases}$$

2.21 The Chicago Redimix Corp. has three warehouses, located at 3300 North and 1400 West, 6500 South and 2900 West, and 3400 North and 9600 West, containing 3300, 6700, and 4600 bags of mix from which to deliver orders of 5000, 3500, and 2500 bags of mix to lumber yards located 600 South and 800 West, 2000 North and 5600 West, and 1200 South and 7200 West. Because of Chicago's street system, the effective distance between (x_1,y_1) and (x_2,y_2), where the positive x axis points north, is $|x_1 - x_2| + |y_1 - y_2|$.

(a) Can you formulate an LP to minimize total effective distance traveled in delivering the orders if we assume that each truck can carry up to 5000 bags of mix, that trucks carry only one order at a time, and that they return to their warehouse of origin?

(b) Suppose that Mr. Clean signs a contract to service the warehouses and the Redimix office located 6800 North and 4800 West. He will work 1 day a week at each warehouse and 2 days a week at the office. He wishes to rent an apartment in a location that will minimize his weekly driving distance between house and work for Redimix Corp. Formulate an LP that will get him started looking for an apartment at an optimal point in terms of distance to travel to work per week.

2.22 A heating and air conditioning manufacturer wishes to have warehouse storage space. Space is available on a monthly, quarterly, semi annual, or annual basis. The following tables provide lists of space required in thousands of square feet and cost of space in dollars per thousand square feet per lease period. Quarterly lease periods begin January 1, April 1, July 1, and October 1. Half-year lease periods begin February 1 and August 1. Formulate an appropriate LP model to lease space at minimal cost for a calendar year.

Month	1	2	3	4	5	6	7	8	9	10	11	12
Space required	10	15	23	32	43	52	50	56	40	25	15	10

Leasing period	Cost per 1000 feet
Month	$ 300
Quarter	500
Half-year	700
Year	1000

2.23 A town must obtain 100,000 gallons of water daily with a polution level less than or equal to 100 parts per million. It can get water from a river and a well. Ample river water is available, and a filtration plant can provide river water with pollution levels of 150 parts per million and 75 parts per million at costs of $10 and $30 per 1000 gallons in sufficient quantity. The well can provide up to 40,000 gallons of water daily with a pollution level of 50 parts per million. Well water can also be processed to provide superwater with a pollution level of only 10 parts per million by an experimental filtering station. Well water costs $40 per 1000 gallons to pump and an additional $15 per 1000 gallons to filter experimentally
(a) Formulate an appropriate LP.
(b) How does your solution change if the 75 parts per million process is shut down for repairs?

2.24 A small manufacturer sells toy truck kits comprised of an assembly packet and the following wooden parts: one body, two axles, and four wheels. Each part must be processed by a saw, a sander, and a sprayer. Time (in minutes) that the machines spend processing a part is tabulated below.

	Saw	Sander	Sprayer
Body	0.8	1.2	0.6
Axle	0.1	0.25	0.3
Wheel	0.5	0.8	0.4

(a) Suppose that the manufacturer has one saw, two sanders, and one sprayer. Formulate an LP to maximize the number of kits that can be made in an 8-hour day.
(b) Suppose that the manufacturer has only two employees. One can operate the saw and a sander and the other can operate a sander and the sprayer. Formulate an LP to maximize the number of kits that the two employees can make in an 8-hour day.

2.25 A computer programmer decides to move to the country, so she buys 40 acres, including a 5-acre homesite and 35 acres of pasture. Then she decides to buy some animals to raise and sell. However, she wishes to keep and not sell at least three horses, one cow, and four goats. Horses, cows, and goats cost $80, $60, and $15 to buy in the spring and will sell for $400, $200, and $35 in the fall. They require 1.5, 0.8, and 0.25 acre to graze. She can also buy feeder calves for $40 in the spring; they will sell for $250 and require 1 acre to graze. She has plenty of money with which to buy animals for grazing.

(a) Formulate an appropriate LP.

(b) Suppose that she could also use up to 10 acres of pasture for a feedlot, which would hold up to 10 feeder calves per acre; however, she would have to pay $110 for feed per calf in the feedlot during the summer. She has $12,000 to invest in animals and feed. Formulate an LP.

2.26 Speedway Toy Company has a standing order to supply 1000 tricycles and 500 wagons per month to Toy World. Tricycles require 17 minutes of time on a shaper and 8 minutes of finishing time; wagons require 14 minutes on a shaper and 6 minutes of finishing time. Speedway is going to shut down for part of a month to remodel. They will have only 20,000 minutes of shaper time and 10,000 minutes of finishing time available that month. Speedway has agreed to give Toy World a $2 discount per tricycle and a $2 discount per wagon for toys delivered late.

(a) Formulate an LP to minimize the total discount.

(b) Suppose that Speedway's usual profit is $11 per tricycle and $5 per wagon and Toy World says: "Forget the discount; it's small potatoes. Just deliver what you can and we'll talk about increasing our order in a few weeks."

2.27 Referring to Exercise 2.6, suppose that there is a storm and only three truckloads of fir and two truckloads of pine logs are available at the Douglas mill. (The rest of the trucks are stuck in the mountains.) The company must pay a penalty of $0.2 per board foot of lumber that is delivered late and $1 per sheet of plywood that is delivered late. What part of the 1000 sheets of plywood and 5000 board feet of lumber order should the Douglas Company fill by using the five truckloads that are available?

2.28 Traction Tire Company gets a new order to supply 10,000 small tractor tires, 10,000 large tractor tires, 100,000 large passenger car tires, 200,000 medium passenger car tires, 300,000 small passenger car tires, and 240,000 truck tires per year for the next 10 years.

Management has decided to buy new presses with which to produce the tires. These presses will be used 24 hours a day, 6 days a week to make tires for their order. Two types of presses are available, costing $40,000 and $80,000 and able to produce tires at the following rates (tires per hour).

	$40,000 press	$80,000 press
Small tractor	1.5	2.5
Large tractor	0.5	1.5
Small passenger car	5.0	8.0
Medium passenger car	3.0	5.0
Large passenger car	2.0	4.0
Truck	1.0	3.0

2.29 Quick Chemical Corporation supplies chemicals to Traction Tire Company. Quick delivers pallets loaded with sacks of three types of chemicals in a truck that carries 11,000 pounds of cargo. The three types of pallets weigh 2000, 2500, and 3000 pounds. Quick requires that its truck travel fully loaded; 50 of the 2000-pound pallets, 100 of the 2500-pound pallets, and 30 of the 3000-pound pallets are in stock. Traction would like to receive as many truckloads as possible. Formulate an LP to determine how many truckloads Quick can deliver with its stock in hand.

2.30 Quick Chemical Corporation expects to have 40 hours of mixing time and 40 hours of packaging time available during the next week. It packages chemicals for its 2000-pound and 3000-pound pallets from five ingredients; it takes 1 and 0.6 hour to mix the chemicals and 0.4 and 1.2 hours to package the chemicals for each 2000- and 3000-pound pallet, respectively. The ingredients contain essential elements A, B, and C in the following percentages.

Ingredient	A	B	C	Cost per ton
1	12	10	2	$600
2	20	8	2	900
3	3	30	0	200
4	10	10	1	750
5	0	0	100	50

Ingredients 3, 4, and 5 can be used to mix the chemical shipped on 2000-pound pallets; it requires at least 5% A, 18% B, 2% C, and

10% ingredient 4. Ingredients 1, 2, and 3 can be used to mixed the chemical shipped on 3000-pound pallets; it requires at least 11% A and 15% B. The 2000-pound pallets sell for $750 and 3000-pound pallets sell for $1500. What should be the composition of the chemicls, and how many pallets of each should be manufactured during the next week to maximize potential income from these two chemicals.

2.31 Classic Oak Company receives an order from National Furniture Company for 1000 oak tables. Each table requires a base, four shafts, and a top. Each part requires processing time on a saw and on a sander.

	Minutes of time per part		
Part	Saw	Sander	Cost per part
Base	10	7	$4.8
Shaft	0.2	0.25	0.3
Top	5	7.5	6.5

Up to 160 hours of saw time and up to 160 hours of sander time are available to process the parts. Classic Oak can buy sanded bases, shafts, and tops from Custom Cabinet Company for $5.50, $0.50, and $7.50. Formulate an appropriate LP.

2.32 An organization has allocated $1.8 million of its budget to purchase cars, vans, and buses. These can be acquired at a unit cost of $12,000, $16,000, and $40,000, respectively. It is decided that at least 20 cars and 20 buses must be acquired. It is also decided not to purchase more than 30 buses. Moreover, the ratio of vans to buses must fall in the range $\frac{1}{4}$ to $\frac{1}{2}$. The ultilities of these types of vehicles are estimated to be 1, 2, and 3. Formulate an appropriate LP and find all six optimal solutions. (*Hint*: The benefit/cost ratio of an object is the ratio of its estimated utility to its cost.)

2.33 Verify the two fundamental facts about standard and symmetric primal–dual pairs stated in Section 2.5.

2.34 A company has a large order to ship. There are four types of products to be shipped. The company transports the products from its factory to a dock. Then the orders are transported from the dock to customers by boat. The boat leaves the dock each Sunday and returns to the lock the following Friday. The company uses two trucks to transport products to the dock. The trucks can hold the following quantities of the four types of products:

	Type of product			
Truck	1	2	3	4
1	25	20	15	10
2	42	37	34	31

It takes 1 hour for a truck to load at the factory, deliver to the dock, unload at the dock, and return to the factory (independent of the type of product being transported). However, the per truckload cost (in U.S. dollars) of transporting the types of products varies as follows:

	Type of product			
Truck	1	2	3	4
1	18	16	12	10
2	26	26	18	16

Products cannot be mixed on a truckload. Each of the trucks can be run up to 40 hours per week.

The order to be shipped is for 7000 of type 1, 6500 of type 2, 7500 of type 3, and 8000 of type 4 products.

(a) How many weeks are needed to transport the order to the dock?

(b) What will it cost to transport the order to the dock in the minimum number of weeks?

(c) Suppose that the trucks could be run a sixth 8-hour day each week at a cost per hour $4 higher than those listed in the table. How much would transportation costs increase if the company were to use some overtime to transport the order to the dock in 12 weeks?

SOLUTIONS TO EXERCISES

2.1 Returning to the original problem, because

$0.25L + 0.1D$ represents jet fuel produced,

$0.21L + 0.55D$ represents fuel oil produced, and

$0.21L + 0.55D = 0.84(0.25L + 0.1D) + 0.466D,$

we see that any combination of light and dark crude that produces

5 million barrels of jet fuel will produce at least 4.2 million barrels of fuel oil. Consequently, we can restrict our attention to combinations of light and dark crude that produce 5 million barrels of jet fuel. Thus it suffices to

$$\min \quad 25L + 17D$$
$$\text{s.t.} \quad 0.25L + 0.1D = 5$$

or, equivalently, $25L + 10D = 500$. However, if $25L + 10D = 500$, then $25L + 17D = 25L + 10D + 7D = 500 + 7D$, which attains its minimum value of 500 when $D = 0$ and $L = 20$.

2.2 Let M = millions of barrels of medium crude to process and solve the linear program

$$\min \quad 25L + 17D + 22M$$
$$\text{s.t.} \quad 0.21L + 0.55D + 0.25M \geq 3$$
$$0.50L + 0.30D + 0.40M \geq 7$$
$$0.25L + 0.10D + 0.30M \geq 5$$
$$L \geq 0, \quad D \geq 0, \quad M \geq 0.$$

Using an LP software package on a PC (personal computer), we find the optimal solution $L = 2$, $D = 0$, $M = 15$ with a corresponding processing cost of $380 million. Consequently, building the pipeline could reduce processing costs by $120 million.

2.3 We can focus on either production or time.

Production Focus

Let P_{IA} = no. of part I produced on machine A, $\quad I = 1,2,3,4$
P_{IB} = no. of part I produced on machine B, $\quad I = 1,2,3,4$
P_{IC} = no. of part I produced on machine C, $\quad I = 1,2,3,4$
s.t.

(1) $\dfrac{1}{300}P_{1A} + \dfrac{1}{250}P_{2A} + \dfrac{1}{200}P_{3A} + \dfrac{1}{100}P_{4A} \leq 50$

(2) $\dfrac{1}{600}P_{1B} + \dfrac{1}{400}P_{2B} + \dfrac{1}{350}P_{3B} + \dfrac{1}{200}P_{4B} \leq 50$

(3) $\dfrac{1}{800}P_{1C} + \dfrac{1}{700}P_{2C} + \dfrac{1}{600}P_{3C} + \dfrac{1}{300}P_{4C} \leq 50$

(4) $P_{1A} + P_{1B} + P_{1C} = 10,000$

(5) $P_{2A} + P_{2B} + P_{2C} = 8000$

$$(6) \quad P_{3A} + P_{3B} + P_{3C} = 6000$$
$$(7) \quad P_{4A} + P_{4B} + P_{4C} = 6000$$

min
30{left side of (1)} + 50{left side of (2)} + 80{left side of (3)}.

The left side of constraint (1) tells you the number of hours used on machine A to process the order. Constraints (2) and (3) give you similar information about machines B and C. Constraints (4) to (7) require that production meet demand. There is no need to produce any surplus.

Time focus

Let T_{IA} = no. of hours used on machine A to produce part I, $I \le 4$
     T_{IB} = no. of hours used on machine B to produce part I, $I \le 4$
     T_{IC} = no. of hours used on machine C to produce part I, $I \le 4$

s.t.

$$(1) \quad T_{1A} + T_{2A} + T_{3A} + T_{4A} \le 50$$
$$(2) \quad T_{1B} + T_{2B} + T_{3B} + T_{4B} \le 50$$
$$(3) \quad T_{1C} + T_{2C} + T_{3C} + T_{4C} \le 50$$
$$(4) \quad 300T_{1A} + 600T_{1B} + 800T_{1C} = 10{,}000$$
$$(5) \quad 250T_{2A} + 400T_{2B} + 700T_{2C} = 8000$$
$$(6) \quad 200T_{3A} + 350T_{3B} + 600T_{3C} = 6000$$
$$(7) \quad 100T_{4A} + 200T_{4B} + 300T_{4C} = 6000.$$

min
$$30(T_{1A} + T_{2A} + T_{3A} + T_{4A}) + 50(T_{1B} + T_{2B} + T_{3B} + T_{4B})$$
$$+ 80(T_{1C} + T_{2C} + T_{3C} + T_{4C}).$$

2.4 (a) Several formulations of a valid LP model for this situation come to mind. For instance, we could focus on land, water, money, or labor. Let us focus on production and let C, W, and S denote the numbers of metric tons of corn, wheat, and soybeans to be grown. Then we have the model

$$\max[10,6,8]\begin{bmatrix} C \\ W \\ S \end{bmatrix} = cx$$

s.t. $\quad Ax = \begin{bmatrix} \frac{1}{10} & \frac{1}{4} & \frac{1}{6} \\ \frac{3}{10} & \frac{1}{4} & \frac{2}{6} \\ \frac{15}{10} & \frac{10}{4} & \frac{12}{6} \\ \frac{10}{10} & \frac{4}{4} & \frac{6}{6} \\ \frac{8}{10} & \frac{6}{4} & \frac{10}{6} \end{bmatrix} x \le \begin{bmatrix} 1000 \\ 2000 \\ 20{,}000 \\ 8000 \end{bmatrix} = b, \quad x \ge 0.$

Now let's focus on land and write the following LP model. Let

C = no. of units of land to plant in corn
W = no. of units of land to plant in wheat
S = no. of units of land to plant in soybeans.

$$\begin{array}{ll}
\max & 100C + 24W + 48S \\
\text{s.t.} & C + W + S \leq 1000 \\
& 3C + W + 2S \leq 2000 \\
& 15C + 10W + 12S \leq 20{,}000 \\
& 8C + 6W + 10S \leq 8000 \\
& C,W,S \geq 0.
\end{array}$$

(b) Suppose that a large food-processing corporation is interested in leasing the resources of the agency. The corporation wishes to be competitive and obtain the use of the resources at the lowest cost in profit units. Suppose that we let L, W, M, and LA denote the number of profit units that the corporation will offer to pay to use a unit of land, water, money, and labor for a crop season. An LP for the corporation's position is

$$\min \quad [L,W,M,LA] \begin{bmatrix} 1000 \\ 2000 \\ 20{,}000 \\ 8000 \end{bmatrix} = yb$$

$$\text{s.t.} \quad yA \geq c, \quad y \geq 0.$$

The constraints tell the agency that it is at least as profitable to rent their resources as it is to grow corn, wheat, or soybeans.

2.6 (a) Let FL = no. of truckloads of fir made into lumber
FP = no. of truckloads of fir made into plywood
PL = no. of truckloads of pine made into lumber
PP = no. of truckloads of pine made into plywood.

$$\begin{array}{ll}
\max & (0.5)(1000)FL + (0.8)(800)PL + 5(140FP + 160PP) \\
\text{s.t.} & 140FP + 160PP \geq 1000 \\
& 1000FL + 800PL \geq 5000 \\
& FL + FP \leq 10 \\
& PL + PP \leq 10, \quad FL,FP,PL,PP \geq 0.
\end{array}$$

2.8 Let A = no. of kilograms of feed 1 to put in diet per day

B = no. of kilograms of feed 2 to put in diet per day

C = no. of kilograms of feed 3 to put in diet per day.

(a) s.t. $160 \leq 36A + 20B + 16C \leq 180$ (100's of cal)

$25A + 35B + 15C \geq 200$ ($\frac{1}{100}$'s of kg protein)

$70A + 40B + 25C \geq 300$ ($\frac{1}{100}$'s of grams vitamins)

min $8A + 6B + 2C$ ($\frac{1}{10}$'s of dollar).

(b) Add two constraints:

$$A \geq 0.2(A + B + C) \Leftrightarrow 0.8A - 0.2B - 0.2C \geq 0$$
$$C \leq 0.5(A + B + C) \Leftrightarrow 0.5A + 0.5B - 0.5C \geq 0.$$

You can multiply these by 10 or 100 if you wish.

2.15 Let C_{IJ} = no. of containers of type I to pack with mix J, $I = 1,2$, $J = 1,2$.

min $15C_{11} + 15C_{12} + 25C_{21} + 25C_{22}$

s.t. $2C_{11} + 2C_{12} + 3C_{21} \qquad\qquad \geq \;\; 78$ (floor)

$4C_{11} + 2C_{12} + 8C_{21} + 8C_{22} \geq 214$ (table)

$4C_{12} \qquad\qquad 12C_{22} \geq 198$ (dresser).

2.16 Let E = no. of earthmover kits to make

R = no. of racing kits to make.

max $15E + 9.5R$

(a)   s.t. $2E + \;\;\; R \leq 215$

$3E + \;\; 2R \leq 525$

$4E + \;\; 2R \leq 440$

$2E + \;\; 3R \leq 560.$

(b) Let E_1 = min{no. of earth moverkits, 60}

$E_2 = E - E_1$

R_1 = min{no. of racing kits, 50}

$R_2 = R - R_1.$

max $15E_1 + 10E_2 + 9.5R_1 + 6.5R_2$

s.t.   all constraints in part (a) plus

$$E \quad - E_1 - E_2 \quad = 0$$
$$R \qquad - R_1 - R_2 = 0$$
$$E_1 \qquad\qquad \leq 60$$
$$R_1 \qquad \leq 50.$$

The objective function will automatically prefer to push E_1 to 600 before E_2 becomes >0, and it will automatically push R_1 to 50 before R_2 becomes >0.

2.17 Put $t = u - v$, where $u \geq 0$ and $v \geq 0$. Consider the case $t > 0$; $u = t + v$, so $u + v = t + 2v$, which is $\geq t$; $u + v$ attains its minimum value, $t = |t|$, when $u = t$ and $v = 0$. Recall that $|t| = -t$ when $t < 0$. Now suppose $t < 0$. Use $v = u - t$ to get $u + v = 2u - t = 2u + |t| \geq |t|$; $u + v$ attains its minimum value, $|t| = -t$, when $u = 0$ and $v = -t$.

2.18 (a) Put $q = qP - qN$, $r = rP - rN$; $qP, qN, rP, rN \geq 0$.

min $\quad 12qP - 12qN + 5rP + 5rN$

s.t. $\qquad qP - \quad qN + 2rP - 2rN \geq 4$: replace q by $(qP - qN)$,
$$\qquad\qquad\qquad\qquad\qquad\qquad\qquad r \text{ by } (rP - rN)$$

$$-5qP + 5qN - 6rP + 6rN \geq -7 \qquad \text{(multiply by } -1)$$

$$\left.\begin{array}{l} 8qP - \ 8qN - 9rP + 9rN \geq 11 \\ -8qP + \ 8qN + 9rP - 9rN \geq -11 \end{array}\right\} see \ below$$

$$8q - 9r = 11 \Leftrightarrow \begin{cases} 8q - 9r \geq 11 \\ 8q - 9r \leq 11 \end{cases} \quad and$$

$$\Leftrightarrow \begin{cases} 8q - 9r \geq \ \ 11 \\ -8q + 9r \geq \ -11 \end{cases} \quad and$$

2.21 (b)

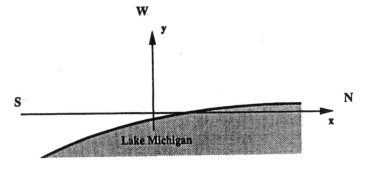

Let (x,y) = coordinates of a place to start looking.
Put $x - 33 = P_1 - N_1$ $y - 14 = P_2 - N_2$
$x + 65 = P_3 - N_3$ $y - 29 = P_4 - N_4$
$x - 34 = P_5 - N_5$ $y - 96 = P_6 - N_4$
$x - 68 = P_7 - N_7$ $y - 48 = P_8 - N_8$.
min $P_1 + N_1 + P_2 + N_2 + P_3 + N_3 + P_4 + N_4 + P_5$
      $+ N_5 + P_6 + N_6 + 2(P_7 + N_7 + P_8 + N_8)$
s.t. $P_1 - N_1 + 33 = P_3 - N_3 - 65 = P_4 - N_5 + 34 =$
      $P_7 - N_7 + 68$
      $P_2 - N_2 + 14 = P_4 - N_4 + 32 = P_6 - N_6 + 96 =$
      $P_8 - N_8 + 48$

2.29 Since $1 \times 2 \qquad\quad + 3 \times 3 = 11,$
        $4 \times 2 \qquad\quad + 1 \times 3 = 11$
        $3 \times 2 + 2 \times 2.5 \qquad = 11, \qquad$ and
        $2 \times 2.5 + 2 \times 3 = 11,$
there are four ways to fully load a truck. Let
L_1 = no. of trucks loaded with one 2000-pound and three 3000-pound pallets
L_2 = no. of trucks loaded with four 2000-pound and one 3000-pound pallets
L_3 = no. of trucks loaded with three 2000-pound and two 2500-pound pallets
L_4 = no. of trucks loaded with two 2500-pound and two 3000-pound pallets.

$$\text{max} \quad L_1 + L_2 + L_3 + L_4$$
$$\text{s.t.} \quad L_1 + 4L_2 + 3L_3 \qquad\quad \leq 50$$
$$2L_3 + 2L_4 \leq 100$$
$$3L_1 + L_2 \qquad + 2L_4 \leq 30.$$

2.31 We wish to minimize the cost of acquiring 1000 bases, 4000 shafts, and 1000 tops. These parts can be purchased from Custom Cabinet Company for the prices indicated in Exercise 2.31, or they can be produced by Classic Oak Company from saw time, sander time, and money. The savings in money by using saw time and sander time to produce the parts, rather than purchase them from Custom Cabinet Company is $0.70 per base, $0.20 per shaft, and $1.00 per top. We will formulate an LP to maximize the savings in dollars due to production rather than purchase. Let
B = no of bases to produce for the order
S = no. of shafts to produce for the order
T = no. of tops to produce for the order.

max $0.7B + 0.2S + T$ (no. of dollars saved)

s.t. $10B + 0.2S + 5T \leq 160 \times 60$ (saw time)

$7B + 0.25S + 7.5T \leq 160 \times 60$ (sander time)

$B \qquad\qquad\qquad \leq 1000$

$S \qquad\qquad \leq 4000$

$T \leq 1000 \qquad (B,S,T \leq 0).$

2.33 We shall see below that fundamental fact 1 about standard and symmetric primal–dual pairs is a consequence of the following simple fact:

Suppose that each of α and β is a number and $\alpha \leq \beta$. If γ is a number and $\gamma \geq 0$, then $\alpha\gamma \geq \beta\gamma$.

Recall that when each of $u = [u_1, \ldots ,u_n]$ and $v = [v_1, \ldots ,v_n]$ is an n-dimensional row vector, $u \leq v$ means $u_i \leq v_i$, $1 \leq i \leq n$. When $u = (u_1, \ldots ,u_n)$ and $v = (v_1, \ldots ,v_n)$ are column vectors, we also defined $u \leq v$ to mean that $u_i \leq v_i$, $1 \leq i \leq n$.

The preceding definition and elementary arithmetic of inequalities give us the following relationships.

1. Suppose that each of u and v is an m-dimensional row vector and $u \leq v$. If w is an m-dimensional column vector and $w \geq 0$, then $uw \leq vw$.

2. Suppose that each of u and v is an n-dimensional column vector and $u \leq v$. If w is an n-dimensional row vector and $w \geq 0$, then $wu \leq wv$.

These two relationships imply that fundamental fact 1 is true.

Fundamental fact 2 follows from fundamental fact 1, which tells us that the set of values of a dual lies to the left on a number line of the set of values of the corresponding primal in both the standard and symmetric cases: No value of the primal can be below $yb = cx$, so cx is the minimum value of the primal. Similarly, no value of the dual can be above $cx = yb$, so yb is the maximum value of the dual.

3

Elementary Properties of the Feasible Set for an LP

This chapter introduces some important properties of the feasible set for an LP and establishes the *fundamental theorem of linear programming*: Given an LP, the feasible set (for that LP) is the set of vectors (equivalently points) in the appropriate n-dimensional space R^n which satisfy the constraints for the given LP.

3.1 BASIC PROPERTIES

The basic properties of the feasible set for an LP can be summarized as follows:

1. The feasible set K is convex.
2. K has a finite number of corners.
3. Finite minimums and maximums of the objective function for the LP occur at corners of K.

Convex sets are defined below and corner points are defined in Exercise 3.9; geometric intuition should suffice to identify the corner points of the feasible sets, which are sketched in this section. Exercise 3.9(b) asks you to show that the corner points of the feasible set for an LP are the basic feasible solutions defined in Section 3.2; in case you get stuck, a sketch of a proof appears among the solutions that follow the exercises.

Definition of a Convex Set. A subset K of R^n is convex if whenever x and y belong to K and $0 \le t \le 1$, then $x + t(y - x) = (1 - t)x + ty$ belongs to K; equivalently, putting $u = 1 - t$, if x and y belong to K and u and t are nonnegative numbers with $u + t = 1$, then $ux + ty$ belong to K.

To illustrate with an example in R^2, put $x = (1,2)$ and $y = (3,5)$; then

$$y \leftarrow (3,5) = (1,2) + 1(2,3), \quad t = 1$$

$$\leftarrow (2,\tfrac{7}{2}) = \tfrac{1}{2}(1,2) + \tfrac{1}{2}(3,5), \quad t = \tfrac{1}{2}, \quad u = \tfrac{1}{2}$$

$$\leftarrow (\tfrac{3}{2},\tfrac{11}{4}) = (1,2) + \tfrac{1}{4}(2,3), \quad t = \tfrac{1}{4}$$

$$x \leftarrow (1,2) = 1(1,2) + 0(2,3), \quad t = 0.$$

Examples of sets in R^2 that are convex and nonconvex are shown in Figs. 1 and 2, respectively.

As an exercise, show that

(a) $K_1 = \{x : Ax = b, x \ge 0\}$ is convex.
(b) $K_2 = \{x : Ax \ge b, x \ge 0\}$ is convex.

Now put $K = \{x \in R^2 : x_1 \ge 0, x_2 \ge 0, x_1 + x_2 \le 5\}$. So K has three corners, $(0,0)$, $(0,5)$, and $(5,0)$ (see Fig. 3). Suppose that we put $A = [1,1]$,

$$x = \begin{bmatrix} x_1 \\ x_2 \end{bmatrix} = (x_1, x_2),$$

and $b = [5] = 5$: b is formally the 1×1 matrix $[5]$, but *we will not distinguish between a 1×1 matrix and its entry*; K is the feasible set for the LP

$$\min \text{ (or max) } cx$$

$$\text{s.t. } Ax \le b, \quad x \ge 0.$$

Suppose that $c = [c_1, c_2] \ne [0,0]$. Then $f(x) = cx = c_1 x_1 + c_2 x_2$ is constant

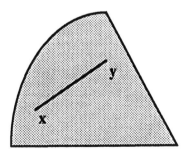

Figure 1. Convex set.

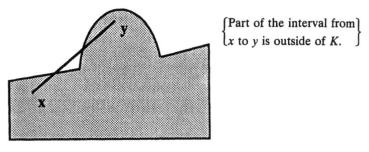

Figure 2. Non-convex set.

on a line that has $c = [c_1, c_2]$ as a normal vector. Consequently, a solution to the LP

$$\min cx = c_1 x_1 + c_2 x_2$$
$$\text{s.t. } Ax = x_1 + x_2 \le 5, \quad x \ge 0$$

occurs at a corner. For example, consider the case $c = [\frac{5}{4}, 3]$ (Fig. 4). cx is constant on lines perpendicular to the vector $c = [\frac{5}{4}, 3]$, and the constant increases as the line moves in the direction of c. Consequently, the minimum value of cx on K occurs at $(0,0)$ and the maximum is attained at $(0,5)$. This numerical example is typical of the case $0 < c_1 < c_2$. When

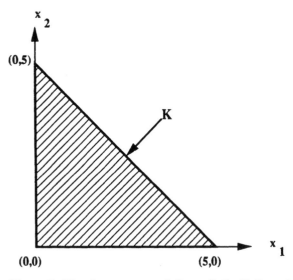

Figure 3. The three corners of K are $(0,0)$, $(0,5)$, and $(5,0)$.

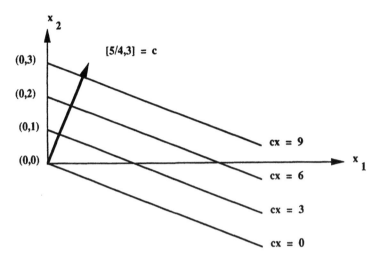

Figure 4.

$0 < c_1 = c_2$, each point on the edge of K from $(0,5)$ to $(5,0)$ is a maximum; however, the maximum does occur at the corners $(0,5)$ and $(5,0)$. If $0 < c_2 < c_1$, the minimum remains at $(0,0)$, but the maximum occurs only at $(5,0)$. If $c_1 < 0 < c_2$, the minimum occurs at $(5,0)$ and the maximum value is $c(0,5) = 5c_2$. You should consider what happens when both c_1 and c_2 are negative.

Introducing a slack variable x_3 gives rise to the set (see Fig. 5)

$$\tilde{K} = \{x \in R^3 : x_1 + x_2 + x_3 = 5, x \geq 0\} \text{ in } R^3.$$

Notice that $\tilde{K}$ has three corners, and if we project $\tilde{K}$ onto K by putting $x_3 = 0$, corners project onto corners.

The next example will be used in several places to illustrate various aspects of the simplex method.

Example 3.1.   Let $K = \{x : Ax \leq b, x \geq 0\}$, where

$$A = \begin{bmatrix} 1 & \frac{2}{3} & 1 \\ \frac{1}{2} & 1 & 0 \end{bmatrix} \quad \text{and} \quad b = \begin{bmatrix} 1 \\ 1 \end{bmatrix}.$$

Thus K is the set of points x in R^3 that satisfy the constraints

$$x_1 + \tfrac{2}{3}x_2 + x_3 \leq 1$$
$$\tfrac{1}{2}x_1 + x_2 \qquad \leq 1$$
$$x = (x_1, x_2, x_3) \geq 0.$$

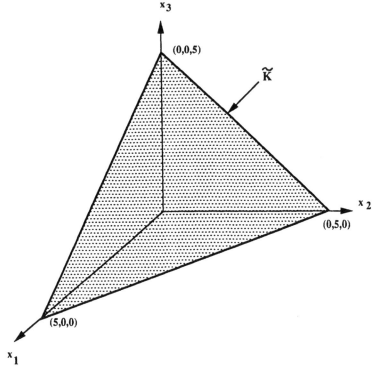

Figure 5. $\tilde{K} = \{x \in R^3 : x_1 + x_2 + x_3 = 5, x \geq 0\}$ in R^3.

Adding slack variables, x_4 and x_5, produces the system

$$x_1 + \tfrac{2}{3}x_2 + x_3 + x_4 \qquad = 1$$
$$\tfrac{1}{2}x_1 + x_2 \qquad\qquad + x_5 = 1$$
$$\tilde{x} = (x_1, x_2, x_3, x_4, x_5) \geq 0,$$

whose corresponding solution set $\tilde{K}$ in R^5 is given by the formula

$$\tilde{K} = \{\tilde{x} : \tilde{A}\tilde{x} = b, \tilde{x} \geq 0\},$$

where

$$\tilde{A} = \begin{bmatrix} 1 & \tfrac{2}{3} & 1 & 1 & 0 \\ \tfrac{1}{2} & 1 & 0 & 0 & 1 \end{bmatrix}.$$

Notice that K has six corners (see Fig. 6).

For exercise, find the points of $\tilde{K}$ that project onto the corners of K.

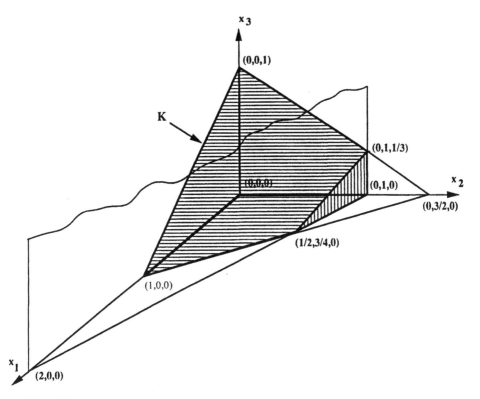

Figure 6.

Observe that if $f(x) = cx = c_1x_1 + c_2x_2 + c_3x_3$, where $c = [c_1,c_2,c_3] \neq [0,0,0]$, a maximum or minimum of f on K occurs at a corner of K because the set of points x for which $f(x)$ has constant value is a plane in R^3 with normal vector c. For example, put $c = [2,3,6]$ (Fig. 7). The minimum value of cx occurs at $(0,0,0)$ and the maximum occurs at $(0,0,1)$. If we change c to $[1,1,1]$, the minimum remains at $(0,0,0)$, but the maximum moves to $(0,1,\frac{1}{3})$. For $c = [2,2,1]$, the maximum is attained at $(\frac{1}{2},\frac{3}{4},0)$.

3.2 BASIC FEASIBLE SOLUTIONS

Now consider the standard form LP, $\min cx$, s.t. $Ax = b$, $x \geq 0$. Suppose that A is a $m \times n$ matrix of rank m; the rank of A can be checked by row reduction on the augmented matrix $[A \mid b]$. If rank $A < \text{rank}[A \mid b]$, there are no solutions to the system $Ax = b$. If $\text{rank}[A \mid b] = \text{rank} A$, there are

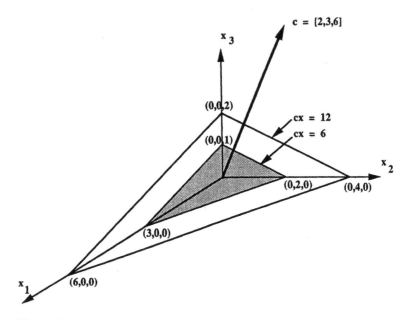

Figure 7.

solutions to the system; some, all, or none of the solutions may be feasible. However, the rows of the row-reduced A matrix are linearly independent, so without loss of generality we begin with a matrix that has independent rows. Let K denote the feasible set for the LP. Strictly speaking, a solution x to the LP should be a point in the feasible set K at which the function $f(x) = cx$ attains its minimum value on K. Nevertheless, for convenience of communication, we shall call such points *optimal solutions*, and we shall call points of K *feasible solutions*. Thus a *solution* x of $Ax = b$ is a feasible solution to the LP if and only if $x \geq 0$. A solution x of $Ax = b$ is called a *basic solution* if the columns of the A matrix corresponding to the nonzero entries in x are linearly independent. A *basic feasible solution* (BFS) is a solution that is basic and feasible. Let us use Example 3.1 to illustrate.

In Example 3,1, the third and fourth columns of

$$A = \tilde{A} = \begin{bmatrix} 1 & \frac{2}{3} & 1 & 1 & 0 \\ \frac{1}{2} & 1 & 0 & 0 & 1 \end{bmatrix}$$

are linearly dependent; any other pair of columns is a linearly independent pair. For instance, the fourth and fifth columns are linearly independent and the slack variable solution, $x = (0,0,0,1,1)$, is a basic feasible solution to $Ax = b$, $x \geq 0$. The first and fourth are independent, and a basic solution $x = (x_1,0,0,x_4,0)$ to $Ax = b$ is determined by solving the linear system

$$\begin{bmatrix} 1 & 1 \\ \frac{1}{2} & 0 \end{bmatrix}\begin{bmatrix} x_1 \\ x_4 \end{bmatrix} = \begin{bmatrix} 1 \\ 1 \end{bmatrix}.$$

Using row reduction to solve this system,

$$\left[\begin{array}{cc|c} 1 & 1 & 1 \\ \frac{1}{2} & 0 & 1 \end{array}\right] \rightarrow \left[\begin{array}{cc|c} 1 & 1 & 1 \\ 0 & -\frac{1}{2} & \frac{1}{2} \end{array}\right] \rightarrow \left[\begin{array}{cc|c} 1 & 1 & 1 \\ 0 & 1 & -1 \end{array}\right] \rightarrow \left[\begin{array}{cc|c} 1 & 0 & 2 \\ 0 & 1 & -1 \end{array}\right].$$

we obtain $x = (2,0,0,-1,0)$, which is basic (the columns corresponding to nonzero entries are independent) but not feasible because $x_4 < 0$.

For exercise, consider a basic solution $x = (x_1,0,x_3,0,0)$; find x_1 and x_3 and locate this point in Fig. 6, the "three-dimensional picture" which includes K.

There are

$$\binom{5}{2} = \frac{5!}{2!\,3!}$$

or 10 pairs of columns in A. Nine of these pairs are independent, and six of the independent pairs lead to basic feasible solutions corresponding to the six corner points in Fig. 6.

Before the fundamental theorem of linear programming is discussed in Section 3.3, Example 3.2 illustrates a technical point called degeneracy that we must deal with later, in Sections 4.7 and 4.10.

Example 3.2.   Consider the standard-form LP

$$\min\ [3,2,1,1]x$$

$$\text{s.t.}\ \begin{bmatrix} 1 & 0 & 0 & 1 \\ 0 & 1 & -2 & 3 \end{bmatrix}x = \begin{bmatrix} 2 \\ 0 \end{bmatrix} = (2,0) = b.$$

Recall that $(2,0,0,0)$ denotes a four-dimensional column vector.

Notice that $x = (2,0,0,0)$ is a basic feasible solution to this problem because the first column is a linearly independent column. There are three different choices for a second column to go with the first column to compose a B-matrix corresponding to the BFS x; each of these three choices leads to a different 2×2 linear system $Bx_B = b$ to solve, and each of the three linear systems provides a different basic feasible solution x_B. The three choices for B, corresponding to choosing the second, third, or fourth to go with the first column, are

$$B = \begin{bmatrix} 1 & 0 \\ 0 & 1 \end{bmatrix}\quad \text{or}\quad \begin{bmatrix} 1 & 0 \\ 0 & -2 \end{bmatrix}\quad \text{or}\quad \begin{bmatrix} 1 & 1 \\ 0 & 3 \end{bmatrix}.$$

The corresponding x_B are $x_B = (x_1, x_2) = (2,0)$, $x_B = (x_1, x_3) = (2,0)$, and $x_B = (x_1, x_4) = (2,0)$. All three of these x_B's correspond to $x = (2,0,0,0)$.

3.3 THE FUNDAMENTAL THEOREM OF LINEAR PROGRAMMING

1. If there are points in the feasible set for a standard-form LP, the LP has basic feasible solutions.
2. If there is an optimal solution x to this LP (i.e., x is feasible and cx is minimal), there is a basic feasible solution that is optimal.

A proof of the fundamental theorem consists of two parts, verifying 1 and 2. To verify 1, suppose that x is in the feasible set K. Then

$$Ax = \sum_{j=1}^{n} x_j a_j = b,$$

where a_j denotes the jth column of A. By relabeling the columns if necessary, we can suppose, without loss of generality, that $x_1 > 0$, $x_2 > 0, \ldots, x_p > 0$, $x_{p+1} = 0, \ldots, x_n = 0$: $x_i > 0$ for $1 \leq i \leq p$ and $x_j = 0$ for $p < i \leq n$. There are two cases to consider.

Case 1.

If $a_1, a_2, \ldots, a_p$ are linearly independent, x is a basic feasible solution and part 1 is completed.

Case 2.

Suppose that $a_1, a_2, \ldots, a_p$ are linearly dependent. Then there exist real numbers $z_1, z_2, \ldots, z_p$ with at least one $z_i > 0$ and $\sum_{i=1}^{p} z_i a_i = 0$. (For exercise, the careful reader should verify the validity of the assertion that we can require that some z_i be positive.)

Put $z = (z_1, \ldots, z_p, 0, \ldots, 0_n)$. Then $Az = \sum_{i=1}^{p} z_i a_i = 0$. Let t denote a positive number and consider $x - tz = (x_1 - tz_1, \ldots, x_p - tz_p, 0, \ldots, 0)$. For small t, $x_i - tz_i$ is positive for $1 \leq i \leq p$. But if z_i is positive, $x_i - tz_i$ is negative for $t > (x_i / z_i)$. So we increase t from zero until at least one $x_i - tz_i = 0$, $1 \leq i \leq p$, and all $x_i - tz_i \geq 0$, $1 \leq i \leq p$. This situation occurs because, by assumption, some z_i is positive. For this choice of t, $x - tz$ is feasible because $A(x - tz) = Ax - tAz = b - tAz = b$, and $x - tz$ has fewer positive entries than x. Cycle through the preceding part of this proof for $x - tz$. Either case 1 applies and $x - tz$ is a basic feasible solution to the LP or we proceed to case 2 and get a new feasible solution $x - tz - t'z'$ with at most $p - 2$ positive entries. After at most p cycles, we arrive at a basic feasible solution to the LP. This completes part 1 of a proof of the fundamental theorem of linear programming.

Part 2 is similar. Suppose that x is an optimal solution. As in part 1, there are two cases. If x is basic, we are done. Otherwise, consider $x - tz$ as in part 1. If we show that $cz = 0$, then $x - tz$ is also optimal and we can continue as in part 1. To show that $cz = 0$, consider $x - tz$ for small positive and negative values of t. Notice that $x - tz$ is feasible for small positive and negative values of t. If cz were > 0, then $c(x - tz) = cx - tcz$ would be less than cx for small positive t; whereas if cz were < 0, then $c(x - tz)$ would be less than cx for small negative t. Neither of the preceding possibilities can occur because cx is minimal due to the supposition that x is optimal. Thus $cz = 0$. Consequently, $c(x - tz) = cx - tcz = cx$. Hence $x - tz$ is optimal and we can cycle as in part 1 and terminate after at most p cycles with a verification of 2 of the fundamental theorem.

Solving Exercise 3.9(b) shows that the basic feasible solutions are the corner points of the feasible set K; there are at most

$$\binom{n}{m} = \frac{n!}{m!\,(n - m)!}$$

basic feasible solutions.

According to the fundamental theorem of linear programming, to find an initial optimal solution it suffices to restrict attention to basic feasible solutions. The simplex method developed in Chapter 4 is an efficient way to examine basic feasible solutions for optimality.

EXERCISES

3.1 Consider the set K of points (x_1, x_2) in R^2 that satisfy the following inequalities:

$$x_1 + 2x_2 \leq 16$$
$$2x_1 + x_2 \leq 12$$
$$x_1 + 2x_2 \geq 2, \qquad x_1, x_2 \geq 0.$$

(a) Sketch K.
(b) List the extreme points (corner points) of K.
(c) Maximize $3x_1 + 5x_2$ s.t. x in K: Graphically solve the LP

$$\max \quad 3x_1 + 5x_2$$

$$\text{s.t.} \quad x_1 + 2x_2 \leq 16$$

$$2x_1 + \ x_2 \leq 12$$

$$x_1 + 2x_2 \geq 2, \qquad x_1, x_2 \geq 0.$$

(d) Repeat part (c) with "maximize" replaced by "minimize."

(e) Repeat part (d) with "$3x_1 + 5x_2$" replaced by "$3x_1$"; find all points at which the minimum is attained.

3.2 Sketch the feasible region for the primal in Example 2.1 and solve the problem graphically.

3.3 Consider the LP

$$\max \quad cx$$

$$\text{s.t.} \quad x_1 + 2x_2 \leq 4$$

$$2x_1 + \ x_2 \leq 4$$

$$2x_1 + 2x_2 \leq 5, \qquad x \geq 0.$$

(a) Sketch the feasible region.

(b) Find the coordinates of all the corner points of the feasible region.

(c) If $c = [3,4]$, find the point where the maximum is attained.

(d) Find all points where the maximum is attained when $c = [2,1]$.

3.4 Sketch the feasible regions and solve the following LPs graphically.

(a)
$$\max \quad 2x_1 + 3x_2$$
$$\text{s.t.} \quad x_1 + \ x_2 \leq 6$$
$$2x_1 - \ x_2 \leq 4$$
$$x_2 \leq 2, \quad x \geq 0.$$

(b)
$$\max \quad 4x_1 - 3x_2$$
$$\text{s.t.} \quad x_1 + \ x_2 \leq 6$$
$$2x_1 - \ x_2 \leq 4$$
$$x_2 \geq 2, \quad x \geq 0.$$

(c)
$$\min \quad -2x_1 + 3x_2$$
$$\text{s.t.} \quad 2x_1 + 3x_2 \leq 12$$
$$-x_1 + \ x_2 \geq -2$$
$$2x_1 - \ x_2 \geq -1, \quad x \geq 0.$$

(d)   max $3x_1 + 7x_2$

   s.t. $x_1 + 2x_2 \leq 6$

   $4x_1 + 5x_2 \leq 20$

   $-2x_1 + x_2 \leq 1, \quad 1 \leq x_2 \leq 2, \quad x \geq 0.$

3.5 (a) Sketch the set of points $x = (x_1, x_2)$ in R^2 satisfying

$$x_1 - x_2 \leq -3$$
$$x_1 + 2x_2 \leq 2$$
$$2x_1 + x_2 \geq -4.$$

(b) Does the linear program

min cx

$$\text{s.t.} \quad \begin{bmatrix} -1 & 1 \\ -1 & -2 \\ 2 & 1 \end{bmatrix} x \geq \begin{bmatrix} 3 \\ -2 \\ -4 \end{bmatrix}, \quad x \geq 0$$

have a feasible x?

(c) Write the dual problem for b in the form min $y(-b)$ s.t. $yA \leq c$.

(d) Write the transpose of the solution to part (c).

(e) Consider the linear program

min $-3y_1 + 2y_2 + 4y_3$

$$\text{s.t.} \quad \begin{bmatrix} -1 & -1 & 2 \\ 1 & -2 & 1 \end{bmatrix} y \leq \begin{bmatrix} 1 \\ -1 \end{bmatrix}, \quad y = (y_1, y_2, y_3) \geq 0.$$

Show that $y = (0, 1, 1)$ is feasible.

3.6 Solve Exercise 2.26 graphically.

3.7 Solve Exercise 2.16a graphically.

3.8 Flatiron Co. has two pits, which we label A and B, and a crusher on some property near the Poudre River. Processing a ton of rock from pit A produces 0.5 ton of smooth stone and 0.5 ton of rough stone, while processing a ton of rock from pit B produces 0.75 ton of smooth stone and 0.25 ton of rough stone. The processing cost for rock from pit B is p times that for rock from pit A. Suppose that Flatiron gets an order for S tons of smooth stone and R tons of rough stone.

(a) For which values of S and R is it possible to fill the order without producing any surplus stone?

(b) Suppose that it is possible to process the order without producing any surplus stone. For which values of p are processing costs less when we permit surplus production?

3.9 *Definition*: A point x of a convex set K in R^n is *not* a corner point of K if there are two points y and z in K and two positive numbers p and q, $p > 0$ and $q > 0$, such that (1) $p + q = 1$ and (2) $x = py + qz$. Corner points are also called extreme points.

(a) Show that each of the following sets K has the set of corner points claimed for it in Section 3.1.

(1) $K = \{x_1 + x_2 \le 5, x_1 \ge 0, x_2 \ge 0\}$ in R^2.

(2) $K = \{x_1 + x_2 + x_3 = 5, x_1 \ge 0, x_2 \ge 0, x_3 \ge 0\}$ in R^3.

(3) K in Example 3.1.

(b) Show that a feasible solution, x, for a LP is a BFS (i.e., is basic) if, and only if, x is a corner point of the feasible set, K, for the LP.

3.10 For $n = 1, 2, \ldots$, consider the unit ball B^n in R^n composed of the points $x = (x_1, x_2, \ldots, x_n)$ in R^n satisfying the inequality $x_1^2 + x_2^2 + \cdots + x_n^2 \le 1$.

(a) Show that B^n is a convex set.

(b) Show that every point of the unit sphere S^n is an extreme point of B^n; x is in S^n if, and only if, $x_1^2 + x_1^2 + \cdots + x_n^2 = 1$.

3.11 Consider the set K of points $p = (x, y, z)$ in R^3 that satisfy the following two inequalities: (1) $0 \le |z| \le 1$ and (2) $(x - |z|)^2 + y^2 \le (1 - |z|)^2$.

(a) Sketch K.

(b) Show that K is a convex set.

(c) Describe the set of extreme points of K.

3.12 Find all basic feasible solutions to Example 3.2.

SOLUTIONS TO EXERCISES

3.1 (a)

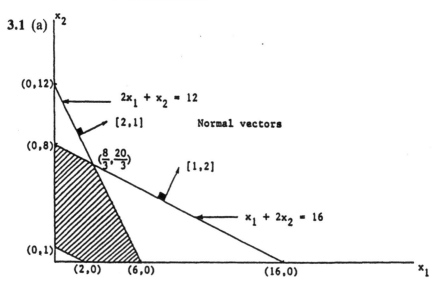

Comments: Included are the normal vectors to two lines; however, they were not required. The point $(\frac{8}{3}, \frac{20}{3})$ was obtained by solving the linear system

$$2x_1 + x_2 = 12$$
$$x_1 + 2x_2 = 16.$$

(b) There are five extreme points: $(0,8)$, $(0,1)$, $(2,0)$, $(6,0)$, and $(\frac{8}{3}, \frac{20}{3})$.

(c) The vector $[3,5]$ is a normal vector to the line $3x_1 + 5x_2 = $ constant. Because $[3,5]$ is a not as steep as $[1,2]$, $3x_1 + 5x_2$ is greater when $x = (x_1, x_2) = (\frac{8}{3}, \frac{20}{3})$ than it is when $x = (0,8)$. On the other hand, $[3,5]$ is steeper than $[2,1]$, so $3x_1 + 5x_2$ is less when $x = (6,0)$ than when $x = (\frac{8}{3}, \frac{20}{3})$. The maximum value of $\frac{124}{3}$ occurs when $x = (\frac{8}{3}, \frac{20}{3})$.

(d) The minimum value of 5 occurs when $x = (0,1)$.

(e) The minimum value of 0 occurs when x is on the interval $(0,1)$ to $(0,8)$: $x = (0, x_2)$, $1 \leq x_2 \leq 8$.

3.5 (a)

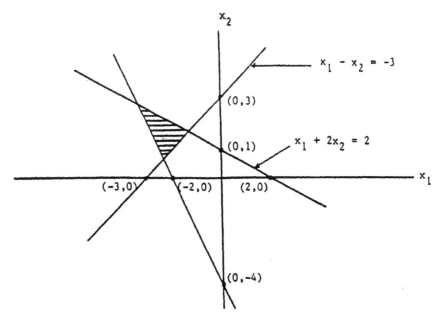

(b) There is no feasible solution.

(c)   max yb

           s.t. $yA \leq c$, $y \geq 0$

or

$$\max \quad 3y_1 - 2y_2 - 4y_3$$
$$\text{s.t.} \quad -y_1 - y_2 + 2y_3 \le c_1$$
$$\phantom{\text{s.t.}} \quad y_1 - 2y_2 + y_3 \le c_2, \quad y \ge 0$$

or

$$(*)\begin{cases} \min \quad -3y_1 + 2y_2 + 4y_3 \\ \text{s.t.} \quad -y_1 - y_2 + 2y_3 \le c_1 \\ \phantom{\text{s.t.}} \quad y_1 - 2y_2 + y_3 \le c_2, \quad y \ge 0. \end{cases}$$

(d) $$\min \quad -b^t y^t$$
$$\text{s.t.} \quad A^t y^t \le c^t.$$

Comment: The dual of the LP (*) has no feasible solutions. In Section 4.8 we will show that the objective values of (*) are unbounded below when $c = [1, -1]$.

3.8 (a) Let a = tons of rock from pit A to process
b = tons of rock from pit B to process.
Put

$$A = \begin{bmatrix} 0.5 & 0.75 \\ 0.5 & 0.25 \end{bmatrix}, \quad x = \begin{bmatrix} a \\ b \end{bmatrix}, \quad \text{and} \quad \bar{b} = \begin{bmatrix} S \\ R \end{bmatrix}.$$

Then

$$A^{-1} = \begin{bmatrix} -1 & 3 \\ 2 & -2 \end{bmatrix} \quad \text{and} \quad A^{-1}\begin{bmatrix} S \\ R \end{bmatrix} = \begin{bmatrix} -S + 3R \\ 2S - 2R \end{bmatrix}.$$

Thus $A^{-1}\bar{b} \ge 0$ when $R \le S \le 3R$.

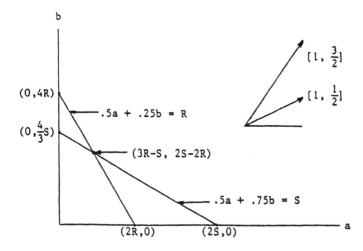

(b) Consider the case $R < S < 3R$. Processing cost is proportional to $a + pb$, so we put $c = [1,p]$. Then $A^{-1}\bar{b}$ is a solution to the LP

$$\min \quad cx$$
$$\text{s.t.} \quad Ax \geq \bar{b}, \qquad x \geq 0$$

if and only if $\frac{1}{2} \leq p \leq \frac{3}{2}$.

3.9 (b) A sketch of a solution follows.

$\Rightarrow$: If x is a BFS, then x is a corner point. As in the proof of the fundamental theorem, we can suppose, without loss of generality, that

$$x = (x_1, \ldots, x_m, 0, \ldots, 0), \qquad \text{where } x_i > 0 \text{ if } i \leq m.$$

1. Show that if x is a proper convex combination of two feasible points, $y = (y_1, \ldots, y_n)$ and $z = (z_1, \ldots, z_n)$, then $y_i = 0$ and $z_i = 0$ if $i > m$: $x = py + qz$, where $p > 0$, $q > 0$ and $p + q = 1$.
2. Show that $y = z = x$. (Recall that $Ay = Az = Ax = b$.)

$\Leftarrow$: If x is not a BFS, then x is not a corner point. As in the proof of the fundamental theorem,

$$\sum_{i=1}^{m} y_i a_i = 0,$$

where at least one y_i is not equal to zero.

3. Put $y = (y_1, \ldots, y_m, 0, \ldots, 0)$ and show that there is a positive number p for which both

$$u = x + py \qquad \text{and}$$
$$v = x - py \qquad \text{both feasible solutions.}$$

4. Observe that $x = \frac{1}{2}u + \frac{1}{2}v$.

4
Introduction to the Simplex Method

The simplex method is essentially an algorithm that systematically examines basic feasible solutions of a standard-form LP for optimality. The discussion begins with some preliminary simplifications and then proceeds to set forth our notation and recall some pertinent algebra. The section headings provide you with a list of the major parts of the method for easy reference. There are many variations of the simplex method and there are newer methods. Exploring the books on the Reading List and the references in those books will expose you to a vast literature on LP; this chapter is merely an introduction.

Preliminary Simplifications

The simplex method assumes that you have an initial basic feasible solution (BFS) with which to begin. Section 4.11 shows how to determine whether a BFS exists and how to find a BFS when one exists. Thus, without incurring any loss of generality, suppose that *we have an initial BFS* with which to begin the simplex method. We can make two other preliminary simplifications without loss of generality: Suppose that the $m \times n$ marix *A has rank m* and suppose that $b \geq 0$.

4.1 NOTATION

Suppose that x is a BFS with $x_i > 0$ for $1 \leq i \leq p$ and $x_i = 0$ for $p < i \leq n$.

Basic Fact

According to basic (pun intended) linear algebra, we can choose $m - p$ columns in the set $\{a_{p+1}, \ldots, a_n\}$ of columns of A and relabel them if necessary, so that they are labeled $a_{p+1}, \ldots, a_m$, and together with $a_1, \ldots, a_p$, compose a set $\{a_1, \ldots, a_p, a_{p+1}, \ldots, a_m\}$ of m linearly independent vectors in R^m. Thus we can label the columns of A so that $A = [a_1, a_2, \ldots, a_m, a_{m+1}, \ldots, a_n] = [B | V]$, where $B = [a_1, \ldots, a_m]$ and $V = [a_{m+1}, \ldots, a_n]$; B has rank m and the columns of B are a basis for R^m. Any vector v in R^m can be written uniquely as a linear combination of $a_1, a_2, \ldots, a_m$; row reduction can be used to determine which combination. For instance, considering the basic feasible solution $x = (2,0,0,0)$ in Example 3.2, $p = 1$ and $m = 2$, so $m - p = 1$; any one of the columns $a_2 = (0,1)$, $a_3 = (0, -2)$, or $a_4 = (1,3)$ can be chosen and relabeled, if necessary, to become the second column in B; B is called a B-matrix.

The notation $A = [B | V]$, where $B = [a_1, \ldots, a_m]$ is chosen to emphasize two basic (for emphasis) points: (1) The columns of B are columns of A, hence the labeling $a_1, \ldots, a_m$ for the columns of B, and (2) the $m \times m$ matrix B composed of these columns is denoted by B to emphasize that these columns correspond to a basic solution (B for basic); the variables corresponding to the columns in B are the only variables that can have nonzero values in the associated basic solution.

4.2 PERTINENT ALGEBRA

Given an $m \times m$ matrix B of rank m and b in R^m, the linear system $Bx = b$ can be solved by row reduction:

$$[B | b] = \begin{bmatrix} a_{11} & \cdots & a_{1m} & b_1 \\ \vdots & & \vdots & \vdots \\ a_{m1} & \cdots & a_{mm} & b_m \end{bmatrix} \to \cdots \to \begin{bmatrix} 1 & 0 & \cdots & & 0 & d_1 \\ 0 & 1 & 0 & \cdots & 0 & \\ \vdots & & & & \cdots & \\ 0 & & \cdots & 0 & 1 & d_m \end{bmatrix}$$

$$= [I | B^{-1}b]$$

or

$$Bx = b \to \cdots \to x = B^{-1}b = d.$$

The same sequence of row operations that changes b to $d = B^{-1}b$ changes v to $B^{-1}v$, where v is a vector in R^m,

V to $B^{-1}V$,   where V is a $m \times k$ matrix,

B to I,

$[B|V|b]$ to $[I|B^{-1}V|d] = [I|W|d]$,   where $W = B^{-1}V$.

The following comment will be very insightful when reduced costs are discussed below.

Comment

(Exercises at the end of Chapter 1 provide examples.) Given $B = [a_1, \ldots, a_m]$, where a_i denotes the ith column of B. Suppose that $w = B^{-1}v$. Then $Bw = \sum_{i=1}^{m} w_i a_i = v$.

4.3 THE SIMPLEX TABLEAU

The constraints $Ax = b$, $x \geq 0$ for the standard-form LP

$$\min cx$$
$$\text{s.t. } Ax = b, \quad x \geq 0$$

can be written as a tableau:

$$[A|b] = [B|V|b] \quad (x \geq 0).$$

The simplex tableau is obtained by adding a row (representing the objective function) to the constraint tableau $[B|V|b]$. There are two natural places to put this cost row. We can add the cost row at the top or at the bottom. Let us label the row $[c|0] = [c_1, \ldots, c_m, c_{m+1}, \ldots, c_n|0] = [c_B|c_V|0]$, where $c_B = [c_1, \ldots, c_m]$ and $c_V = [c_{m+1}, \ldots, c_n]$, put the row at the bottom, and thus obtain the *simplex tableau*

$$\left[\begin{array}{c|c} A & b \\ \hline c & 0 \end{array}\right] = \left[\begin{array}{c|c|c} B & V & b \\ \hline c_B & c_V & 0 \end{array}\right]$$

for the standard-form LP.

Apply row reduction to $[B|V|b]$ so that $[B|V|b] \to \cdots \to [I|B^{-1}V|B^{-1}b] = [I|W|d]$. Then continue row reduction to replace c_B by zero. Thus

$$\left[\begin{array}{c|c|c} B & V & b \\ \hline c_B & c_V & 0 \end{array}\right] \to \cdots \to \left[\begin{array}{c|c|c} I & B^{-1}V & B^{-1}b \\ \hline c_B & c_V & 0 \end{array}\right] = \left[\begin{array}{c|c|c} I & W & d \\ \hline c_B & c_V & 0 \end{array}\right] \to \cdots$$

$$\to \left[\begin{array}{c|c|c} I & W & d \\ \hline 0 & \bar{c}_V & -c_B d \end{array}\right] = \left[\begin{array}{c|c|c} I & W & d \\ \hline 0 & c_V - c_B W & -c_B d \end{array}\right]$$

$$= \begin{bmatrix} I & B^{-1}V & B^{-1}b \\ \hline 0 & c_V - c_B B^{-1}V & -c_B B^{-1}b \end{bmatrix},$$

where $\tilde{c}_V = [\tilde{c}_{m+1}, \ldots, \tilde{c}_n]$, $\tilde{c}_j = c_j - c_B B^{-1}a_j$, and $W = B^{-1}V$. The cost row will be analyzed in subsequent sections.

According to the preceding discussion, $x_B = B^{-1}b$ determines a basic solution $x = (x_B, 0)$ to the system $Ax = b$. We take the liberty of calling x_B a *basic solution*; x_B is a *basic feasible solution* when $x_B \geq 0$. In Example 3.2, three different basic feasible solutions x_B correspond to the same basic feasible solution $x = (2,0,0,0)$. Remember that

$$\sum_{k=1}^{m} w_{kj}a_k = a_j, \quad \text{where } w_j = (w_{1j}, \ldots, w_{mj}) = B^{-1}a_j;$$

these equations underlie reduced costs.

4.4 REDUCED COSTS

The entries in $\tilde{c}_V$ are called *reduced costs* relative to x_B; $\tilde{c}_j = c_j - c_B B^{-1}a_j$ is called the *reduced cost* of x_j relative to $x_1, \ldots, x_m$.

Example 2.2 provides an intuitive example to help understand reduced costs. Think of the a_j's as output lists and b as a list of nutritional requirements. Focusing on the illustration in Section 2.2, put $m = 4$ so that we are focusing on four nutrients. For a numerical example, suppose that

$$B = [a_1, a_2, a_3, a_4] = \begin{bmatrix} 1 & 3 & 3 & 6 \\ 0 & 1 & 4 & 0 \\ 1 & 2 & 3 & 6 \\ 6 & 2 & 2 & 0 \end{bmatrix}$$

is composed of the nutritional lists for units of green beans, soybeans, cottage cheese, and Twinkie. Suppose that $b = (16,8,14,13)$ is the minimal acceptable nutrient level for the diet. According to Exercise 1.4, a meal composed of 1 unit of green beans, 2 units of soybeans, $1\frac{1}{2}$ units of cottage cheese, and $\frac{3}{4}$ unit of Twinkie meets these minimal requirements exactly, at a cost of $1.32. One unit of f_7, veggie supreme cheese pizza, also meets the minimal requirements exactly, but at a cost of $2.49. Because a unit of f_7 can be replaced by its nutritional equivalent, 1 unit of f_1 + 2 units of f_2 + $1\frac{1}{2}$ units of f_3 + $\frac{3}{4}$ unit of f_4, which costs $1.32, f_7 will not be used in the diet. The reduced cost of f_7, relative to f_1, f_2, f_3, and f_4, is $2.49 - 1.32 = 1.17$. The cost of 1 unit of f_7 would have to be reduced at least \$1.27 for f_5 to be considered for the diet. Suppose that f_6, candy, and f_7, spaghetti, have nutritional vectors

$$a_6 = \begin{bmatrix} 0 \\ 0 \\ 7 \\ 0 \end{bmatrix} \quad \text{and} \quad a_7 = \begin{bmatrix} 16 \\ 8 \\ 0 \\ 13 \end{bmatrix}.$$

Then 1 unit of spaghetti and 2 units of candy will also meet the nutritional requirements exactly. If spaghetti and candy cost \$0.99 and \$0.05 per unit, the reduced cost of f_5 relative to f_6 and f_7 would seem to be \$2.49 − 1.09 = \$1.40; however, for technical reasons it is not defined. Because $m = 4$ in this example, reduced costs are only defined relative to four foods with linearly independent nutritional vectors. To put the illustration in Section 2.2 in standard form, we must introduce four surplus variables. Keep this example in mind as an illustration while you read about reduced costs relative to a BFS x_B below.

Suppose that x_B is feasible and we consider c_j to represent the cost of a_j. Then $b = \sum_{k=1}^m x_k a_k$, so the cost of b as a combination of the columns $a_1, \ldots, a_m$ of B is $\sum_{k=1}^m c_k x_k = c_B x_B = c_B(B^{-1}b) = c_B B^{-1}b$. Similarly, the cost of a_j as a combination of $a_1, \ldots, a_m$ is $c_B w_j = c_B B^{-1} a_j$. If $c_j > c_B B^{-1} a_j$, the direct cost c_j of a_j is more than the cost $c_B B^{-1} a_j$ of a_j as a combination of columns of B; if $c_j = c_B B^{-1} a_j$, the costs are equal, and if $c_j < c_B B^{-1} a_j$, the direct cost of a_j is less. The numbers $\bar{c}_j = c_j - c_B B^{-1} a_j$ are called reduced costs, $\bar{c}_V = [\bar{c}_{m+1}, \ldots, \bar{c}_n]$. When $c_j - c_B B^{-1} a_j < 0$, it would be cheaper to use a_j directly rather than as the combination $\sum_{k=1}^m w_{kj} a_k$ of $a_1, \ldots, a_m$. In the next section it will be shown that a basic feasible solution x_B is optimal if all the reduced costs relative to x_B are nonnegative.

4.5 CONDITIONS FOR OPTIMALITY

The second fundamental fact about primal–dual pairs can be used to give a sufficient condition for a basic feasible solution x_B to be optimal as follows. Simply put

$$\lambda_B = c_B B^{-1}.$$

Since $cx = c_B x_B = c_B B^{-1}b = \lambda_B b$, where $x = (x_B, 0)$ is feasible for a standard-form (primal) LP, the second fundamental fact about standard primal–dual pairs tells us that x and λ_B are optimal for the primal and dual, respectively, if λ_B is feasible for the dual problem, that is, if $\lambda_B A \leq c$.

However,

$$\lambda_B A = \lambda_B[B|V] = c_B B^{-1}[B|V] = [c_B B^{-1}B|c_B B^{-1}V]$$
$$= [c_B|c_B B^{-1}V].$$

Hence we have

$$\lambda_B A = [c_B|c_B B^{-1}V] \leq c = [c_B, c_V]$$

if and only if $c_B B^{-1}V \leq c_V$ or, equivalently, if and only if $\tilde{c}_V = c_V - c_B B^{-1}V \geq 0$. Consequently, $x = (x_B, 0)$ is optimal if $\tilde{c}_V \geq 0$, or, to repeat in slightly different terminology, x_B is optimal if all the reduced costs are nonnegative. When $\tilde{c}_j > 0$, the reduced cost $\tilde{c}_j$ indicates how much c_j must decrease before a_j becomes a candidate for a basic column. If $\tilde{c}_j < 0$, then a_j is a candidate to become a basic column.

Example 3.2 is used for an illustration. Example 3.2 has two basic feasible solutions, $x = (2,0,0,0)$ and $x = (0,0,3,2)$; the first is degenerate and the second is nondegenerate.

Corresponding to $x = (2,0,0,0)$, we will choose

$$B = [a_1, a_2] = \begin{bmatrix} 1 & 0 \\ 0 & 1 \end{bmatrix}.$$

Then

$$c_B = [c_1, c_2] = [3,2], \quad \lambda_B = c_B B^{-1} = [3,2]\begin{bmatrix} 1 & 0 \\ 0 & 1 \end{bmatrix} = [3,2],$$

and

$$\lambda_B A = [3,2]\begin{bmatrix} 1 & 0 & 0 & 1 \\ 0 & 1 & -2 & 3 \end{bmatrix} = [3,2,-4,9],$$

which is not less than or equal to $c = [3,2,1,1]$: the fourth column is a candidate to replace the second column in B.

Corresponding to $x = (0,0,3,2)$,

$$B = [a_3, a_4] = \begin{bmatrix} 0 & 1 \\ -2 & 3 \end{bmatrix}.$$

Then

$$c_B = [c_3, c_4] = [1,1], \quad B^{-1} = (\tfrac{1}{2})\begin{bmatrix} 3 & -1 \\ 2 & 0 \end{bmatrix}, \quad \lambda_B = c_B B^{-1} = (\tfrac{1}{2})[5,-1];$$

$$\lambda_B A = (\tfrac{1}{2})[5,-1]\begin{bmatrix} 1 & 0 & 0 & 1 \\ 0 & 1 & -2 & 3 \end{bmatrix} = (\tfrac{1}{2})[5,-1,2,2] = [\tfrac{5}{2},-\tfrac{1}{2},1,1] \leq c.$$

Thus $x = [0,0,3,2]$ is an optimal BFS for Example 3.2 with $cx = 5$. Since there are only two basic feasible solutions to this problem and $c(2,0,0,0) = 6$, we see by inspection that $x = (0,0,3,2)$ is the unique optimal solution.

4.6 THE OBJECTIVE FUNCTION

The objective function appears as the bottom row in the simplex tableau in Section 4.3, where it is modified in form by row operations involving the constraints; the reduced costs discussed in Section 4.4 appear explicitly in the final modified form of the simplex tableau. Section 4.5 established the fact that a basic feasible solution x_B is optimal if the associated reduced costs are all nonnegative. When there are negative reduced costs, the simplex method systematically examines column vectors with negative reduced costs. The simplex method algorithmically modifies x_B by replacing one of the variables in x_B by a nonbasic variable with a negative reduced cost to produce a new basic feasible solution $x_{\overline{B}}$. When x_B is nondegenerate, a term that will be defined later, $x_{\overline{B}}$ has a lower cost than x_B: $c_{\overline{B}}x_{\overline{B}} < c_B x_B$.

To expose the idea behind the simplex method, let us look at the value cy of the objective function at an arbitrary solution y of $Ay = b$. Put $y = [y_B, y_V]$. Then

$$Ay = [B|V]\begin{bmatrix} y_B \\ y_V \end{bmatrix} = By_B + Vy_V = b,$$

so

$$B^{-1}[By_B + Vy_V] = y_B + B^{-1}Vy_V = B^{-1}b,$$

which implies that

$$y_B = B^{-1}b - B^{-1}Vy_V.$$

Thus

$$\begin{aligned}
cy &= c_B y_B + c_V y_V \\
&= c_B[B^{-1}b - B^{-1}Vy_V] + c_V y_V \\
&= c_B B^{-1}b - c_B B^{-1}Vy_V + c_V y_V \\
&= c_B B^{-1}b + [c_V - c_B B^{-1}V]y_V \\
&= c_B x_B + \tilde{c}_V y_V,
\end{aligned}$$

where $x = (x_B, 0)$, the basic solution corresponding to B, is feasible but not necessarily optimal. If $\tilde{c}_V \geq 0$, then x_B is optimal. But if some $\tilde{c}_j < 0$ and we can replace one of $a_1, \ldots, a_m$ by a_j and get a basic feasible solution with $y_j > 0$, the objective function will decrease. This is the idea behind the simplex method.

A Second View of the Objective Function

To help explain what is occurring in the cost row, let us look at the objective function from another point of view. Put $f = f(y) = cy$ and re-write $f = cy$ in the form $-f + cy = 0$. Then consider f to be additional variable and write a tableau for the system

$$Ay = b$$
$$-f + cy = 0$$

in the form

$$0 + Ay = b$$
$$-f + cy = 0.$$

Equivalently,

$$\begin{bmatrix} 0 \\ \vdots & A \\ 0 \\ \hline -1 & c \end{bmatrix} \begin{bmatrix} f \\ \hline y \end{bmatrix} = \begin{bmatrix} b \\ 0 \end{bmatrix},$$

which corresponds to the expanded tableau

$$\begin{bmatrix} 0 & & & \\ \vdots & B & V & b \\ 0 & & & \\ \hline -1 & c_B & c_V & 0 \end{bmatrix}.$$

Apply row reduction to get

$$\begin{bmatrix} 0 & & & \\ \vdots & I & B^{-1}V & B^{-1}b \\ 0 & & & \\ \hline -1 & 0 & c_V - c_B B^{-1}V & -c_B B^{-1}b \end{bmatrix}.$$

Because of its form, the first column remains constant; since it does not change, stop writing it. The first column was introduced to relate the cost row to a form of the objective function that was considered earlier in this discussion: namely,

$$f = cy = \tilde{c}_V y_V + c_B B^{-1} b \quad \text{or} \quad -f + \tilde{c}_V y_V = -c_B B^{-1} b,$$

which is the bottom row of the preceding tableau.

Example 4.1. To illustrate, we return to Example 3.1 and introduce the objective function $cx = x_1 + x_2 - x_3$, which we shall minimize subject to

$$\begin{bmatrix} 1 & \frac{2}{3} & 1 \\ \frac{1}{2} & 1 & 0 \end{bmatrix} \begin{bmatrix} x_1 \\ x_2 \\ x_3 \end{bmatrix} \leq \begin{bmatrix} 1 \\ 1 \end{bmatrix}, \quad x_1, x_2, x_3 \geq 0.$$

The associated standard-form LP is

$$\min [1,1,-1,0,0]x = cx$$

$$\text{s.t.} \begin{bmatrix} 1 & \frac{2}{3} & 1 & 1 & 0 \\ \frac{1}{2} & 1 & 0 & 0 & 1 \end{bmatrix} x = \begin{bmatrix} 1 \\ 1 \end{bmatrix}, \quad x = (x_1, x_2, x_3, x_4, x_5) \geq 0,$$

with expanded (to include f) tableau

$$\begin{bmatrix} 0 & 1 & \frac{2}{3} & 1 & 1 & 0 & 1 \\ 0 & \frac{1}{2} & 1 & 0 & 0 & 1 & 1 \\ -1 & 1 & 1 & -1 & 0 & 0 & 0 \end{bmatrix} = \begin{bmatrix} 0 & & & \\ 0 & V & B & b \\ -1 & c_V & c_B & 0 \end{bmatrix}$$

$$= \begin{bmatrix} 0 & & & \\ 0 & W & I & B^{-1}b \\ -1 & \tilde{c}_V & 0 & -c_B B^{-1} b \end{bmatrix},$$

since $B = I$ and $c_B = 0$.

The slack variable solution $x_4 = 1$, $x_5 = 1$ is a basic feasible solution. There is one negative entry in $\tilde{c}_V = [1,1,-1]$, so $j = 3$, and we wish to bring

$$a_3 = \begin{bmatrix} 1 \\ 0 \end{bmatrix}$$

into the basic solution. We need to replace one of

$$a_4 = \begin{bmatrix} 1 \\ 0 \end{bmatrix} \quad \text{and} \quad a_5 = \begin{bmatrix} 0 \\ 1 \end{bmatrix}$$

by a_3. If we were to replace a_5, we would have

$$[a_4 a_3] = \begin{bmatrix} 1 & 1 \\ 0 & 0 \end{bmatrix},$$

which would not be a rank-two matrix, so we must replace a_4; put

$$B = [a_3, a_5] = \begin{bmatrix} 1 & 0 \\ 0 & 1 \end{bmatrix}.$$

Now $c_B = [c_3, c_5] = [-1, 0]$. We can replace c_3 by zero in the tableau by adding the first row to the bottom row to get the expanded tableau

$$\begin{bmatrix} 0 & 1 & \frac{2}{3} & 1 & 1 & 0 & 1 \\ 0 & \frac{1}{2} & 1 & 0 & 0 & 1 & 1 \\ \hline -1 & 2 & \frac{5}{3} & 0 & 1 & 0 & 1 \end{bmatrix}.$$

Thus $x = (0,0,1,0,1)$ is a basic feasible solution to the standard-form version of Example 4.1, and the objective function is in the form

$$f = f(y) = 2y_1 + \tfrac{5}{3}y_2 + y_4 - 1.$$

Because all the reduced costs of the nonbasic variables are >0, $(0,0,1)$ is the unique solution to the original LP:

$$\min \quad x_1 + x_2 - x_3$$
$$\text{s.t.} \quad x_1 + \tfrac{2}{3}x_2 + x_3 \leq 1$$
$$\tfrac{1}{2}x_1 + x_2 \quad\quad \leq 1$$
$$x_1, x_2, x_3 \geq 0.$$

We have solved this very simple problem by inspection.

In the next several sections we discuss an algorithm that proceeds automatically and can be programmed.

4.7 SIMPLEX METHOD PIVOTING

Example 4.1 will be used to set up a pivoting situation. The first two columns of the matrix A are linearly independent, so let us see if the basic solution corresponding to those columns is feasible by using row reduction:

$$\begin{bmatrix} 1 & \frac{2}{3} & 1 & 1 & 0 & 1 \\ \frac{1}{2} & 1 & 0 & 0 & 1 & 1 \\ 1 & 1 & -1 & 0 & 0 & 0 \end{bmatrix} = \begin{bmatrix} B & V & b \\ \hline c_B & c_V & 0 \end{bmatrix}$$

$$\xrightarrow[\;-(1/2)(1)\text{ to }(2)\;]{}\quad
\left[\begin{array}{ccccc|c}
1 & \tfrac{2}{3} & 1 & 1 & 0 & 1\\
0 & \tfrac{2}{3} & -\tfrac{1}{2} & -\tfrac{1}{2} & 1 & \tfrac{1}{2}\\ \hline
1 & 1 & -1 & 0 & 0 & 0
\end{array}\right]$$

$$\xrightarrow[(3/2)(2)]{}\quad
\left[\begin{array}{ccccc|c}
1 & \tfrac{2}{3} & 1 & 1 & 0 & 1\\
0 & 1-\tfrac{3}{4} & -\tfrac{3}{4} & \tfrac{3}{2} & \tfrac{3}{4}\\ \hline
1 & 1-1 & 0 & 0 & 0
\end{array}\right]
\xrightarrow[\;-(2/3)(2)\text{ to }(1)\;]{}
\left[\begin{array}{ccccc|c}
1 & 0 & \tfrac{3}{2} & \tfrac{3}{2} & -1 & \tfrac{1}{2}\\
0 & 1 & -\tfrac{3}{4} & -\tfrac{3}{4} & \tfrac{3}{2} & \tfrac{3}{4}\\ \hline
1 & 1 & -1 & 0 & 0 & 0
\end{array}\right]$$

$$\xrightarrow[\substack{-(1)\text{ to }(3)\\-(2)\text{ to }(3)}]{}\quad
\left[\begin{array}{ccccc|c}
1 & 0 & \tfrac{3}{2} & \tfrac{3}{2} & -1 & \tfrac{1}{2}\\
0 & 1 & -\tfrac{3}{4} & -\tfrac{3}{4} & \tfrac{3}{2} & \tfrac{3}{4}\\ \hline
0 & 0 & -\tfrac{7}{4} & -\tfrac{3}{4} & -\tfrac{1}{2} & -\tfrac{5}{4}
\end{array}\right].$$

We see that $x_B = [\tfrac{1}{2},\tfrac{3}{4}]$ is a basic feasible solution and the corresponding $x = [\tfrac{1}{2},\tfrac{3}{4},0,0,0]$. The form of the objective function is now

$$f = f(y) = -\tfrac{7}{4}y_3 - \tfrac{3}{4}y_4 - \tfrac{1}{2}y_5 + \tfrac{5}{4},$$

where y denotes any solution of $Ay = b$;

$$\bar{c}_V = [\bar{c}_3, \bar{c}_4, \bar{c}_5] = [-\tfrac{7}{4}, -\tfrac{3}{4}, -\tfrac{1}{2}].$$

The Pivot Column

We will bring *column j* into B, where $\bar{c}_j$ *is a most negative entry in* $\bar{c}_V$. (In the case of ties, choose one systematically.) This choice will cause the objective function to decrease as fast as is possible by changing one basic variable. Continuing with Example 4.1, we see that column 3 is the column to be brought in because $-\tfrac{7}{4} \le -\tfrac{3}{4}$ and $-\tfrac{7}{4} \le -\tfrac{1}{2}$.

The Pivot Point

We will replace one column in the $m \times m$ identity matrix I by the jth (pivot) column. The column of I that we replace has one nonzero entry, a 1. If that entry occurs in the ith place, we *call (i,j) the pivot point*. We will use row operations to change w_{ij} to 1 and change all the other entries in w_j to zero. We must choose the pivot point so that the new basic solution is feasible. We will use the idea introduced in the proof of the fundamental theorem of LP to decide which column of I to replace by the jth (pivot) column.

To focus on the method, we leave Example 4.1 temporarily and consider the standard-form LP

$$\min cx$$
$$\text{s.t. } Ax = b, \quad x \geq 0,$$

in tableau form

$$\left[\begin{array}{c|c} A & b \\ \hline c & 0 \end{array}\right] = \left[\begin{array}{c|c|c} B & V & b \\ \hline c_B & c_V & 0 \end{array}\right] \rightarrow \cdots \rightarrow \left[\begin{array}{c|c|c} I & W & d \\ \hline 0 & \tilde{c}_V & -c_B B^{-1} b \end{array}\right].$$

Suppose that $\tilde{c}_j < 0$ is a smallest entry in $\tilde{c}_V$. Let us see if we can put x_j into a basic feasible solution by replacing some x_i, where $1 \leq i \leq m$, and improve the value of the objective function. We need $d_i > 0$ for the objective function to decrease. The current BFS $x_B = d$, and we want the objective function to decrease. The objective function will decrease if $x_B > 0$, in which case we say that x_B is nondegenerate: x_B is *nondegenerate* if every entry in x_B is > 0; otherwise, x_B is *degenerate*. An example follows.

In Example 3.2, the three basic feasible solutions x_B corresponding to the basic feasible solution $x = (2,0,0,0)$ are degenerate. The rank m of the matrix A for Example 3.2 is two and x has fewer than two nonzero entries. When a BFS x has fewer than m ($m = 2$ in Example 3.2) positive entries, x is said to be a *degenerate* BFS.

Dealing with Degeneracy—Assumption of Nondegeneracy

When x is degenerate, the "Basic Fact" in Section 4.1 can be used to find a corresponding B matrix, and an anticycling feature can be incorporated in the algorithm to prevent cycling forever in a loop of B's that correspond to the same BFS x. An anticycling feature assures that each basic matrix B is used at most once in the simplex algorithm; consequently, in due course, the algorithm chooses a basic matrix B corresponding to a BFS distinct from x. Thus we can deal with degeneracy; these complications are avoided in the discussion below by the following assumption: *Every BFS that is encountered henceforth in this section is nondegenerate.*

Looking at the column $w_j = (w_{1j}, w_{2j}, \ldots, w_{ij}, \ldots, w_{mj})$ in W corresponding to x_j, we see that we need $w_{ij} > 0$ in order to replace x_i by x_j. We want to choose i with $w_{ij} > 0$ such that the new basic solution will be feasible. We will use row operations to replace x_i by x_j and then discover a rule for choosing i so that the new BFS is feasible. The row operations we use correspond to applying row reduction to column j with respect to the pivot point (i,j). We use row operations to change w_{ij} to 1 and w_{kj} to zero for $k \neq i$. The first change is effected by multiplying row i by w_{ij}^{-1} and the other changes are effected by subtracting w_{kj} times the modified ith row from the kth row, $k \neq i$. These modifications change w_j to $\bar{w}_j = (0_1, \ldots, 0_{i-1}, 1_i, 0_{i+1}, \ldots, 0_m)$ and d to $\bar{d} = (\bar{d}_1, \ldots, \bar{d}_m)$, where $\bar{d}_i = d_i/w_{ij}$ and $\bar{d}_k = d_k - w_{kj} (d_i/w_{ij})$, $k \neq i$. The new basic solution is $x_k = \bar{d}_k$, where $k \leq m$ and $k \neq i$, and $x_j = \bar{d}_i > 0$. The change in the objective function is $\bar{c}_j x_j$. We want the new basic solution to be feasible, so we want $\bar{d} \geq 0$; for $1 \leq k \leq m$, we want

$$d_k - w_{kj}\frac{d_i}{w_{ij}} \geq 0 \quad \text{or} \quad d_k \geq w_{kj}\frac{d_i}{w_{ij}}.$$

This requirement is satisfied if $w_{kj} \leq 0$. Thus we choose i so that $d_i/w_{ij} \leq d_k/w_{kj}$ whenever $w_{kj} > 0$. To state this choice slightly differently, consider the vector ratio $d/w_j = (d_1/w_{1j}, d_2/w_{2j}, \ldots, d_m/w_{mj})$ and *choose i to correspond to a smallest positive entry* in this vector. If there is only one choice for x_i, then $\bar{d} > 0$. But if there is a tie, $\bar{d}$ has a zero entry and the new BFS is degenerate. The assumption of nondegeneracy for this section implies that no ties occur here. The tableau now has the form

column→	1	2	·	·	·	i	·	·	·	·	m	···	j	········	RHS
row															
1	1	0	·	·	·	0	·	·	·	·	0		0		$\bar{d}_1$
2	0	1	0	·	·	0	·	·	·	·	0		0		·
·		·	0	1	0	·	0	·	·	·	·	0		0	·
·			·	·	0		·				·		·		
·						0					·		0		
i	0	·	·	·	0	1	0	·	·		0		1		$\bar{d}_i$
·		·				0					·		0		·
·		·							·		·		·		
·		·						0			·		·		
m	0	·	·	·	·	0	·	·	·		1		0		$\bar{d}_m$
	0	·	·	·	·	0	·	·	·	·	0		$\bar{c}_j$		$-c_B x_B$

We can permute the positions of columns i and j and subtract $\bar{c}_j$ times row i from the objective function row to get a tableau of the form

$$\left[\begin{array}{c|c|c} I & \bar{W} & \bar{d} \\ \hline 0 & \bar{c}_W & -c_B x_B - \bar{c}_j \bar{d}_i \end{array}\right].$$

On the other hand, if we permute a_i and a_j in the original A matrix to get a matrix $\bar{A} = [\bar{B}|\bar{V}]$ and continue

$$\left[\begin{array}{c|c} \bar{A} & b \\ \hline \bar{c} & 0 \end{array}\right] = \left[\begin{array}{c|c|c} \bar{B} & \bar{V} & b \\ \hline c_{\bar{B}} & c_{\bar{V}} & 0 \end{array}\right] \to \cdots \to \left[\begin{array}{c|c|c} I & \bar{B}^{-1}\bar{V} & \bar{d} \\ \hline 0 & \bar{c}_V & -c_{\bar{B}}\bar{d} \end{array}\right].$$

we obtain the same tableau.

The preceding discussion boils down to an algorithm that terminates with optimal basic solution x_B when $\bar{c}_V \geq 0$ or stops at a point where some reduced cost $\bar{c}_j < 0$ and $w_j \leq 0$. Section 4.8 verifies that the objective function is unbounded below in the latter case.

Let us apply this process to Example 4.1.

Example 4.1 (cont.). From the tableau

$$\left[\begin{array}{cc|ccc|c} 1 & 0 & \frac{3}{2} & \frac{3}{2} & -1 & \frac{1}{2} \\ 0 & 1 & -\frac{3}{4} & -\frac{3}{4} & \frac{3}{2} & \frac{3}{4} \\ \hline 0 & 0 & -\frac{7}{4} & -\frac{3}{4} & -\frac{1}{2} & -\frac{5}{4} \end{array}\right], \qquad \bar{c}_V = [-\tfrac{7}{4}, -\tfrac{3}{4}, -\tfrac{1}{2}].$$

Because $-\frac{7}{4} < -\frac{3}{4}$ and $-\frac{7}{4} < -\frac{1}{2}$, $j = 3$; thus

$$w_j = \left[\begin{array}{c} \frac{3}{2} \\ -\frac{3}{4} \end{array}\right].$$

Also,

$$d = \left[\begin{array}{c} \frac{1}{2} \\ \frac{3}{4} \end{array}\right],$$

so

$$\frac{d}{w_j} = \left[\begin{array}{c} \frac{1}{2}/\frac{3}{2} \\ \frac{3}{4}/-\frac{3}{4} \end{array}\right] = \left[\begin{array}{c} \frac{1}{3} \\ -1 \end{array}\right].$$

The smallest positive ratio in d/w_j occurs in the position corresponding to the first column, so the third column will replace the first column. Consequently, (1,3) is the pivot point. The tableaux corresponding to pivoting at (1,3) are

$$\xrightarrow{(2/3)(1)} \left[\begin{array}{cc|ccc|c} \frac{2}{3} & 0 & 1 & 1 & -\frac{2}{3} & \frac{1}{3} \\ 0 & 1 & -\frac{3}{4} & -\frac{3}{4} & \frac{3}{2} & \frac{3}{4} \\ \hline 0 & 0 & -\frac{7}{4} & -\frac{3}{4} & -\frac{1}{2} & -\frac{5}{4} \end{array}\right] \xrightarrow[\substack{(3/4)(1)\ \text{to}\ (2)\\(7/4)(1)\ \text{to}\ (3)}]{} \left[\begin{array}{cc|ccc|c} \frac{2}{3} & 0 & 1 & 1 & -\frac{2}{3} & \frac{1}{3} \\ \frac{1}{2} & 1 & 0 & 0 & 1 & 1 \\ \hline \frac{7}{6} & 0 & 0 & 1 & -\frac{10}{6} & -\frac{2}{3} \end{array}\right].$$

The latter tableau tells us that $x = (0,1,\frac{1}{3},0,0)$ is a basic feasible solution with objective value $\frac{2}{3}$ and also that

$$f = f(x) = \tfrac{7}{6}x_1 + x_4 + -\tfrac{10}{6}x_5 + \tfrac{2}{3}$$

is a valid form of the objective function. Consequently, we wish to bring in the fifth column. By comparing ratios we learn that the second column is to be replaced by the fifth column. Pivoting at position (2,5) corresponds to the tableau

$$\xrightarrow[\substack{(2/3)(2)\ \text{to}\ (1)\\(5/3)(2)\ \text{to}\ (3)}]{} \left[\begin{array}{ccccc|c} 1 & \frac{2}{3} & 1 & 1 & 0 & 1 \\ \frac{1}{2} & 1 & 0 & 0 & 1 & 1 \\ \hline 2 & \frac{5}{3} & 0 & 1 & 0 & 1 \end{array}\right]$$

We have arrived at the optimal solution $x = (0,0,1,0,1)$, which was obtained by inspection in Section 4.6. If you look back to that discussion, you will see that the simplex method was applied there, beginning with the slack variable BFS, without saying so. Notice that the incoming and outgoing columns are conceptually not physically interchanged as the simplex method is applied.

An important observation made earlier about this solution is repeated here for emphasis: Because the coefficients of all the nonbasic variables in the current form $f = 2x_1 + \frac{5}{3}x_2 + x_4 - 1$ of the objective function are positive, if we were to use any of these variables at a positive level, the objective function would increase. Consequently, the optimal solution $x = (0,0,1,0,1)$ is unique.

The simplex method was presented in this section under the assumption that only nondegenerate basic feasible solutions were encountered. Section 4.10 deals with degeneracy after you are shown in Section 4.9 that there are multiple optimal solutions when a nondegenerate BFS has some zero-level reduced costs. Section 4.8 shows that the objective function is unbounded below when the case $c_j < 0$, $w_j \leq 0$ is encountered.

4.8   WHEN NO OPTIMAL SOLUTION EXISTS

We now consider the case where $\tilde{c}_j < 0$ and $w_j \leq 0$. Recall that

$$\sum_{i=1}^{m} w_{ij}a_i = a_j$$

or

$$\sum_{i=1}^{m} (-w_{ij})a_i + a_j = 0.$$

Put $h_i = -w_{ij} \geq 0$ and let $p > 0$. Then

$$\sum_{i=1}^{m} h_i a_i + a_j = 0$$

so

$$\sum_{i=1}^{m} (ph_i)a_i + pa_j = 0.$$

Put

$$y = (ph_1, ph_2, \ldots, ph_m, 0_{m+1}, \ldots, 0_{j-1}, p_j, 0_{j+1}, \ldots, 0_n).$$

Then

$$Ay = 0,$$

so

$$A(x + y) = b.$$

But $c(x + y) = c_B x_B + p\bar{c}_j$, which is less than any preassigned number if p is sufficiently large. Consequently, if $\bar{c}_j < 0$ and $w_j \leq 0$, the objective function is unbounded below and there is no optimal solution. For example, referring to Exercise 3.5, we know that the dual of the linear program

$$\min \quad -3y_1 + 2y_2 + 4y_3$$
$$\text{s.t.} \quad -y_1 - y_2 + 2y_3 + y_4 \quad = 1$$
$$y_1 - 2y_2 + y_3 \quad + y_5 = -1, \quad y \geq 0,$$

has no feasible solution to block values of the objective function, so observing that $y = (0,\frac{1}{2},0,\frac{3}{2},0)$ is a basic feasible solution, we apply the simplex method to change the fourth and second columns to the identity.

$$
\begin{array}{cc}
x_2 \quad x_4 & \\
\end{array}
\begin{bmatrix}
-1 & -1 & 2 & 1 & 0 & 1 \\
1 & -2 & 1 & 0 & 1 & -1 \\
\hline
-3 & 2 & 4 & 0 & 0 & 0
\end{bmatrix}
\xrightarrow{-(1/2)(2)}
\begin{bmatrix}
-1 & -1 & 2 & 1 & 0 & 1 \\
-\frac{1}{2} & 1 & -\frac{1}{2} & 0 & -\frac{1}{2} & \frac{1}{2} \\
\hline
-3 & 2 & 4 & 0 & 0 & 0
\end{bmatrix}
\xrightarrow{(2)\ \text{to}\ (1)}
$$

$$
\begin{array}{cc}
x_2 \quad x_4 & \\
\end{array}
\begin{bmatrix}
-\frac{1}{3} & 0 & \frac{3}{2} & 1 & -\frac{1}{2} & \frac{3}{2} \\
-\frac{1}{2} & 1 & -\frac{1}{2} & 0 & -\frac{1}{2} & \frac{1}{2} \\
\hline
-3 & 2 & 4 & 0 & 0 & 0
\end{bmatrix}
\xrightarrow{-(2)\ \text{to}\ (3)}
\begin{bmatrix}
-\frac{3}{2} & 0 & \frac{3}{2} & 1 & -\frac{1}{2} & \frac{3}{2} \\
-\frac{1}{2} & 1 & -\frac{1}{2} & 0 & -\frac{1}{2} & \frac{1}{2} \\
\hline
-2 & 0 & 5 & 0 & 1 & -1
\end{bmatrix}.
$$

The first column has a negative reduced cost, -2, and

$$w_1 = \begin{bmatrix} -\frac{3}{2} \\ -\frac{1}{2} \end{bmatrix}$$

has all negative entries. Thus we focus on the representation of the first column as a linear combination of the second and fourth columns. Referring to Exercise 1.3 for amplification,

$$-\frac{1}{2}\begin{bmatrix} -1 \\ -2 \end{bmatrix} - \frac{3}{2}\begin{bmatrix} 1 \\ 0 \end{bmatrix} = \begin{bmatrix} -1 \\ 1 \end{bmatrix} \quad \text{or} \quad \begin{bmatrix} -1 \\ 1 \end{bmatrix} + \frac{1}{2}\begin{bmatrix} -1 \\ -2 \end{bmatrix} + \frac{3}{2}\begin{bmatrix} 1 \\ 0 \end{bmatrix} = 0;$$

consequently, for each $p \geq 0$, $(0,\frac{1}{2},0,\frac{3}{2},0) + p(1,\frac{1}{2},0,\frac{3}{2},0)$ is a feasible solution with value $1 - 2p$.

The ray $\{y_0 + py_1 : y_0 = (0,\frac{1}{2},0,\frac{3}{2},0), \ y_1 = (1,\frac{1}{2},0,\frac{3}{2},0), \ p \geq 0\}$ is in the feasible set. It is a half-line emanating from y_0 in the direction y_1. As a point on the ray moves away from y_0, the value of the objective function at the point decreases.

Notice that if $\tilde{c}_j$ were equal to zero, the objective function would remain constant on the ray. Multiple optimal solutions are discussed next.

4.9 MULTIPLE SOLUTIONS

The presence of zero reduced costs for nonbasic variables in a final tableau is a signal that there may be multiple optimal solutions to the LP. For example, consider the LP

$$\min \ x_1 + x_2 + x_3 + x_4$$

$$\text{s.t.} \ \ x_1 + x_2 \qquad\quad = 3$$

$$x_2 + x_3 \quad\ \ = 1$$

$$x_3 + x_4 = 1, \qquad x \geq 0.$$

Your solution to the linear system in Exercise 1.5 shows you that the feasible set for the LP given here has two extreme points, $(3,0,1,0)$ and $(2,1,0,1)$. The first is a degenerate BFS and the second is a nondegenerate BFS. Let us focus on $(3,0,1,0)$, put the first three columns of the coefficient matrix in B, and perform the simplex method operations that will change B to the identity matrix on the following tableau:

$$\begin{bmatrix} 1 & 1 & 0 & 0 & 3 \\ 0 & 1 & 1 & 0 & 1 \\ 0 & 0 & 1 & 1 & 1 \\ 1 & 1 & 1 & 1 & 0 \end{bmatrix} \xrightarrow[(-3) \text{ to } (2)]{} \begin{bmatrix} 1 & 1 & 0 & 0 & 3 \\ 0 & 1 & 0 & -1 & 0 \\ 0 & 0 & 1 & 1 & 1 \\ 1 & 1 & 1 & 1 & 0 \end{bmatrix}$$

$$\xrightarrow[-(2) \text{ to } (1)]{} \begin{bmatrix} 1 & 0 & 0 & 1 & 3 \\ 0 & 1 & 0 & -1 & 0 \\ 0 & 0 & 1 & 1 & 1 \\ 1 & 1 & 1 & 1 & 0 \end{bmatrix} \xrightarrow[\substack{-(1) \text{ to } (4) \\ -(2) \text{ to } (4) \\ -(3) \text{ to } (4)}]{} \begin{bmatrix} 1 & 0 & 0 & 1 & 3 \\ 0 & 1 & 0 & -1 & 0 \\ 0 & 0 & 1 & 1 & 1 \\ 0 & 0 & 0 & 0 & -4 \end{bmatrix}.$$

So $(3,0,1,0)$ is an optimal basic feasible solution and the value of the objective function at $(3,0,1,0)$ is 4. However, since x_4 has a reduced cost of zero, we might have other optimal solutions involving x_4. To try to find a BFS involving x_4, look at the vector ratio d/w_4, where $w_4 = (1,-1,1)$ is the fourth column in the preceding tableau and $d = (3,0,1)$ is the last column. The smallest positive ratio in $d/w_4 = (3,0,1)$ occurs in the third position. Consequently, we bring x_4 into a BFS by pivoting at the $(3,4)$

position. The simplex method tableau operations follow:

$$\begin{array}{c} -\text{(3) to (1)} \\ \text{(3) to (2)} \\ \longrightarrow \end{array} \left[\begin{array}{cccc|c} 1 & 0 & -1 & 0 & 2 \\ 0 & 1 & 1 & 0 & 1 \\ 0 & 0 & 1 & 1 & 1 \\ \hline 0 & 0 & 0 & 0 & -4 \end{array}\right].$$

This pivot produces the optimal BFS, $x = (2,1,0,1)$. The set of optimal solutions to a LP is a convex set. Consequently, for each number t between zero and 1, $x_t = t(3,0,1,0) + (1 - t)(2,1,0,1) = (2 + t, 1 - t, t, 1 - t)$ is an optimal solution.

In this example we have a degenerate optimal BFS $(3,0,1,0)$ and a zero reduced cost $\tilde{c}_4 = 0$. We were able to pivot in the fourth column and get another optimal BFS $(2,1,0,1)$. However, if we look at the LP with simplex tableau

$$\left[\begin{array}{ccccc} 1 & 0 & 1 & 1 & 0 \\ 0 & 1 & 0 & 2 & 1 \\ \hline 0 & 0 & 0 & 3 & 0 \end{array}\right] = \left[\begin{array}{c|c} A & b \\ \hline c & 0 \end{array}\right]$$

and put

$$B = [a_1, a_2] = \begin{bmatrix} 1 & 0 \\ 0 & 1 \end{bmatrix},$$

then $x_B = (0,1)$ is an optimal BFS and the reduced cost $\tilde{c}_3 = 0$. But we cannot get another optimal BFS because the optimal BFS is really $x = (0,1,0,0)$, which is the unique optimal solution to this LP. The point here is that in the presence of degeneracy, zero reduced costs may or may not lead to multiple solutions. If $\tilde{c}_j = 0$ and $w_j \leq 0$, then, as was observed at the end of Section 4.8, there is a ray of optimal solutions; multiple optimal solutions may or may not occur if w_j has at least one positive entry. However, if x_B is a nondegenerate optimal BFS and $\tilde{c}_j = 0$, then (cf. Exercise 4.4c) there is a ray of optimal solutions when $w_j \leq 0$ and the simplex method provides another optimal BFS when w_j has a positive entry. To conclude, if a nondegenerate optimal BFS has a zero reduced cost, there are multiple optimal solutions. Degeneracy is discussed further below.

4.10 DEGENERACY

We have seen that a nondegenerate optimal BFS has nonnegative reduced costs and that a zero reduced cost then leads to multiple optimal solutions.

We also know that the simplex method produces a degenerate BFS from a nondegenerate BFS when a tie occurs in the minimum positive ratio. When degeneracy occurs, there are subroutines that prevent cycling in a loop through a sequence of choices for B which correspond to the same BFS for the LP. Applying such a subroutine will either cause the simplex method to show that the LP has no minimum or will produce an optimal BFS and a corresponding B matrix for which the reduced costs are all nonnegative. The latter point is technical, but it is important. For example, suppose that we are given an optimal BFS x for an LP. We can find B so that x corresponds to x_B. If x is nondegenerate, B is uniquely determined and the reduced costs are all nonnegative. But as the example below shows, if x is degenerate, x_B may have negative reduced costs. However, we apply the simplex method, beginning at x_B, to find $\bar{B}$ with $x_{\bar{B}}$ corresponding to x and all reduced costs ≥ 0; $x_{\bar{B}}$ will correspond to x rather than some other optimal BFS because of the way the cycling subroutine works (Exercise 4.6). We will use $x_{\bar{B}}$ in Chapter 6.

Example 4.2. Consider the LP with standard-form simplex tableau

$$\left[\begin{array}{ccccc|c} 1 & 0 & 0 & 1 & 3 & 0 \\ 0 & 1 & 0 & 1 & 4 & 1 \\ 0 & 0 & 1 & 1 & 5 & 2 \\ \hline 0 & 1 & 1 & 1 & 7 & 0 \end{array}\right].$$

Applying the simplex method for the degenerate BFS $x_B = (x_1,x_2,x_3) = (0,1,2)$:

$$\begin{array}{c} {}^{-(2)\text{ to }(4)}_{-(3)\text{ to }(4)} \\ \longrightarrow \end{array} \left[\begin{array}{ccccc|c} 1 & 0 & 0 & 1 & 3 & 0 \\ 0 & 1 & 0 & 1 & 4 & 1 \\ 0 & 0 & 1 & 1 & 5 & 2 \\ \hline 0 & 0 & 0 & -1 & -2 & -3 \end{array}\right],$$

we see that there are negative reduced costs. But pivoting now at (1,4),

$$\begin{array}{c} {}^{-(1)\text{ to }(2)}_{-(1)\text{ to }(3)} \\ {}^{(1)\text{ to }(4)} \\ \longrightarrow \end{array} \left[\begin{array}{ccccc|c} 1 & 0 & 0 & 1 & 3 & 0 \\ -1 & 1 & 0 & 0 & 1 & 1 \\ -1 & 0 & 1 & 0 & 2 & 2 \\ \hline 1 & 0 & 0 & 0 & 1 & -3 \end{array}\right],$$

we find that $x_{\bar{B}} = (x_4,x_2,x_3) = (0,1,2)$ is the unique optimal solution to the

LP. However, x_B and $x_{\bar{B}}$ both correspond to the unique solution $x = (0,1,2,0,0)$ to the LP. Consequently, x_B is an optimal BFS—even though there are negative reduced costs!

Stability of solutions of a LP under small changes of the parameters A, b, and c is discussed in Section 5.9, where it is shown that *if* the following three conditions are satisfied: (1) the rank of A is n, (2) x_B is a nondegenerate BFS, and (3) all the reduced costs relative to x_B are positive [so that $x = (x_B,0)$ is the unique solution to the LP], *then* small changes in the parameters A, b, or c result in new LPs with unique optimal solutions. Moreover, the solutions to the new LPs involve the same variables.

Now let us return to the simplex method. In order to solve an LP, the simplex method needs an initial BFS with which to begin. Section 4.11 presents phase 1, which takes a standard-form LP and either asserts that the LP has no feasible solution or provides an initial BFS with which to begin the simplex method.

4.11 PHASE 1

Sometimes a standard-form LP has an obvious initial BFS; for instance, if a given LP has constraints $Ax \le b$, $x \ge 0$ and $b \ge 0$, the slack variables provide an initial BFS to the associated standard-form LP. When an initial BFS to a standard-form LP is not obvious, phase 1 solves a different standard-form LP which has an obvious initial BFS. If there is no solution to the original LP, the solution to the new LP informs you of that fact; otherwise, phase 1 provides an initial BFS with which to begin the simplex method. The following example explains how phase 1 works.

Consider the standard-form LP

$$\min 4x_1 + x_2 + x_3$$
$$\text{s.t. } 2x_1 + x_2 + 2x_3 = 4$$
$$3x_1 + 3x_2 + x_3 = 3, \qquad x \ge 0.$$

Phase 1 consists of introducing a nonnegative *artificial variable* for each of the two constraints and solving the associated LP:

$$\min \qquad x_4 + x_5$$
$$\text{s.t. } 2x_1 + x_2 + 2x_3 + x_4 \qquad = 4$$
$$3x_1 + 3x_2 + x_3 \qquad + x_5 = 3, \qquad x_1,x_2,x_3,x_4,x_5 \ge 0.$$

Because the right-hand side of the constraints is ≥ 0 [i.e., $b = (4,3) \ge (0,0)$], the artificial variables compose an initial basic feasible solution to the new problem. The value of the objective function for the phase 1 problem is simply the sum of the artificial variables. Since there are feasible solutions to the phase 1 problem and the values of the objective function are all greater than or equal to zero, there is an optimal BFS $(x,\bar{x})$ to the phase 1 problem. As in the example, where $x = (x_1,x_2,x_3)$ and $\bar{x} = (x_4,x_5)$, x represents the original variables and $\bar{x}$ represents the artificial variables. There are two cases to consider.

Case 1: $\bar{x} = 0$

In this case x is an initial basic feasible solution to the original problem: If $(x,\bar{x})$ is a nondegenerate solution to the phase 1 problem, all of its basic variables correspond to entries in x, so x is a BFS to the original problem. If $(x,\bar{x})$ is a degenerate solution to the phase 1 problem, some of its basic variables can correspond to entries in $\bar{x}$; however, x is a BFS to the original LP because the columns corresponding to basic variables in x are linearly independent (see Section 4.1).

Case 2: $\bar{x} \neq 0$

In this case the minimum value of the objective function for the phase 1 problem is greater than zero (i.e., $x_4 + x_5 > 0$). Consequently, there is no BFS y to the original problem because if y were a BFS to the original problem, $(y,0)$ would be a BFS to the phase 1 problem with objective function value equal to zero, which is less than the minimum value $x_4 + x_5$. The simplex method solution to the phase 1 problem follows in tableau format:

$$
\begin{array}{ccccc}
& & x_4 & x_5 & \\
\end{array}
\left[\begin{array}{ccccc|c}
2 & 1 & 2 & 1 & 0 & 4 \\
3 & 3 & 1 & 0 & 1 & 3 \\
\hline
0 & 0 & 0 & 1 & 1 & 0
\end{array}\right]
\xrightarrow[\substack{-(1)\text{ to }(3)\\-(2)\text{ to }(3)}]{}
\begin{array}{ccccc}
& & x_4 & x_5 & \\
\end{array}
\left[\begin{array}{ccccc|c}
2 & 1 & 2 & 1 & 0 & 4 \\
3 & 3 & 1 & 0 & 1 & 3 \\
\hline
-5 & -4 & -3 & 0 & 0 & -7
\end{array}\right]
$$

$$
\xrightarrow[\substack{-2(2)\text{ to }(1)\\5(2)\text{ to }(3)}]{(1/3)(2)}
\begin{array}{ccccc}
x_1 & & & x_4 & \\
\end{array}
\left[\begin{array}{ccccc|c}
0 & -1 & \frac{4}{3} & 1 & -\frac{2}{3} & 2 \\
1 & 1 & \frac{1}{3} & 0 & \frac{1}{3} & 1 \\
\hline
0 & 1 & -\frac{4}{3} & 0 & \frac{5}{3} & -2
\end{array}\right]
$$

$$
\xrightarrow[\substack{-(1/3)(1)\text{ to }(2)\\(4/3)(1)\text{ to }(3)}]{(3/4)(1)}
\begin{array}{ccccc}
x_1 & & x_3 & & \\
\end{array}
\left[\begin{array}{ccccc|c}
0 & -\frac{3}{4} & 1 & \frac{3}{4} & -\frac{1}{2} & \frac{3}{2} \\
1 & \frac{5}{4} & 0 & -\frac{1}{4} & \frac{1}{2} & \frac{1}{2} \\
\hline
0 & 0 & 0 & 1 & 1 & 0
\end{array}\right] .
$$

So $x = [\frac{1}{2},0,\frac{3}{2}]$ is an initial basic feasible solution for the original LP.

Phase 1 either develops an initial BFS for a standard-form LP or indicates that the LP has no feasible solutions. When phase 1 provides an initial BFS, *phase 2* consists of using the simplex method and the initial basic feasible solution obtained in phase 1 to solve the original problem. Phase 2 for the example follows:

$$
\begin{matrix} x_1 & & x_3 \end{matrix} \qquad \begin{matrix} x_1 & & x_3 \end{matrix} \qquad \begin{matrix} x_1 & & x_3 \end{matrix}
$$

$$
\begin{bmatrix} 2 & 1 & 2 & | & 4 \\ 3 & 3 & 1 & | & 3 \\ \hline 4 & 1 & 1 & | & 0 \end{bmatrix} \xrightarrow[\text{phase 1}]{} \begin{bmatrix} 0 & -\frac{3}{4} & 1 & | & \frac{3}{2} \\ 1 & \frac{5}{4} & 0 & | & \frac{1}{2} \\ \hline 4 & 1 & 1 & | & 0 \end{bmatrix} \rightarrow \begin{bmatrix} 0 & -\frac{3}{4} & 1 & | & \frac{3}{2} \\ 1 & \frac{5}{4} & 0 & | & \frac{1}{2} \\ \hline 0 & -\frac{13}{4} & 0 & | & -\frac{7}{2} \end{bmatrix}
$$

$$
\begin{matrix} x_2 & & x_3 \end{matrix}
$$

$$
\rightarrow \begin{bmatrix} \frac{3}{5} & 0 & 1 & | & \frac{9}{5} \\ \frac{4}{5} & 1 & 0 & | & \frac{2}{5} \\ \hline \frac{13}{5} & 0 & 0 & | & -\frac{11}{5} \end{bmatrix}.
$$

Thus we obtain the optimal solution $x = [0, \frac{2}{5}, \frac{9}{5}]$ to the original problem. The minimum value is $\frac{11}{5}$ and the solution is unique because the reduced cost of the nonbasic variable x_1 is positive. A variation of the simplex method called the revised simplex method is sketched below.

4.12 THE REVISED SIMPLEX METHOD

The revised simplex method (RSM) has computational advantages which make it a popular choice for LP software; it also provides a handy way to check answers to LPs that you have solved by hand. A flowchart for the RSM follows.

Given a B that corresponds to a basic feasible solution:
compute B^{-1}
compute $B^{-1}b$ $(=d)$
compute $c_B B^{-1}$
compute $(c_B B^{-1})V$
compute $c_V - c_B B^{-1}V$ $(=\tilde{c}_V)$
compute j: $\tilde{c}_j \le \tilde{c}_k$, $\tilde{c}_j$ and $\tilde{c}_k$ in $\tilde{c}_V$;
if $\tilde{c}_j \ge 0$, then compute $c_B d$ and stop; else

compute $B^{-1}a_j$ $(=w_j)$
compute i: $w_{ij} > 0$ and d_i/w_{ij} is a smallest positive ratio to obtain the column to be replaced in B by c_j; repeat what precedes for the updated B, which is described below.

Updating B

Focus on the set of basic variables corresponding to the columns of B; replace the variable corresponding to the outgoing ith column of B by the variable x_j to get an updated set of basic variables; then go back to the original A matrix and choose the updated B to be composed of the columns of A that correspond to the updated basic variables.

Alternative Way to Update B and Compute (update B)$^{-1}$

Recall that the list of row operations that changes B to I also changes a_j to $B^{-1}a_j$; with the pivot point determined to be at the (i,j) position, replace the jth row in the $m \times m$ identity matrix by $B^{-1}a_j$ to obtain (update B) in a form that can be used to compute (update B)$^{-1}$ easily.

Example 4.3. The revised simplex method can be applied to Exercise 4.3(e) as follows. After changing the maximize to a minimize by multiplying the objective function by negative 1 and adding the slack variables, we have a standard-form primal LP with corresponding initial tableau

$$\begin{bmatrix} 1 & 2 & 2 & 1 & 0 & | & 1 \\ 1 & 3 & 2 & 0 & 1 & | & 3 \\ \hline -2 & -1 & -3 & 0 & 0 & | & 0 \end{bmatrix}.$$

By inspection we observe that the slack variable solution is feasible, that the third column pivots in, and that the fourth column pivots out; let us begin the revised simplex method with

$$B = \begin{bmatrix} 2 & 0 \\ 2 & 1 \end{bmatrix},$$

corresponding to the third and fifth columns:

$$[B \mid I] = \begin{bmatrix} 2 & 0 & | & 1 & 0 \\ 2 & 1 & | & 0 & 1 \end{bmatrix} \rightarrow \begin{bmatrix} 2 & 0 & | & 1 & 0 \\ 0 & 1 & | & -1 & 1 \end{bmatrix}$$

$$\rightarrow \begin{bmatrix} 1 & 0 & | & \frac{1}{2} & 0 \\ 0 & 1 & | & -1 & 1 \end{bmatrix} = [I \mid B^{-1}]$$

$$B^{-1}b = \begin{bmatrix} \frac{1}{2} & 0 \\ -1 & 1 \end{bmatrix}\begin{bmatrix} 1 \\ 3 \end{bmatrix} = \begin{bmatrix} \frac{1}{2} \\ 2 \end{bmatrix}$$

$$c_B B^{-1} = [-3,0]\begin{bmatrix} \frac{1}{2} & 0 \\ -1 & 1 \end{bmatrix} = -[\tfrac{3}{2},0]$$

$$c_B B^{-1} V = -[\tfrac{3}{2},0] \begin{bmatrix} 1 & 2 & 1 \\ 1 & 3 & 0 \end{bmatrix} = -[\tfrac{3}{2},3,\tfrac{3}{2}]$$

$$c_V - c_B B^{-1} V = [-2,-1,0] + [\tfrac{3}{2},3,\tfrac{3}{2}] = [-\tfrac{1}{2},2,\tfrac{3}{2}] \quad \text{(first column pivots in)}$$

$$B^{-1} a_1 = \begin{bmatrix} \tfrac{1}{2} & 0 \\ -1 & 1 \end{bmatrix} \begin{bmatrix} 1 \\ 1 \end{bmatrix} = \begin{bmatrix} \tfrac{1}{2} \\ 0 \end{bmatrix}$$

$$\frac{B^{-1}b}{B^{-1}a_1} = \begin{bmatrix} 1 \\ \infty \end{bmatrix} \quad \text{(third column pivots out)}.$$

Update Applied

The set of basic variables changes from {3,5} to {1,5}, so the first and fifth columns of the A matrix compose the updated B:

$$B = \begin{bmatrix} 1 & 0 \\ 1 & 1 \end{bmatrix} \quad \text{(corresponding to the first and fifth columns)}$$

$$[B \mid I] = \begin{bmatrix} 1 & 0 & 1 & 0 \\ 1 & 1 & 0 & 1 \end{bmatrix} \rightarrow \begin{bmatrix} 1 & 0 & 1 & 0 \\ 0 & 1 & -1 & 1 \end{bmatrix} = [I \mid B^{-1}].$$

Alternative Update Applied

$$B^{-1} a_1 = \begin{bmatrix} \tfrac{1}{2} \\ 2 \\ 0 \end{bmatrix}$$

replaces the first column in the 2×2 identity matrix. Now

$$\begin{bmatrix} \tfrac{1}{2} & 0 \\ 0 & 1 \end{bmatrix}$$

is the updated B and we compute

$$[\text{update } B \mid B^{-1}] = \begin{bmatrix} \tfrac{1}{2} & 0 & \tfrac{1}{2} & 0 \\ 0 & 1 & -1 & 1 \end{bmatrix} \rightarrow \begin{bmatrix} 1 & 0 & 1 & 0 \\ 0 & 1 & -1 & 1 \end{bmatrix}$$

$$= [I \mid (\text{update } B)^{-1}]$$

to find that

$$\begin{bmatrix} 1 & 0 \\ -1 & 1 \end{bmatrix}$$

is the updated B^{-1}. Continuing with the example, we have

$$B^{-1}b = \begin{bmatrix} 1 & 0 \\ -1 & 1 \end{bmatrix} \begin{bmatrix} 1 \\ 3 \end{bmatrix} = \begin{bmatrix} 1 \\ 2 \end{bmatrix}$$

$$c_B B^{-1} = [-2,0] \begin{bmatrix} 1 & 0 \\ -1 & 1 \end{bmatrix} = -[2,0]$$

$$c_B B^{-1}V = -[2,0] \begin{bmatrix} 2 & 2 & 1 \\ 3 & 2 & 0 \end{bmatrix} = -[4,4,2]$$

$$c_V - c_B B^{-1}V = [-1,-3,0] + [4,4,2] = [3,1,2]$$

$$c_B d = [-2,0] \begin{bmatrix} 1 \\ 2 \end{bmatrix} = -2 \quad \text{(stop)}.$$

Consequently, the maximum value of 2 occurs at $x = (1,0,0)$; moreover, the solution is unique.

Using RSM to Check Answers

To check your answer to a problem that you have solved by hand, choose the initial B for the RSM to be composed of the columns corresponding to the basic variables in your optimal BFS for the problem.

EXERCISES

Save your solutions to Exercise 4.1 until you finish Exercise 6.1.
4.1 Use the simplex method to solve the following LPs.

(a) 　　　　　max $4x_1 + 2x_2 + 6x_3$

　　　　　　s.t. $3x_1 + 2x_2 \quad\quad \leq 10$

　　　　　　　　　　　$x_2 + 2x_3 \leq 8$

　　　　　　　　$2x_1 + x_2 + x_3 \leq 8, \quad x \geq 0.$

(b) 　　　　　min $x_1 + x_2 + x_3 + x_4$

　　　　　　s.t. $x_1 \quad\quad + x_3 \quad\quad = 3$

　　　　　　　　　　　$x_2 + x_3 + 3x_4 = 2$

　　　　　　　　　　　　　　$x_3 + x_4 = 1, \quad x \geq 0.$

[*Hint:* $x = (2,1,1,0)$ is a basic feasible solution.]

(c) 　　　　　min $4x_1 + x_2 + 3x_3 + x_4$

　　　　　　s.t. $x_1 + x_2 + x_3 \quad\quad = 3$

　　　　　　　　　$x_1 \quad\quad + x_3 + 3x_4 = 2$

　　　　　　　　　　　$x_2 + x_3 + x_4 = 2, \quad x \geq 0.$

[*Hint:* $x = (1,1,1,0)$ is a basic feasible solution.]

(d) $\quad$ min $\quad x_1 \quad + 3x_3 + 2x_4 + x_5$

$\quad$ s.t. $\quad x_1 + x_2 + \ x_3 + 5x_4 - x_5 = 3$

$$2x_1 + x_2 + 5x_3 + 2x_4 + x_5 = 4, \qquad x \geq 0.$$

[*Hint:* $x = (1,2,0,0,0)$ is a basic feasible solution.]

(e) $\qquad$ max $\ 2x_1 + \ x_2 + \ x_3$

$\qquad$ s.t. $\quad x_1 + 4x_2 + 2x_3 \leq 10$

$$2x_1 + 3x_2 + 4x_3 \leq \ 8, \qquad x \geq 0.$$

(f) $\qquad$ max $\ x_1 + \ x_2 + 2x_3$

$\qquad$ s.t. $\quad x_1 + 2x_2 + 2x_3 \leq 2$

$$x_1 + 4x_2 + 2x_3 \leq 4, \qquad x \geq 0.$$

(Find all solutions.)

(g) $\qquad$ min $\qquad 2x_3 - x_4$

$\qquad$ s.t. $\quad x_1 - 4x_3 + x_4 + 3x_5 = 1$

$$x_2 + 6x_3 - x_4 \qquad = 2, \qquad x \geq 0.$$

(h) $\qquad$ max $\ 4x_1 + 2x_2 + \ x_3$

$\qquad$ s.t. $\quad x_1 + 4x_2 + 2x_3 \leq 8$

$$2x_1 + 3x_2 + 4x_3 \leq 4, \qquad x \geq 0.$$

(i) $\qquad$ min $\ -2x_1 - \ x_2 - 2x_3$

$\qquad$ s.t. $\quad 2x_1 + \ x_2 + \ x_3 \leq 2$

$$x_1 + 2x_2 + 3x_3 \leq 5, \qquad x \geq 0.$$

[*Hint:* Note that if $x = (0,1,1)$, then $[2,1,1]x = 2$ and $[1,2,3]x = 5$.)

(j) $\qquad$ min $\qquad\qquad x_2 + 3x_4$

$\qquad$ s.t. $\quad x_1 + 2x_2 + \ x_3 - \ x_4 = 1$

$$x_2 + 3x_3 + \ x_4 \geq 4, \qquad x \geq 0.$$

[*Hint:* $x = (5,0,0,4)$ is an initial basic feasible solution.]

4.2 Use the two-phase method to solve the following LPs.

(a) $\qquad$ min $\ 2x_1 + x_2 + x_3$

$\qquad$ s.t. $\quad 2x_1 - x_2 + x_3 = 3$

$$x_1 + x_2 + x_3 = 2, \qquad x \geq 0.$$

(b) $\qquad$ max $\ x_1 + 2x_2 - x_3$

$\qquad$ s.t. $\quad x_1 + \ x_2 + x_3 = 1$

$$x_1 - \ x_2 + x_3 = 2, \qquad x \geq 0.$$

(c)   min $3x_2 + x_3$

      s.t. $x_1 + x_2 + 2x_3 \geq 3$

          $x_1 - 2x_2 + x_3 \leq 2, \quad x \geq 0.$ (Find all solutions.)

4.3 Use the revised simplex method to solve the following LPs.

(a) Exercise 4.1(f).

(b)   max $2x_1 + x_2 + x_3$

         s.t. $x_1 + 4x_2 + 2x_3 \leq 10$

             $2x_1 + 3x_2 + 4x_3 \leq 12, \quad x \geq 0.$

(c)   max $x_1 + 2x_2 + 3x_3$

          s.t. $x_1 + 2x_2 + 2x_3 \leq 2$

             $x_1 + 4x_2 + 2x_3 \leq 3, \quad x \geq 0.$

(d)   max $2x_1 + 3x_2 + x_3$

          s.t. $2x_1 + x_2 + 2x_3 \leq 4$

             $2x_1 + 3x_2 + 4x_3 \leq 9, \quad x \geq 0.$

(e)   min $-2x_1 - x_2 - 3x_3$

          s.t. $x_1 + 2x_2 + 2x_3 \leq 1,$

             $x_1 + 3x_2 + 2x_3 \leq 3, \quad x \geq 0.$

4.4 Consider the LP

         max $x_1 \qquad - 3x_3$

         s.t. $x_1 - 2x_2 - x_3 + x_4 = 1$

             $-x_1 + 4x_2 - x_3 \qquad \leq 2$

             $-x_1 - 3x_2 + x_3 \qquad \leq 3, \quad x \geq 0.$

(a) Use the simplex method to find a solution.

(b) Use the revised simplex method to check your answer.

(c) Look at your final simplex tableau in part (a); then find all solutions to this LP.

4.5 A farmer has 500 acres of land and wishes to determine the acreage allocated to the following three crops: wheat, corn, and soybeans. The worker-days, preparation cost, and profit per acre of the three crops are summarized below.

Crop	Worker-days	Preparation cost	Profit
Wheat	6	$100	$ 60
Corn	8	150	100
Soybeans	10	120	80

Suppose that the maximum number of worker-days available are 5000 and that the farmer has $60,000 for preparation.

(a) Write an appropriate LP model.

(b) Use the simplex method to solve your model.

(c) What is the profit from planting as many acres as possible in corn?

(d) What is the profit from planting as many acres as possible in soybeans?

(e) Find all optimal solutions.

(f) Suppose that a neighbor wants to rent some (or all) of the farmer's land for $90 per acre for a growing season. What should the farmer do? (Why?)

(g) Suppose that instead of renting the land, the farmer can borrow money for preparation cost at 20% interest for a season. What would you do if you were the farmer? (Why?)

(h) Suppose that the farmer cannot rent his land and cannot borrow money and that his labor cost increases $1 per worker-day. What should the farmer do now?

4.6 Use the anticycling method of R. G. Bland, "New Finite Pivoting Rules for the Simplex Method," *Mathematics of Operations Research*, 2 (1977), pages 103–107, to verify the assertion made in Section 4.10 that every optimal BFS x has a corresponding B such that $x = (x_B, 0)$ and all the reduced costs with respect to x_B are nonnegative.

4.7 Referring to Exercise 2.6(a), let $x = (FL, PL, FP, PP) = (0, 6.25, 10, 3.75)$.

(a) Write the standard-form simplex tableau for Exercise 2.6(a), use the simplex method to find the basic feasible solution corresponding to x, and conclude that x is an optimal solution to Exercise 2.6(a).

(b) There is a nonbasic variable with zero reduced cost in your final tableau for part (a). Pivot in that column and find another optimal basic feasible solution, (5,0,5,10), to Exercise 2.6(a).

(c) Find all solutions to Exercise 2.6(a).

4.8 Discuss the computer solution to Exercise 2.27 (minimizing the penalty); find all optimal solutions to the problem.

SOLUTIONS TO EXERCISES

4.1 (a)

$$
\begin{bmatrix}
3 & 2 & 0 & 1 & 0 & 0 & 10 \\
0 & 1 & ② & 0 & 1 & 0 & 8 \\
2 & 1 & 1 & 0 & 0 & 1 & 8 \\
-4 & -2 & -6 & 0 & 0 & 0 & 0
\end{bmatrix}
\rightarrow
\begin{bmatrix}
3 & 2 & 0 & 1 & 0 & 0 & 10 \\
0 & \frac{1}{2} & 1 & 0 & \frac{1}{2} & 0 & 4 \\
2 & \frac{1}{2} & 0 & 0 & -\frac{1}{2} & 1 & 4 \\
-4 & 1 & 0 & 0 & 3 & 0 & 24
\end{bmatrix}
$$

$$\rightarrow \begin{bmatrix} 0 & \frac{1}{2} & 1 & 0 & \frac{1}{2} & 0 & 4 \\ 1 & \frac{1}{4} & 0 & 0 & -\frac{1}{4} & \frac{1}{2} & 2 \\ \hline 0 & 2 & 0 & 0 & 2 & 2 & 32 \end{bmatrix}.$$

Tableau 1 is the standard-form tableau which we get after we add slack variables and change the mass to a min by multiplying the objective function by -1.

Discussion: In tableau 1 we see that the slack variable solution $x = (0,0,0,10,8,8)$ is an initial BFS with which to begin the simplex method. The most negative reduced cost is -6, in the third column, so x_3 pivots in. The smallest positive ratio is $\frac{8}{2}$, so x_5 pivots out. The pivot point is circled. Going on to tableau 2, we see the system after one pivot step; $x = (0,0,4,10,0,4)$ is a BFS. The first column has a negative reduced cost, so we pivot again and arrive in tableau 3 at the optimal solution $x = (2,0,4)$ to the original problem. The first line in tableau 3 is not completed because the value of the slack variable x_4 is superfluous in this case. The solution is unique because all the nonbasic reduced costs are positive.

(b) Using the hint, we choose B to be composed of the first three columns in the A matrix in the tableau

$$\left[\begin{array}{c|c} A & b \\ \hline c & 0 \end{array}\right] = \begin{bmatrix} 1 & 0 & 1 & 0 & 3 \\ 0 & 1 & 1 & 3 & 2 \\ 0 & 0 & 1 & 1 & 1 \\ \hline 1 & 1 & 1 & 1 & 0 \end{bmatrix} = \left[\begin{array}{c|c|c} B & V & b \\ \hline c_B & c_V & 0 \end{array}\right].$$

The computation $\rightarrow \cdots \rightarrow \left[\begin{array}{c|c|c} I & B^{-1}V & B^{-1}b \\ \hline 0 & c_V - c_B B^{-1}V & -c_B B^{-1}b \end{array}\right]$ follows.

$$\begin{array}{c} -(3) \text{ to } (2) \\ -(3) \text{ to } (1) \\ \longrightarrow \end{array} \begin{bmatrix} 1 & 0 & 0 & -1 & 2 \\ 0 & 1 & 0 & 2 & 1 \\ 0 & 0 & 1 & 1 & 1 \\ \hline 1 & 1 & 1 & 1 & 0 \end{bmatrix}$$

$$\begin{array}{c} (-1) \text{ to } (4) \\ (-2) \text{ to } (4) \\ -(3) \text{ to } (4) \\ \longrightarrow \end{array} \begin{bmatrix} 1 & 0 & 0 & -1 & 2 \\ 0 & 1 & 0 & 2 & 1 \\ 0 & 0 & 1 & 1 & 1 \\ \hline 0 & 0 & 0 & -1 & -4 \end{bmatrix}$$

From the preceding tableau, we see that $x = (2,1,1,0)$ is a BFS with value 4 and that x_4 has a negative reduced cost. We make a pivot.

$$\underset{(1/2)(2)}{\longrightarrow} \begin{bmatrix} 1 & 0 & 0 & -1 & 2 \\ 0 & \frac{1}{2} & 0 & 1 & \frac{1}{2} \\ 0 & 0 & 1 & 1 & 1 \\ 0 & 0 & 0 & -1 & -4 \end{bmatrix} \quad \begin{array}{l} +(2) \text{ to } (1) \\ -(2) \text{ to } (3) \\ +(2) \text{ to } (4) \end{array} \quad \longrightarrow \begin{bmatrix} 1 & \frac{1}{2} & 0 & 0 & \frac{5}{2} \\ 0 & \frac{1}{2} & 0 & 1 & \frac{1}{2} \\ 0 & -\frac{1}{2} & 1 & 0 & \frac{1}{2} \\ 0 & \frac{1}{2} & 0 & 0 & -\frac{7}{2} \end{bmatrix},$$

and see that the optimal BFS $x = (\frac{5}{2},0,\frac{1}{2},\frac{1}{2})$ has objective value $\frac{7}{2}$ and is unique because the nonbasic reduced cost $\frac{1}{2}$ is greater than 0.

(c) $x = (\frac{5}{4},\frac{7}{4},0,\frac{1}{4})$, 7, unique.
(d) $x = (0,\frac{7}{2},0,0,\frac{1}{2})$, $\frac{1}{2}$, unique.
(e) $x = (4,0,0)$, 8, unique.
(f)

$$\begin{bmatrix} 1 & 2 & ② & 1 & 0 & 2 \\ 1 & 4 & 2 & 0 & 1 & 4 \\ -1 & -1 & -2 & 0 & 0 & 0 \end{bmatrix} \quad \begin{array}{l} -(1) \text{ to } (2) \\ (1) \text{ to } (3) \\ \\ (1/2)(1) \end{array} \quad \longrightarrow \begin{bmatrix} \frac{1}{2} & 1 & 1 & \frac{1}{2} & 0 & 1 \\ 0 & 2 & 0 & -1 & 1 & 2 \\ 0 & 1 & 0 & 1 & 0 & 2 \end{bmatrix}$$

tell us that $x = (0,0,1)$ is an optimal BFS with objective value 2. However, x_1 has a reduced cost of 0, so $x_1 \neq 0$ may be part of another optimal BFS. The smallest positive ratio in the first column is 2, so let us try pivoting at the (1,1) entry:

$$\underset{2(1)}{\longrightarrow} \begin{bmatrix} 1 & 2 & 2 & 1 & 0 & 2 \\ 0 & 2 & 0 & -1 & 1 & 2 \\ 0 & 1 & 0 & 1 & 0 & 2 \end{bmatrix}.$$

Thus $x = (2,0,0)$ is also an optimal BFS. Consequently, any convex combination

$$x_t = t(2,0,0) + (1 - t)(0,0,1) = (2t,0,1 - t), \qquad 0 \le t \le 1$$

is optimal. But only the endpoints $(2,0,0)$ and $(0,0,1)$ are basic solutions.

(g) $x = (0,0,\frac{3}{2},7,0)$, -4, unique.
(h) $x = (2,0,0)$, 8, unique.
(i) $x = (\frac{1}{3},0,\frac{8}{3})$, $\frac{-18}{5}$, unique.
(j) $x = (0,0,\frac{5}{4},\frac{1}{4})$, $\frac{3}{4}$, unique.

4.2 (a)

$$\begin{bmatrix} 2 & -1 & 1 & 1 & 0 & 3 \\ 1 & 1 & 1 & 0 & 1 & 2 \\ 0 & 0 & 0 & 1 & 1 & 0 \end{bmatrix}$$

$$\xrightarrow[\substack{-(1)\text{ to }(3) \\ -(2)\text{ to }(3)}]{} \begin{bmatrix} 2 & -1 & 1 & 1 & 0 & 3 \\ 1 & 1 & 1 & 0 & 1 & 2 \\ -3 & 0 & -2 & 0 & 0 & -5 \end{bmatrix}$$

$$\xrightarrow[]{(1/2)(1)} \begin{bmatrix} 1 & -\frac{1}{2} & \frac{1}{2} & \frac{1}{2} & 0 & \frac{3}{2} \\ 1 & 1 & 1 & 0 & 1 & 2 \\ -3 & 0 & -2 & 0 & 0 & -5 \end{bmatrix}$$

$$\xrightarrow[3(1)\text{ to }(3)]{-(1)\text{ to }(2)} \begin{bmatrix} 1 & -\frac{1}{2} & \frac{1}{2} & \frac{1}{2} & 0 & \frac{3}{2} \\ 0 & \frac{3}{2} & \frac{1}{2} & -\frac{1}{2} & 1 & \frac{1}{2} \\ 0 & -\frac{3}{2} & -\frac{1}{2} & \frac{3}{2} & 0 & -\frac{1}{2} \end{bmatrix}$$

$$\xrightarrow[(2/3)(2)]{+(2)\text{ to }(3)} \begin{bmatrix} 1 & -\frac{1}{2} & \frac{1}{2} & \frac{1}{2} & 0 & \frac{3}{2} \\ 0 & 1 & \frac{1}{3} & -\frac{1}{3} & \frac{2}{3} & \frac{1}{3} \\ 0 & 0 & 0 & 1 & 1 & 0 \end{bmatrix}$$

$$\xrightarrow[]{(1/2)(2)\text{ to }(1)} \begin{bmatrix} 1 & 0 & \frac{2}{3} & & & \frac{5}{3} \\ 0 & 1 & \frac{1}{3} & & & \frac{1}{3} \\ 0 & 0 & 0 & 1 & 1 & 0 \end{bmatrix} .$$

Thus $x = (\frac{5}{3}, \frac{1}{3}, 0)$ is an initial BFS for the given LP. Moreover, we can use what we did in phase 1;

$$\begin{bmatrix} 2 & -1 & 1 & 3 \\ 1 & 1 & 1 & 2 \\ 2 & 1 & 1 & 0 \end{bmatrix} \xrightarrow[]{\text{phase 1}} \begin{bmatrix} 1 & 0 & \frac{2}{3} & \frac{5}{3} \\ 0 & 1 & \frac{1}{3} & \frac{1}{3} \\ 2 & 1 & 1 & 0 \end{bmatrix}$$

$$\xrightarrow[-(2)\text{ to }3]{-2(1)\text{ to }3} \begin{bmatrix} 1 & 0 & \frac{2}{3} & \frac{5}{3} \\ 0 & 1 & \frac{1}{3} & \frac{1}{3} \\ 0 & 0 & -\frac{2}{3} & -\frac{11}{3} \end{bmatrix}$$

$$\begin{array}{c} \frac{-2(2) \text{ to } (1)}{2(2) \text{ to } (3)} \\ \xrightarrow{\hspace{2cm}} \\ 3(2) \end{array} \begin{bmatrix} 1 & -2 & 0 & | & 1 \\ 0 & 3 & 1 & | & 1 \\ 0 & 2 & 0 & | & -3 \end{bmatrix}.$$

Thus we see that the optimal BFS $x = (1,0,1)$ has objective value 3 and is unique.

(b) No feasible solution.

(c) From phase 1, an initial BFS is $x = (0,0,\frac{3}{2},0,\frac{1}{2})$. The final tableau for phase 2 is

$$\begin{bmatrix} 0 & 3 & 1 & -1 & -1 & | & 1 \\ 1 & -5 & 0 & 1 & 2 & | & 1 \\ 0 & 0 & 0 & 1 & 1 & | & -1 \end{bmatrix}.$$

Thus $x = (1,0,1)$ is an optimal BFS for the original problem. However, x_2 has a reduced cost of zero. Pivot at the (1,2) position and find that $(\frac{8}{3},\frac{1}{3},0)$ is also an optimal BFS. Consequently, $x_t = t(\frac{8}{3},\frac{1}{3},0)+(1-t)(1,0,1)=(1+\frac{5}{3}t,\frac{1}{3}t,1-t)$ is optimal for $0 \le t \le 1$.

4.3 (b) $x = (6,0,0)$, 12, unique.

(c) $x = (0,0,1)$, 3, unique.

(d) $x = (0,3,0)$, 9, not unique: $(\frac{3}{4},\frac{5}{2},0)$ is also an optimal BFS.

(e) $x = (1,0,0)$, 2, unique.

4.4 (a)

$$\begin{bmatrix} 1 & -2 & -1 & 1 & 0 & 0 & | & 1 \\ -1 & 4 & -1 & 0 & 1 & 0 & | & 2 \\ -1 & -3 & 1 & 0 & 0 & 1 & | & 3 \\ -1 & 0 & 3 & 0 & 0 & 0 & | & 0 \end{bmatrix}$$

$$\begin{bmatrix} 1 & -2 & -1 & 1 & 0 & 0 & | & 1 \\ 0 & 2 & -2 & 1 & 1 & 0 & | & 3 \\ 0 & -5 & 0 & 1 & 0 & 1 & | & 4 \\ 0 & -2 & 2 & 1 & 0 & 0 & | & 1 \end{bmatrix}$$

$$\begin{bmatrix} 1 & 0 & -3 & 2 & 1 & 0 & | & 4 \\ 0 & 1 & -1 & \frac{1}{2} & \frac{1}{2} & 0 & | & \frac{3}{2} \\ 0 & 0 & -5 & \frac{7}{2} & \frac{5}{2} & 1 & | & \frac{23}{2} \\ 0 & 0 & 0 & 2 & 1 & 0 & | & 4 \end{bmatrix}$$

max of 4 at $x = (4,\frac{3}{2},0,0)$.

(b) From the tableau for part (a),

$$B^{-1} = \begin{bmatrix} 2 & 1 & 0 \\ \frac{1}{2} & \frac{1}{2} & 0 \\ \frac{7}{2} & \frac{5}{2} & 1 \end{bmatrix}$$

$$B^{-1}b = (4, \tfrac{3}{2}, \tfrac{23}{2})$$

$$c_B B^{-1} = [-1,0,0]B^{-1} = [-2,-1,0]$$

$$c_V - c_B B^{-1}V = [3,0,0] + [2,1,0]\begin{bmatrix} -1 & 1 & 0 \\ -1 & 0 & 1 \\ 1 & 0 & 0 \end{bmatrix}$$

$$= [3,0,0] + [-3,2,1] = [0,2,1] \geq 0$$

$$c_B B^{-1}b = [-2,-1,0]\begin{bmatrix} 1 \\ 2 \\ 3 \end{bmatrix} = -4 \leftarrow \text{negative of value of solution.}$$

(c) The reduced cost of x_3 is zero, but there is no positive ratio. However, from the tableau,

$$-3\begin{bmatrix} 1 \\ -1 \\ -1 \end{bmatrix} - \begin{bmatrix} -2 \\ 4 \\ -3 \end{bmatrix} - 5\begin{bmatrix} 0 \\ 0 \\ 1 \end{bmatrix} = \begin{bmatrix} -1 \\ -1 \\ 1 \end{bmatrix}$$

$$\text{or} \quad A\begin{bmatrix} 3 \\ 1 \\ 1 \\ 0 \\ 0 \\ 5 \end{bmatrix} = 0.$$

Consequently, $x_t = (4, \tfrac{3}{2}, 0, 0) + t(3,1,1,0) = (4 + 3t, \tfrac{3}{2} + t, t, 0)$ is an optimal solution for all $t \geq 0$.

4.7 Let X_5 and X_6 denote surplus variables for the first two constraints, and let X_7 and X_8 denote slack variables for the second two constraints. Then $(FL,PL,FP,PP,X_5,X_6,X_7,X_8)$ displays the decision variables corresponding to the entries in a feasible vector for the associated standard form LP for which a tableau follows.

$$
\begin{bmatrix}
0 & 0 & 14 & 16 & -1 & 0 & 0 & 0 & 100 \\
10 & 8 & 0 & 0 & 0 & -1 & 0 & 0 & 50 \\
1 & 0 & 1 & 0 & 0 & 0 & 1 & 0 & 10 \\
0 & 1 & 0 & 1 & 0 & 0 & 0 & 1 & 10 \\
\hline
-5 & -6.4 & -7 & -8 & 0 & 0 & 0 & 0 & 0
\end{bmatrix}.
$$

You are given that $(0, 50/8, 10, 30/8)$ is a BFS for the original LP; computing the values of the slack and surplus variables, we find the BFS $(0, 50/8, 10, 30/8, 100, 0, 0, 0)$ to the corresponding standard-form LP; applying the appropriate row operations to the preceding simplex tableau will produce the simplex tableau corresponding to this BFS:

$$
\begin{array}{r}
\\
1/8(2) \\
-(2)\ to\ (4) \\
\\
\\
\end{array}
\begin{bmatrix}
0 & 0 & 14 & 16 & -1 & 0 & 0 & 0 & 100 \\
10/8 & 1 & 0 & 0 & 0 & -1/8 & 0 & 0 & 50/8 \\
1 & 0 & 1 & 0 & 0 & 0 & 1 & 0 & 10 \\
-10/8 & 0 & 0 & 1 & 0 & 1/8 & 0 & 1 & 30/8 \\
\hline
-5 & -6.4 & -7 & -8 & 0 & 0 & 0 & 0 & 0
\end{bmatrix}
$$

(Fill in the next tableau)

$-14(3)$ to (1)
$-16(4)$ to (1)

$$
\begin{array}{r}
\\
-(1) \\
6.4(2)\ to\ (5) \\
7(3)\ to\ (5) \\
8(4)\ to\ (5) \\
\end{array}
\begin{bmatrix}
-6 & 0 & 0 & 0 & 1 & 2 & 14 & 16 & 100 \\
10/8 & 1 & 0 & 0 & 0 & -1/8 & 0 & 0 & 50/8 \\
1 & 0 & 1 & 0 & 0 & 0 & 1 & 0 & 10 \\
-10/8 & 0 & 0 & 1 & 0 & 1/8 & 0 & 1 & 30/8 \\
\hline
0 & 0 & 0 & 0 & 0 & 0.2 & 7 & 8 & 140
\end{bmatrix}.
$$

From the latter tableau, we see that the nonbasic variable X_1 corresponding to FL has zero reduced cost; thus you can pivot in the first column and obtain another optimal BFS. Please do the computations!

5
Topics in LP and Extensions

In this chapter we continue the theme of Chapter 2; by the end of the chapter you should be able to use LP to model several more types of problems and be acquainted with some types of integer, network, and dynamic programming models. The text presentations in Sections 5.1 to 5.6 are deliberately brief; their purpose is to help you get started using the ideas to formulate models to solve problems. Do the problems as they are suggested in these sections. Solutions to several of the problems appear at the end of this chapter, and many more are discussed in Appendix 3. Try to solve a problem before you look at my solution. But after you have solved a problem, look at my solution because useful ideas are introduced in some of the discussions. Section 5.7 introduces networks and discusses graphs, cycles, trees, the node-incidence matrix, and the principle of induction in the process of relating LP and network formulations of transportation problems. After discussing the transportation problem, in Section 5.7 we present LP formulations for two important network problems, the minimum cost flow problem and the maximum flow problem, and show you the maximum flow–minimum cut theorem. After an introduction to dynamic programming in Section 5.8, in Section 5.9 we return to LP and discuss stability and sensitivity of solutions to LPs. A variety of exercises are given at the end of the chapter. Section 5.1 begins by showing you how to put some objective functions that do not appear to be LP-type objective functions into an LP form.

114

5.1 EXAMPLES THAT FIT INTO LP FORMAT

Two examples are used to introduce you to some important and frequently occurring types of objectives that fit into LP format. The first example goes back to Example 2.1 and considers the following scenario.

Example 5.1. Suppose that the group of neighbors approaches the organization with a proposal. They offer to sell fuel oil, gasoline, and jet fuel to the organization for $32, $50, and $100 a barrel; they offer to buy surplus fuel oil, gasoline, or jet fuel from the organization for $10 a barrel. How should the organization respond? Recall that we have let

L = millions of barrels of light crude to process
D = millions of barrels of dark crude to process
F = millions of barrels of surplus fuel oil produced
G = millions of barrels of surplus gasoline produced
J = millions of barrels of surplus jet fuel produced.

Now, in addition, we let

BF = millions of barrels of surplus of fuel oil to buy
BG = millions of barrels of surplus of gasoline to buy
BJ = millions barrels of surplus of jet fuel to buy.

The constraints

(1) $\qquad 0.21L + 0.55D - 3 = F - BF$
(2) $\qquad 0.5L + \ 0.3D - 7 = G - BG$
(3) $\qquad 0.25L + \ 0.1D - 5 = J - BJ$
$\qquad\qquad x = (L,D,F,G,J,BF,BG,BJ) \geq 0$

apply; the left sides of constraints (1) to (3) tell us how production of fuel oil, gasoline, and jet fuel compare with requirements. If there are surpluses, we can sell them; if there are shortages, we can buy what we need from the group. The corresponding cost of x is given by the formula

$$25L + 17D - 10F - 10G - 10J + 32BF + 50BG + 100BJ = Cx,$$

where

$$C = [25,17,-10,-10,-10,32,50,100]$$

and

$$x = (L,D,F,G,J,BF,BG,BJ).$$

Because the cost of buying fuel oil, gasoline, or jet fuel from the group is more than the income from selling the products to them, an optimal solution will *automatically* have at least one of each of the pairs, F and BF, G and BG, J and BJ, equal to zero; this idea applies to the exercises! The key facets in this example are:

1. The real decision variables are still L and D.
2. Production of fuel oil, gasoline, and jet fuel can be above or below required levels.
3. Overproduction and underproduction have different consequences.
4. The objective function does not encourage overproduction because the cost of producing or buying fuel oil, gasoline, or jet fuel exceeds the income obtained by selling it.

An LP model for the organization follows.

$$\text{min } 25L + 17D - 10F - 10G - 10J + 32BF + 50BG + 100BJ$$
$$\begin{aligned}
\text{s.t. } \quad 0.21L + 0.55D - F \quad\quad\quad + BF \quad\quad\quad\quad &= 3 \\
0.5L + \quad 0.3D \quad - G \quad\quad\quad + BG \quad\quad &= 7 \\
0.25L + \quad 0.1D \quad\quad\quad - J \quad\quad\quad\quad + BJ &= 5 \\
L,D,F,G,J,BF,BG,BJ &\geq 0.
\end{aligned}$$

Example 5.2. Referring to the example in Section 4.9, we have optimal solutions

$$x_t = (2 + t, 1 - t, t, 1 - t), \quad 0 \leq t \leq 1.$$

Suppose that $x_t = (x_1(t), x_2(t), x_3(t), x_4(t))$, where $x_I(t)$ represents the number of units of a product to be produced in town I, $I \leq 4$. The corner point BFS $(3,0,1,0)$ calls for no production in towns 2 and 4, while the other corner-point solution $(2,1,0,1)$ calls for no production in town 3. Implementing either BFS results in shutting down a manufacturing facility. Suppose that management would prefer an optimal solution with all the entries equal. The answer to the question "Which optimal solutions come closest to fulfilling this additional desire?" depends on how you measure closeness. The average value of the entries in x_t is 1, independent of the value of t, and $x_t - (1,1,1,1) = (1 + t, -t, t - 1, -t)$. Three common measures of closeness of x_t to $(1,1,1,1)$ follow; LP will be used to model the first two of them below.

1. *Sum of deviations*: $d_1(t) = |1 + t| + |-t| + |t - 1| + |-t|$
2. *Largest deviation*: $d_m(t) = \max\{|1 + t|, |-t|, |t - 1|, |-t|\}$
3. *Sum of squares of deviations*: $d_2(t) = (1 + t)^2 + (-t)^2 + (1 - t)^2 + (-t)^2$

Case 1

Put

$$1 + t = P_1 - N_1$$
$$-t = P_2 - N_2$$
$$t - 1 = P_3 - N_3.$$

According to Exercise 2.17, $|1 + t|$ is the minimum value of the objective function of the LP

$$\min \ P_1 + N_1$$
$$\text{s.t.} \ \ 1 + t = P_1 - N_1, \quad (P_1, N_1) \geq 0,$$

and similar representations are available for $|-t|$ and $|1 - t|$ in terms of P_2, N_2, P_3, and N_3. Moreover,

$$t = P_1 - N_1 - 1 = N_2 - P_2 = P_3 - N_3 + 1.$$

Consequently, the minimum value of $d_1(t)$ is the objective function value of an optimal solution to the LP

$$\min \ P_1 + N_1 + 2P_2 + 2N_2 + P_3 + N_3$$
$$\text{s.t.} \ \ P_1 - N_1 - 1 = N_2 - P_2$$
$$P_3 - N_3 + 1 = N_2 - P_2$$
$$0 \leq N_2 - P_2 \leq 1$$

$(P_1, N_1, P_2, N_2, P_3, N_3 \geq 0.)$ Deviations of type d_1 occur in many situations (cf. Exercises 5.1 to 5.4).

Case 2

Continuing the discussion for case 1, introduce an additional nonnegative variable z, which is required to be greater than or equal to each of the numbers $|1 + t|$, $|-t|$, and $|t - 1|$. The minimum value of z is equal to the minimum value of $d_m(t)$, so it suffices to solve the LP

$$\min \ z$$
$$\text{s.t.} \ \ \text{all the constraints of case 1 and also}$$
$$P_1 + N_1 \leq z$$
$$P_2 + N_2 \leq z$$
$$P_3 + N_3 \leq z \quad (z \geq 0).$$

Case 3

This case does not fit into an LP form. However, $d_2(t) = (1 + t)^2 + 2(-t)^2 + (t - 1)^2 = 4t^2 + 2$, so the minimum value of 2 occurs when $t = 0$.

The ideas presented in the two preceding examples will be used in subsequent sections.

5.2 INFEASIBILITY

An old adage asserts that necessity is the mother of change. If you are confronted with a situation that has no feasible solution, you can simply quit, give up, and leave; or you can decide to modify the situation: Consider suitable related situations that have feasible solutions. The following example is an illustation; I urge you to do Exercises 5.6 to 5.11.

Example 5.3. An agricultural mill manufactures feed for cattle and chickens from corn, soybeans, limestone, and fishmeal. These ingredients contain vitamins, protein, calcium, and fat. The units of the nutrients in each kilogram of the ingredients are tabulated below.

Ingredient	Vitamins	Protein	Calcium	Fat
Corn	8	10	6	8
Limestone	6	5	10	6
Soybeans	10	12	6	6
Fishmeal	4	8	6	9

The mill has 5 (metric) tons of corn, 8 tons of limestone, 4 tons of soybeans, and 3 tons of fishmeal available to process an order for 10 tons of cattle feed and 8 tons of chicken feed. The current price of the ingredients is 200, 120, 240, and 120 dollars per ton. The units of nutrients in a kilogram of feed in the order are required to be constrained as indicated below.

Feed	Vitamins Min.	Vitamins Max.	Protein Min.	Protein Max.	Calcium Min.	Calcium Max.	Fat Min.	Fat Max.
Cattle	6		6		7		4	8
Chicken	4	6	6		6		4	6

Upon running its LP model for this problem, the mill found that there was no feasible solution. The mill operators wish to explain the problem to the prospective buyer and, better yet, offer some options. Suppose they call us and ask for help. We examine the tables and observe that they can use a mix of $\frac{1}{2}$ corn and $\frac{1}{2}$ limestone to make an acceptable cattle feed. Hence we look closely at the requirements for chicken feed. The maximum fat limitation on chicken feed implies that only limestone and soybeans can be used in chicken feed; however, the maximum vitamin limitation implies that only limestone can be used. But limestone does not supply enough protein. Consequently, there is no way to mix an acceptable chicken feed using the available ingredients. Our solution by inspection worked for this simple problem. We could have made an LP model for each type of feed and then run it to find that it was not feasible to make chicken feed. At this point we know that the imposed maximums on vitamins and fat cause the infeasibility. Moreover, we can use the tools that were developed in Section 5.1 to make some suggestions. For example, if we allow up to z units of vitamins and up to z units of fat in a kilo of feed, a solution to the following LP tells us how large z must be in order that a feasible mix exist. Let

C = (metric) tons of corn to use in a ton of chicken feed
L = (metric) tons of limestone to use in a ton of chicken feed
S = (metric) tons of soybeans to use in a ton of chicken feed
F = (metric) tons of fishmeal to use in a ton of chicken feed.

Then

$$\min \ z$$
$$\text{s.t.} \quad C + L + S + F = 1$$
$$8C + 6L + 10S + 4F \le z \quad \text{(vitamins)}$$
$$10C + 5L + 12S + 8F \ge 6 \quad \text{(protein)}$$
$$6C + 10L + 6S + 6F \ge 6 \quad \text{(calcium)}$$
$$8C + 6L + 6S + 9F \le z \quad \text{(fat)}$$
$$C,L,S,F \ge 0.$$

In Exercise 5.6(a) you are asked to modify this LP to incorporate the limitations on supplies of ingredients available to fill the order.

5.3 MULTIPERIOD PROBLEMS

Two examples will be used to illustrate how to set up LP models for production scheduling problems.

Example 5.4. Commercial Appliance Division produces cooktops, dishwashers, ovens, and refrigerators. Estimated quarterly sales for the coming year follow.

		Quarter		
Product	1	2	3	4
Cooktops	100	100	50	100
Dishwashers	100	200	150	250
Ovens	50	50	70	50
Refrigerators	150	100	200	120

Cooktops, dishwashers, ovens, and refrigerators require 4, 3, 4, and 2 hours of production time; 1800 hours of production time are available each quarter. There is no current inventory and management has decided to require an estimated inventory of 10 cooktops, 15 dishwashers, 5 ovens, and 15 refrigerators at the end of each quarter. An item in inventory at the end of a quarter incurs an average interest cost of $50. We shall formulate an LP to plan quarterly production in order to meet expected demands and inventory requirements while minimizing expected inventory cost. Let

C_I = no. of cooktops to manufacture in quarter I, $I = 1, 2, 3, 4$
D_I = no. of dishwashers to manufacture in quarter I, $I = 1, 2, 3, 4$
O_I = no. of ovens to manufacture in quarter I, $I = 1, 2, 3, 4$
R_I = no. of refrigerators to manufacture in quarter I, $I = 1, 2, 3, 4$.

$$C_1 \qquad\qquad\quad \geq 100 + 10$$
$$C_1 + C_2 \qquad\quad\; \geq 200 + 10$$
$$C_1 + C_2 + C_3 \quad\; \geq 250 + 10$$
$$C_1 + C_2 + C_3 + C_4 = 350 + 10.$$

You should write the corresponding production-level requirements for dishwashers, ovens, and refrigerators. Production time constraints are

$$4C_I + 3D_I + 4O_I + 2R_I \leq 1800 \qquad I = 1, 2, 3, 4.$$

The estimated inventory interest in dollars due to cooktop production is 50 times:

$$(C_1 - 100) + (C_1 + C_2 - 200) + (C_1 + C_2 + C_3 - 250)$$
$$+ (C_1 + C_2 + C_3 + C_4 - 350) = 4C_1 + 3C_2 + 2C_3 + C_4 - 900.$$

Write corresponding inventory interest costs for dishwashers, ovens, and refrigerators. After eliminating constants, the objective function can be written in the form

$$\min 4(C_1 + D_1 + O_1 + R_1) + 3(C_2 + D_2 + O_2 + R_2)$$
$$+ 2(C_3 + D_3 + O_3 + R_3) + (C_4 + D_4 + O_4 + R_4)$$

or, because $\sum_{I=1}^{4} C_I + D_I + O_I + R_I = 1885$,

$$\min 3(C_1 + D_1 + O_1 + R_1) + 2(C_2 + D_2 + O_2 + R_2)$$
$$+ (C_3 + D_3 + O_3 + R_3).$$

Example 5.5. A company produces two products, which we denote by P and Q. During the next four months, the company wishes to produce the following numbers of products P and Q:

	Month			
Product	1	2	3	4
P	4000	5000	6000	4000
Q	6000	6000	7000	6000

The company has decided to install new machines on which to manufacture product Q. These machines will become available at the end of month 2. The company wishes to plan production during a 4-month changeover period. Maximum monthly production of product Q is 6000 during months 1 and 2, and 7000 during months 3 and 4; maximum total monthly production is 11,000 units during months 1 and 2, and 12,000 units per month during months 3 and 4. Moreover, manufacturing costs per unit of product Q are 15 dollars less per unit during months 3 and 4. Also, during the changeover period, product Q can be delivered late at a penalty of $10 per unit per month. Monthly inventory costs are $10 per unit per month for product P and $8 per unit per month for product Q. We will formulate an LP to minimize production, penalty, and inventory costs during the changeover period. Let

P_I = thousands of product P produced during month I, $I = 1, 2, 3, 4$
Q_I = thousands of product Q produced during month I, $I = 1, 2, 3, 4$
X_I = thousands of product Q in inventory at end of month I
Y_I = thousands of product Q backlogged at end of month I.

Constraints follow:

$$P_1 \qquad \geq \quad 4 \qquad \text{(production requirement for month 1)}$$

$P_1 + P_2$	≥ 9	(production requirement for months 1–2)
$P_1 + P_2 + P_3$	≥ 15	(production requirement for months 1–3)
$P_1 + P_2 + P_3 + P_4$	$= 19$	(production requirement for months 1–4)
Q_1, Q_2	≤ 6	(production capacity)
Q_3, Q_4	≤ 7	(production capacity)
$P_1 + Q_1 \leq 11$		(production capacity)
$P_2 + Q_2 \leq 11$		(production capacity)
$P_3 + Q_3 \leq 12$		(production capacity)
$P_4 + Q_4 \leq 12$		(production capacity)
$Q_1 - 6$	$= X_1 - Y_1$	(effect of production during month 1)
$Q_1 + Q_2 - 12$	$= X_2 - Y_2$	(effect of production during months 1–2)
$Q_1 + Q_2 + Q_3 - 19 = X_3 - Y_3$		(effect of production during months 1–3)
$Q_1 + Q_2 + Q_3 + Q_4 = 25$		(production requirement for months 1–4).

The objective (in thousands of dollars) is to minimize

$$10(3P_1 + 2P_2 + P_3) \quad \text{(inventory cost for } P)$$
$$+ \; 8(X_1 + X_2 + X_3) \quad \text{(inventory cost for } Q)$$
$$+10(Y_1 + Y_2 + Y_3) \quad \text{(penalty cost for } Q)$$
$$-15(Q_3 + Q_4) \quad \text{(reduced cost of producing } Q \text{ on new machines)}.$$

Because both X_I and Y_I incur a positive cost, a solution will automatically put at least one of each pair, X_I and Y_I, equal to zero. Please do Exercises 5.12 and 5.13.

5.4 MORE OBJECTIVES

One encounters many types of thresholds in applications. For example, if you are going to manufacture a product, you may have initial production costs for machinery and so on; or it may be more economical to buy larger quantities of an item. Some types of thresholds can be incorporated in an LP because, like the examples in Section 5.1, their structure is such that LP solutions will automatically model the thresholds accurately. Other types of thresholds can be modeled by a set of LPs; zero–one integer-valued variables (i.e., variables whose values are either zero or 1) can be used to incorporate such a set of LPs into one problem. Example 5.6

shows you how to deal with some common types of thresholds; zero–one integer variables, also called 0–1 integer variables, will be applied in Section 5.5.

Example 5.6. The Speedway Toy Company manufactures tricycles and wagons, which require 17 and 8 minutes to shape and 14 and 6 minutes to finish. Tricycles require one large wheel and two small wheels; wagons require four small wheels. Speedway has decided to allocate up to 520 hours of shaping time and up to 400 hours of finishing time to these two products during the next month. Speedway buys wheels from the Rollright Corporation, which sells large wheels for $5 and small wheels for $1. Speedway has been selling its entire output of these products to Toy World, Inc. at rates of 1000 and 500 units per month at a profit of $11 and $5 per unit. Speedway must decide how many tricycles and wagons to manufacture during the following month.

An LP model to maximize potential profit follows. Let

T = no. of tricycles to manufacture next month
W = no. of wagons to manufacture next month.

(1) $$\max\ 11T + 5W$$
$$\text{s.t.}\quad 17T + 8W \le 520(60)$$
$$14T + 6W \le 400(60) \qquad T, W \ge 0.$$

What assumptions are implicit in this model?
Consider the following four possibilities:

a. Rollright can supply only 1000 large wheels, and Speedway's alternative source of supply charges $6 per large wheel.
b. Toy World will buy wagons in excess of 600 only if given a $1 discount per wagon.
c. Rollright will sell small wheels in excess of 6000 for $0.75 each.
d. Toy World is willing to pay a $2 bonus for tricycles in excess of 1200.

Possibilities a and b are easy to incorporate into the model. The following models incorporate them separately and together.

To deal with possibility a, Speedway must decide how many large wheels to purchase from each supplier. Let

LR = no. of large wheels to purchase from Rollright
LA = no. of large wheels to purchase from the alternative source.

The profit decreases $1 for each large wheel purchased from the alternative supplier; this adjustment to the model can be incorporated in several ways. The following LP model works.

(1,a)   max $11T + 5W$ $- LA$
                         s.t. $17T + 8W$ $\leq 31{,}200$
                               $14T + 6W$ $\leq 24{,}000$
                               $T \quad - LR - LA =$ 0
                               $LR \quad \leq 1{,}000.$

Because large wheels purchased from Rollright do not cause a loss of profit, the model will automatically prefer to purchase large wheels from Rollright.

Considered alone, possibility b does not change the two basic decisions of how many tricycles and wagons to manufacture, but a decrease in profit per unit occurs when we cross the threshold of 600 wagons because wagons in excess of 600 have a decreased profit of one dollar caused by the discount. Thus we introduce decision variables to quantify production below and above the threshold level. Let

WI = min {no. of wagons produced, 600}
WE = no. of wagons produced in excess of 600.

(1,b)   max $11T + 5W$ $- WE$
                         s.t. $17T + 8W$ $\leq 31{,}200$
                               $14T + 6W$ $\leq 24{,}000$
                               $W - WI - WE =$ 0
                               $WI \quad \leq \quad 600.$

As in possibility a, this model automatically prefers to increase WI, so $WE = 0$ unless $WI = 600$. Notice that we could eliminate W from the model. Also notice that we can model possibility a this way, too. We can let TI = no. of tricycles that use Rollright large wheels and TE = no. of tricycles that use other large wheels. Then TI = min{no. of tricycles produced, 1000} and $T - TI - TE = 0$; the objective is to maximize $11TI + 10TE + 5W = 11T + 5W - TE$.

The following model includes both possibilities a and b.

(1,a,b)   max $11T + 5W$ $- LA$ $- WE$
                s.t. $17T + 8W$ $\leq 31{,}200$
                      $14T + 6W$ $\leq 24{,}000$
                      $T \quad - LR - LA \quad = \quad 0$
                      $W \quad \quad - WI - WE = \quad 0$
                      $LR \quad \leq \quad 1{,}000$
                      $WI \quad \leq \quad 600.$

Possibilities c and d require more effort because Speedway's natural preferences are to produce only bonus "trikes" and use only $0.75 small wheels. We will consider these cases individually. Considering c first, the reduction in price begins when S increases beyond 6000. We will write an LP model for each of the cases ($S \leq 6000$) and ($S \geq 6000$); comparing their solutions leads to a decision. Let S = no. of small wheels to purchase. Then

$(1,\text{c}, S \leq 6000)$

$$\text{max } 11T + 5W$$
$$\begin{aligned} \text{s.t.} \quad 17T + 8W &\leq 31{,}200 \\ 14T + 6W &\leq 24{,}000 \\ 2T + 4W - S &= 0 \\ S &\leq 6{,}000. \end{aligned}$$

Let SE = no. of small wheels to purchase in excess of 6000. Then

$(1,\text{c}, S \geq 6000)$

$$\text{max } 11T + 5W + .25SE$$
$$\begin{aligned} \text{s.t.} \quad 17T + 8W &\leq 31{,}200 \\ 14T + 6W &\leq 24{,}000 \\ 2T + 4W - S &= 0 \\ S - SE &= 6{,}000. \end{aligned}$$

(The conditions $S - SE = 6000$ and $SE \geq 0$ imply that $S \geq 6000$.) In Section 5.5 a zero–one integer variable will be introduced to switch automatically between the cases $S \leq 6000$ and $S \geq 6000$.

Possibility d is similar to c; we can make a decision by comparing solutions to the following two LPs.

$(1,\text{d}, T \leq 1200)$

$$\text{max } 11T + 5W$$
$$\begin{aligned} \text{s.t.} \quad 17T + 8W &\leq 31{,}200 \\ 14T + 6W &\leq 24{,}000 \\ T &\leq 1200. \end{aligned}$$

Let TE = no. of tricycles produced in excess of 1200. Then

$(1,\text{d}, T \geq 1200)$

$$\text{max } 11T + 5W + 2TE$$
$$\begin{aligned} \text{s.t.} \quad 17T + 8W &\leq 31{,}200 \\ 14T + 6W &\leq 24{,}000 \\ T - TE &= 1200. \end{aligned}$$

In Section 5.5 a 0–1 integer variable will be used to combine these two cases. Do Exercises 5.14 and 5.15.

5.5 INTEGER VARIABLES

Example 5.6 continues below. First, a 0–1 integer variable Y is used to switch between the cases $S \leq 6000$ and $S \geq 6000$ in possibility c. Let $SI = \min\{S, 6000\}$, $SE = S - SI$, and Y be a zero–one integer variable: The value of the variable Y is either the integer zero or the integer 1.

We want to incorporate Y into our model so that (1) SE must be zero when $Y = 0$ and (2) SE can be positive when $Y = 1$; this is accomplished by the inequality

$$SE \leq 14,000Y.$$

If $Y = 0$, then $SE \leq 0$, but $SE \geq 0$ automatically in an LP model, so $SE = 0$ when $Y = 0$. The constant 14,000 was chosen as follows. The inequality

$$14T + 6W \leq 24,000$$

assures us that

$$T \leq \frac{24,000}{14} < 2000$$

and

$$W \leq \frac{24,000}{6} = 4000;$$

thus, when $S \geq 6000$,

$$SE = S - 6000$$
$$= (2T + 4W) - 6000$$
$$\leq 2(2000) + 4(4000) - 6000$$
$$= 14,000.$$

Now we need a linear inequality that will force Y to be equal to zero when $SI \leq 6000$; the inequality

$$6000Y \leq SI$$

works because if $Y = 1$, then $6000 \leq SI$. The following model incorporates these inequalities.

(1c, integer)   max $11T + 5W + 0.25SE$

s.t. $17T + 8W$ $\leq 31,200$

$14T + 6W$ $\leq 24,000$

$2T + 4W - S$ $=$ 0

$$S - SI - SE \quad\quad = \quad 0$$
$$SI \quad\quad\quad\quad \le \quad 6000$$
$$SI \quad - \; 6000Y \ge \quad 0$$
$$SE - 14{,}000Y \le \quad 0.$$

If $SI = 6000$, then Y can be 1; if $Y = 1$, then SE can go up to 14,000, which is more than Speedway would wish to purchase.

A 0–1 integer variable Z is applied below to combine the two LPs used to deal with possibility d. Let Z be a zero–one integer variable, let $TI = \min\{T,1200\}$, and let $TE = T - TI$. The inequality

$$TE \le 800Z$$

forces TE to be zero when $Z = 0$; 800 suffices as an upper limit on TE because, referring to the discussion for possibility c, when $T \ge 1200$,

$$TE = T - 1200 \le 2000 - 1200 \le 800.$$

The inequality

$$1200Z \le TI$$

forces Z to be equal to zero when $TI \le 1200$. Consequently, possibility d can be modeled as follows.

(1d, integer)   max $11T + 5W + 2TE$

$$
\begin{array}{lll}
\text{s.t.} & 17T + 8W & \le 31{,}200 \\
 & 14T + 6W & \le 24{,}000 \\
 & T \quad - TI - TE & = \quad 0 \\
 & TI & \le \quad 1200 \\
 & TI \quad - 1200Z \ge & 0 \\
 & TE - \; 800Z \le & 0.
\end{array}
$$

The best way for you to begin to learn how to use 0–1 integer variables is to practice by using them to model simple situations; Exercises 5.17 to 5.24 give you opportunities to practice. To provide you with feedback, solutions to Exercises 5.17, 5.18(b), and 5.18(c), 5.19, 5.21, and 5.22 appear among the solutions at the end of Chapter 5. Example 5.7 below discusses some types of costs that can be modeled nicely by 0–1 variables.

Example 5.7. Consider the following situation. There is a set of N supply sources for a product. There is a sequence of K consecutive time periods and a corresponding list $(D_1, \ldots, D_k)$ of quantities of the product required to be supplied by the sources during the periods. Source I has a capacity of S_I units per period, $I \le N$.

Four types of costs will be considered:

1. A fixed-cost A_I that is incurred if source I is used to supply any quantity of the product. Examples include opening a factory, airport, army base, summer camp, lumber camp, and so on, bringing a ship out of mothballs, and providing an administrative team for a supply source.
2. A fixed-cost B_I for each period in which source I is used: for instance, the cost of a production crew to staff a supply source during a time period in which the source is being used to produce the product.
3. A fixed-cost C_I each time source I is activated from inactivity; for example, consider the cost of "firing up" a power plant.
4. A production cost P_I per unit of product supplied by source I.

The case $N = K = 3$ will be used to show how 0–1 variables can be applied to model the first three types of costs. Numerical data follow the discussion below; Exercise 5.22 is another numerical example. Let

X_{IJ} = no. of units of the product produced by source I in period J, $I \le 3$, $J \le 3$.

$X_I = X_{I1} + X_{I2} + X_{I3}$, $I \le 3$.

$Q_J = X_{1J} + X_{2J} + X_{3J}$, $J \le 3$.

Type 4 Costs

The following LP models type 4 costs:

$$\min P_1 X_1 + P_2 X_2 + P_3 X_3$$
$$\text{s.t.} \quad Q_J \ge D_J, \quad J \le 3$$
$$X_{IJ} \le S_I, \quad I \le 3, \quad J \le 3.$$

Type 1 and Type 4 Costs

To model the type 1 costs, introduce 0–1 integer variables as follows: Let U_I be a 0–1 integer variable that is equal to 1 if source I is used at all during the K time periods. The following LP models this situation:

$$\min P_1 X_1 + P_2 X_2 + P_3 X_3 + A_1 U_1 + A_2 U_2 + A_3 U_3$$
$$\text{s.t.} \quad Q_J \ge D_J, \quad J \le 3$$
$$X_{IJ} \le S_I U_I, \quad I \le 3, \quad J \le 3.$$

Type 1, Type 2, and Type 4 Costs

Here, in addition to the type 1 and type 4 costs modeled above, there are also type 2 costs (B_1, B_2, B_3) to consider; 0–1 integer variables can be used to model these costs, too. Let V_{IJ} be a 0–1 integer variable that is equal to 1 if source I is used in period J, $I \le 3$, $J \le 3$. Let $V_I = V_{I1} + V_{I2} + V_{I3}$, $I \le 3$.

The following LP models this situation:

$$\min\ P_1 X_1 + P_2 X_2 + P_3 X_3 + A_1 U_1 + A_2 U_2 + A_3 U_3 + B_1 V_1$$
$$+ B_2 V_2 + B_3 V_3$$

$$\text{s.t.}\quad Q_J \ge D_J, \qquad J \le 3$$
$$X_{IJ} \le S_I V_{IJ}, \qquad I \le 3, \quad J \le 3$$
$$V_I \le 3U_I, \qquad I \le 3.$$

Type 1, Type 2, Type 3, and Type 4 Costs

To incorporate the effect of type 3 costs into the preceding model, it suffices to make the following additions to that model. For $1 \le I \le 3$ and $1 < J \le 3$, additional variables P_{IJ} and Q_{IJ} are defined below. These variables do not need to be specified to be integer variables; however, in effect, $P_{IJ} = 1$ if source I gets turned on at the beginning of period J, and $P_{IJ} = 0$ otherwise. Add the following constraints:

$$V_{IJ} - V_{I(J-1)} = P_{IJ} - Q_{IJ}, \qquad I \le 3, \quad J = 2,3.$$
$$W_I = V_{I1} + P_{I2} + P_{I3}, \qquad I \le 3.$$

Add $C_1 W_1 + C_2 W_2 + C_3 W_3$ to the objective function.

Numerical data

Costs (thousands of dollars)

Type	Source 1	2	3
A	1500	750	980
B	160	65	110
C	0	800	270
P	1.8	1.55	1.4

Requirements

Type	Source (capacity) – period (demand) 1	2	3
Capacity	40	90	35
Demand	125	50	70

5.6 TRANSPORTATION PROBLEMS

Transportation problems were introduced in Section 2.3. You should reread the first two paragraphs in that section before proceeding to Example 5.8, which shows how to deal with some frequently occuring variations on a standard transportation problem. You can find lots of examples in the news. In 1989 they include transportation in the Mideast, garbage in the eastern United States, nuclear waste in the western United States, and so on. Example 5.8 deals with blueberries, which can be healthy: nontoxic and nonhazardous.

Example 5.8. The tableau

		Warehouse				
		Calif.	Ariz.	Colo.	N. Mex.	Supply
	Washington	460	550	650	720	100
Orchard	Oregon	350	450	560	620	170
	Michigan	990	920	500	540	120
	Demand	175	100	80	35	390

contains a list of orchards that produce blueberries and a list of distribution centers to which truckloads of blueberries are to be shipped. The supplies and demands are given in truckloads of blueberries and the cost of shipping are given in dollars per truckload. The special structure of transportation problems permits us to assert that *if the quantities entered in the tableau are integers*, then an optimal (basic feasible) solution will be composed of integers; to repeat, *a solution will automatically be integer valued.* Thus we do not need to consider what to do if we were to get an answer that told us to ship fractions of a truckoad to some destinations. Of course, if there were multiple optimal solutions, then, as in Example 5.2, we might choose to consider noninteger solutions.

After using Example 5.8 to show you how to incorporate three simple variations on a transportation problem into a transportation model, I will discuss an interesting variation that occurs in many settings. First, suppose that Michigan could supply only 100 truckloads; then, to balance supply and demand, we would introduce an artificial supply point with an artificial supply of 20 truckloads and artificial costs of zero, because we are not really going to ship from the artificial supply. Similarly, if there were less total demand than total supply, we could introduce an artificial demand point to balance supply and demand. Again, because we are not actually going to ship to the dummy demand, the costs could be put equal to zero;

actually, any constant value for these costs will suffice. The third simple variation considers the case where, for some reason, it becomes necessary to prohibit shipment from, for example, Oregon to California; this constraint can be incorporated in the tableau model by changing the corresponding shipping cost to a large number, traditionally denoted by M ("big-M"), which will cause the optimal solution to bypass that route. Now we are ready to consider an interesting variation: suppose that supply remains constant but demand grows to 300, 150, 100, and 40 truckloads at the warehouses. Moreover, suppose it is decided to ship at least 150, 75, 50, and 30 truckloads to the warehouses. To incorporate these new constraints into a transportation model, we can introduce an artificial supply A and split the demands into guaranteed minimal demands MCA, MAZ, MCO, MNM, which will be met by shipping from the actual supplier at the listed costs, and optional demands OCA, OAZ, OCO, ONM, which can either be met by actual supplies at the listed costs or not be met by the artificial supply at zero cost. To set the level of the artificial supply, note that total actual supply is 390, guaranteed shipments add to 305, and optional desired shipments sum to 285. Since desired shipments add to 590, we can put the dummy supply level at 590–390 = 200 to balance supply and demand. The modified tableau follows; as above, *M denotes a large number.*

		Warehouse								
		MCA	OCA	MAZ	OAZ	MCO	OCO	MNM	ONM	Supply
	Washington	460	460	550	550	650	650	720	720	100
Orchard	Oregon	350	350	450	450	560	560	620	620	170
	Michigan	990	990	920	920	500	500	540	540	120
	A	M	O	M	O	M	O	M	O	200
	Demand	150	150	75	75	50	50	30	10	590

You should do Exercises 5.25 and 5.26; formulating other exercises is also a good exercise. At this point we know LP and tableau formulations for a transportation problem. In Section 5.7 a modified LP formulation and a network formulation are introduced for transportation problems, and relationships between the various formulations are discussed.

5.7 INTRODUCTION TO NETWORKS

We begin this section by looking at the transportation problem (TP) from a network perspective and slightly modifying the form of an LP model for a TP. Then several basic network concepts, including graphs, digraphs, cycles, trees, and the node-incidence matrix, are introduced and used to analyze the TP by relating structures that occur in the various models as follows: (1) a basic feasible solution in the LP formulation corresponds to a spanning tree in the network formulation; (2) considering a nonbasic variable for inclusion in a BFS corresponds to introducing a cycle into the associated spanning tree; and (3) the reduced cost of the nonbasic variable corresponds to a "cycle cost" assigned to this cycle. Negative cost cycles correspond to BFSs, which have lower objective function value. Cycles in the network formulation can also be visualized as cycles on the transportation tableau introduced in Section 5.6, and one can do variations of the simplex algorithm for a TP right on the tableau. You will see the correspondence between cycles in the two contexts, but computations will not be done on the tableau. After discussing the TP, you will be shown LP formulations of two important network problems, the minimum cost flow problem and the maximum flow problem. Next, you will see a very nice interpretation of the dual of the maximum flow problem. Section 5.7 concludes by introducing two useful functions which are used in Section 5.8 and in Chapter 9, where some network algorithms are discussed.

Transportation Problem

The transportation problem can be viewed advantageously from a network perspective. The act of shipping a unit of material from supply point i to demand point j can be viewed as an activity that produces a unit flow out of the supply point i and into the demand point j. For an m-supply-point and n-demand-point transportation, label the m supply points $N_1, N_2, \ldots, N_m$ and the n demand points $N_{m+1}, \ldots, N_{m+n}$. In Example 5.8, $m = 3$ and $n = 4$, so N_1 corresponds to Washington, N_2 corresponds to Oregon, N_3 corresponds to Michigan, N_4 corresponds to California, $\ldots, N_7$ corresponds to New Mexico; $N_1, \ldots, N_7$ are called *nodes*. A network is a finite set of nodes with some conditions relating nodes in that network.

Here, I have arbitrarily decided to assign positive numbers to represent flows out of a node and negative numbers to represent flows into a node. This convention changes the A matrix and b vector in Example 2.3; the 1's in the last 2300 rows in A become -1's, and the d_j's in b become $-d_j$'s. Column $a_{n(i-1)+j}$ in A corresponds to a unit flow from N_i to N_{m+j}, and $x_{n(i-1)+j}$ corresponds to the number of tons of potatoes shipped from N_i to N_{m+j}; A is called a node-incidence matrix.

In Example 5.8, $x_{n(i-1)+j}$ represents the number of truckloads of blueberries shipped from N_i to N_{m+j}. The A matrix and b vector for Example 5.8 follow (only nonzero entries are shown in A):

$$
A = \begin{bmatrix}
1 & 1 & 1 & 1 & & & & & & & & \\
& & & & 1 & 1 & 1 & 1 & & & & \\
& & & & & & & & 1 & 1 & 1 & 1 \\
-1 & & & & -1 & & & & -1 & & & \\
& -1 & & & & -1 & & & & -1 & & \\
& & -1 & & & & -1 & & & & -1 & \\
& & & -1 & & & & -1 & & & & -1
\end{bmatrix}
$$

$$
b = \begin{bmatrix}
100 \\
170 \\
120 \\
-175 \\
-100 \\
-80 \\
-35
\end{bmatrix}.
$$

The special structure of A permits various simplifications. For example, it is easy to find initial basic feasible solutions. However, the rank of A is not equal to seven. The rank of A was defined in Chapter 1; in fact, the rank of A is equal to the number of rows of A in a maximal set of linearly independent rows of A. Since the sum of the rows of A is equal to the twelve-dimensional zero row vector, the rank of A is less than seven. However, any six rows of A are linearly independent, so the rank of A is equal to six. Let us verify that rows 1 to 6 of A are linearly independent:

Suppose that $v = (v_1, v_2, \ldots, v_{12}) = y_1 r_1 + y_2 r_2 + \cdots + y_6 r_6 = 0$. Then $v_4 = y_1 = 0$, $v_8 = y_2 = 0$, $v_{12} = y_3 = 0$ and $v_1 = y_1 - y_4 = -y_4 = 0$; also $v_2 = y_1 - y_5 = -y_5 = 0$ and similarly, $v_3 = -y_6 = 0$. Hence $r_1, r_2, \ldots, r_6$ are linearly independent. Thus the rank of the column space of A is six; sets of six linearly independent columns of A correspond to basic feasible solutions x_B. (The a priori constraint assures us that b is in the column space of A.) Notice that B is a 7×6 matrix in this example, so B^{-1} is not defined; but there is a unique solution to the 7×6 linear system $Bx_B = b$. To characterize sets of six linearly independent columns in A, it will be convenient to consider a network representation of Example 5.8. To do that, notation that will enable us to discuss networks with some facility is introduced below.

Notation

A *link* is an undirected path between two nodes; realizations include a connection between computers, a pipeline between junctions, a piece of freeway between interchanges, a long-distance telephone line between cities, and a railraod line between stations. A link will be denoted by $(i,j) = (j,i)$, where i and j are the nodes (ends) of the link. A link is defined by specifying its ends; it is a set of two nodes.

A *route* from node i to node j, denoted by $[i,j]$, is a directed link, the ends i and j of a route are distinguished, the route $[i,j]$ has origin node i and destination (terminal) node j, and the route $[i,j]$ goes from its origin node i to its terminal node j. In contrast with links, $[i,j] \neq [j,i]$.

A *graph* G is a set of k nodes, labeled $1,2,\ldots,k$, and a set of links (i,j) between a node i and a node j in G.

A *subgraph* of a graph G is composed of a subset S of nodes in G and a subset of the links in G which have both ends in S.

A *directed graph: digraph* G is a set of k nodes, labeled $1,2,\ldots,k$, and a set of routes $[i,j]$ from a node i to a node j in G.

Two links or two routes are connected if they have a common node, called their *transfer node*.

The following definitions apply to graphs; by replacing "link" with "route" and "graph" with "digraph" in appropriate places, you can get corresponding definitions for digraphs.

A *simple path* from node a to another node b in a graph G is a sequence of links $L_1, L_2, \ldots, L_n$ in G satisfying the following: (1) a is in L_1, (2) b is in L_n, (3) each adjacent pair of links is connected, and (4) if t_i denotes the transfer node between links L_i and L_{i+1}, then $a, t_1, t_2, \ldots, t_{(n-1)}, b$ is a sequence of $n + 1$ distinct nodes. (In a graph the later sequence determines the simple path, but in a digraph the directions of the routes must be taken into consideration.)

A *cycle* is a simple path from a node i to a node j with a final link (i,j) or (j,i) added. A cycle in a digraph is called an *order cycle* if all of its routes are aligned in the same direction. An *acyclic* graph is a graph without cycles; a *NOcycle* digraph is a digraph containing *No Order cycles*. The digraph representation of a transportation problem contains cycles, but it is a NOcycle digraph; for instance, referring to Example 5.8, the cycle composed of the ordered list of nodes 2,6,3,5,2 and the four routes [2,6], [3,6], [3,5] and [2,5] contains two *forward routes*, [2,6] and [3,5], aligned with the path and two *backward routes*, [3,6] and [2,5], with direction opposite to the path.

A *connected graph* is a finite set of nodes and a set of links such that there is a simple path between any two nodes.

A *path* from node i to node j in a graph G is a list $L_1, \ldots, L_n$ of links in G satisfying the following three properties.

1. i is in L_1,
2. every pair of adjacent links has a common node,
3. j is in L_n.

A simple path is a path which contains no cycles.

Two nodes i and j are in the same *component* of a graph G if there is a path from i to j in G. A graph with one component is connected. The components of G are maximal connected subgraphs of G.

A *tree* is a connected, acyclic graph (or digraph). A *spanning tree* is a tree in a graph (or digraph) G which contains every node in G; if G contains n nodes, then a spanning tree contains $n - 1$ links. A node i in a tree T is an *end* of the tree T if there is only one link in T which contains i.

Mathematical Induction

Mathematical induction, also called the principle of induction, is a simple, intuitive, and very useful principle that is introduced informally in the following example before being stated formally.

Example 5.9. Suppose that you are given a list $S_1, S_2, \ldots, S_n, \ldots$, of all the settlements in a vast desert, jungle, or swamp and you are asked to determine whether or not you can travel to the region and then travel about the region, visiting each of the settlements in the order listed exactly once. To determine if such a journey is possible, it suffices to know two things:

1. You can reach S_1 without touching $S_2, S_3, \ldots$.
2. Suppose that you can reach S_n; then there is a route from S_n to S_{n+1}.

Principle of Induction

Given a sequence, statement 1, statement 2, $\ldots$, statement n, $\ldots$ of statements; all of the statements in the sequence are true if the following two conditions are satisfied:

1. Statement 1 is true.
2. Suppose that statement n is true; then statement $n + 1$ is true.

The supposition that statement n is true is called the *induction hypothesis*. Verifying condition 2 is called the *inductive step*.

The principle of induction is applied in Lemma 5.1 below to show that a tree has at least two ends; statement 1 says that the simplest kind of tree, a single link or route, has two ends, the statement n says that trees with no more than $n + 1$ nodes have at least two ends. The inductive steps consists of verifying that if trees with fewer than $n + 2$ nodes have at least two ends, trees with $n + 2$ nodes have at least two ends.

Lemma 5.1 A non-trivial tree T has at least two ends.

Proof (proof for graphs using the principle of induction): Statement n:

The lemma is true for all trees that contain $\leq n + 1$ nodes. Statement 1 is clearly true: T is composed of nodes 1 and 2 with one link, (1,2); so we are ready for the inductive step that follows.

Inductive Step

Suppose that statement n is true and suppose that T has $n + 2$ nodes. Choose a node i in T. Remove node i and all links that contain i. Look at the components of what remains. If i is an end of T, we have removed one link and what remains is a tree with $n + 1$ nodes. By the induction hypothesis, this new trees has at least two ends; at least one of the ends of the new tree is a second end of the original tree. If i is not an end of T, there are at least two components in what remains; to each component, add i and the links containing i and a node in that component (in fact, there is only one such link). Figures 1 to 4 are interjected below to illustrate. Each subtree that we obtain by this process satisfies the induction hypothesis and, consequently, contains an end distinct from i. The ends of the subtrees that are distinct from i are ends of T. Thus statement $n + 1$ is also true and we are done with a proof of the lemma.

The digraph corresponding to Example 5.8 has seven nodes and 12 routes. It is a connected NOcycle digraph. The corresponding matrix A is called the *node-incidence matrix* for the digraph. Sets of independent columns in A are characterized by the following:

Independence Theorem. Given a digraph G; the columns in a node-incidence matrix for G are linearly independent if, and only if, the digraph G is acyclic.

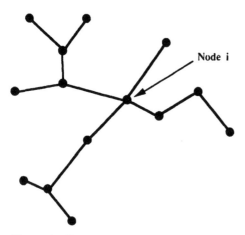

Node i

Figure 1   Tree with node i marked.

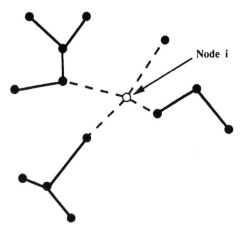

Figure 2 Tree with node i and connecting links highlighted.

Proof: Part 1. Suppose that G is acyclic. I will verify that the columns are linearly independent by applying the principle of induction below under the assumption that G is connected and leave the case where G has more than one component for you to verify. Suppose that the connected digraph G is acyclic: suppose that G is a tree.

Statement n: The result is valid for all connected digraphs with $\leq n + 1$ nodes.

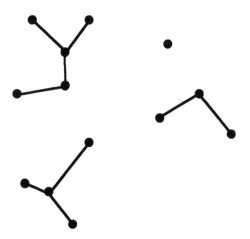

Figure 3 Node i and connecting links removed; four components remain.

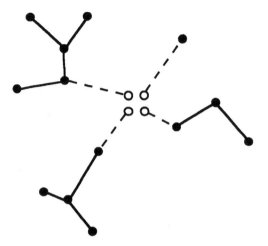

Figure 4   Four components with node i and appropriate connecting link attached.

Statement 1 is clearly true; suppose that statement n is true and that G has $n + 2$ nodes. Relabeling two nodes, if necessary, we may assume that node $n + 2$ is an end of G. Remove node $n + 2$ and the link containing that node. What remains satisfies the inductive hypothesis; consequently, the corresponding columns in the node-incidence matrix are linearly independent. The column corresponding to the link containing node $n + 2$ is clearly independent of the others: Verify this assertion. Then establish the case where G is not connected by applying this result to each component. This completes a proof of the first half of the independence theorem; the second half is verified below.

Part 2.   Suppose that G contains a cycle a, $t_1, t_2, \ldots, t_k$, a. We define coefficients to establish dependence of an appropriate set of columns of a node-incidence matrix via the following correspondences:

If $[a, t_1]$ is in the cycle and this route corresponds to a_{i_1}, then $y_{i_1} = 1$.
If $[t_1, a]$ is in the cycle and this route corresponds to a_{i_1}, then $y_{i_1} = -1$.
If $[t_j, t_{j+1}]$ is in the cycle and it corresponds to $a_{i_{j+1}}$, then $y_{i_{j+1}} = 1$.
If $[t_{j+1}, t_j]$ is in the cycle and it corresponds to $a_{i_{j+1}}$, then $y_{i_{j+1}} = -1$.
If $[t_k, a]$ is in the cycle and it corresponds to $a_{i_{k+1}}$, then $y_{i_{k+1}} = 1$.
If $[a, t_k]$ is in the cycle and it corresponds to $a_{i_{k+1}}$, then $y_{i_{k+1}} = -1$.

Thus $y_{i_1} a_{i_1} + y_{i_2} a_{i_2} + \cdots + y_{i_{k+1}} a_{i_{k+1}} = 0$ and we have completed a proof of the independence theorem.

Spanning Trees

A *spanning tree* in a graph, or digraph, G is a tree in G that contains all k of the nodes of G. Going back to Example 5.8, we are interested in *linearly independent spanning sets for the column space of A*; these are the maximal sets of linearly independent column vectors in A. In two steps below, these maximal sets are shown to correspond to *spanning trees in the associated digraph of routes*.

Step 1

If the digraph of routes corresponding to a set of linearly independent columns in a node-incidence matrix is not connected, the set of column vectors is not maximal because a route can be added to connect two components in the digraph. The added route corresponds to a column vector that is added to the set. The new digraph contains no cycles, so the augmented set of vectors is linearly independent.

Step 2

If the tree corresponding to a maximal set S of linearly independent columns were not a spanning tree, we could choose a node not in the tree and add a route which contained that node. The resulting digraph would be a tree; consequently, the set S would not be maximal (why not?), and we would have a contradiction. Thus we see that a basic matrix B is composed of a set of arcs that corresponds to a spanning tree for the associated digraph.

Routes in the node-incidence matrix for a transportation problem correspond to positions in the tableau representation for that problem. It turns out that cycles play a central role in solving transportation problems efficiently. Cycles correspond to a cyclic configuration on the tableau model for a transportation problem; one can thus define cycles on the tableau and do appropriate hand computations on the tableau. You can find many pages of those kinds of hand computations in various textbooks; such computations will not be discussed here. The methods are easy to explain in a network flow context; Example 5.8 will be used below to do that. Let us begin with an allocation tableau on which an initial BFS is described by assigning flows lexicographically: Blueberries are shipped, preferring lower-numbered nodes.

Allocation Tableau

	4	5	6	7	Supply
1	100				100
2	75	95			170
3		5	80	35	120
Demand	175	100	80	35	

A network display of this BFS follows:

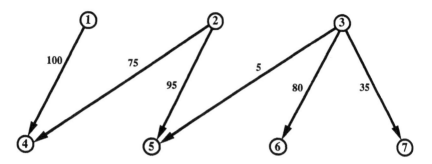

Cycles

Now we will modify this BFS by introducing a flow from node 2 to node 6. Adding route [2,6] to routes in the BFS introduces a cycle on nodes 2, 3, 5, and 6:

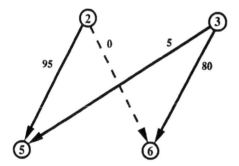

The current flow on this cycle can be augmented within the limits displayed below.

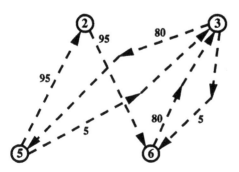

The augmentation

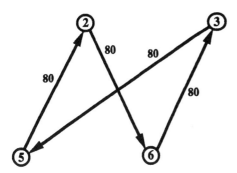

reduces the flow on route [3,6] to zero and results in the flow

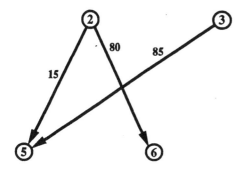

on nodes 2, 3, 5, and 6; the full BFS that results from this augmentation follows:

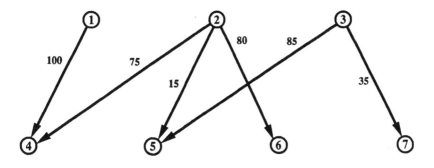

Note the cost of transporting one truckload around the cycle:

+560: 2 to 6 cost
−500: 6 to 3 cost reduction
+920: 3 to 5 cost
−450: 5 to 2 cost reduction

+530: net cost (number of dollars required) to transport one truckload around the cycle.

Consequently, the new BFS costs 80 × 530 = $42,400 more than the first BFS. If the net cost to transport one unit around the cycle were negative, we would call the cycle a *negative cost cycle*. Augmenting a flow by a positive flow on a negative cost cycle reduces the total cost.

Cycles on the Allocation Tableau

Routes correspond to positions on the allocation tableau and nodes correspond to vertical or horizontal line intervals between adjacent locations. A flow on a cycle in the network representation of a transportation problem also appears as a flow on a cycle in the allocation tableau representation of the problem, but the nodes and routes in the two representations have quite different interpretations. The allocation tableau representation for the cycle on nodes 2, 3, 5, and 6 which we considered above is shown in Fig. 5.

The transportation problem is an example of a minimum cost flow problem. LP formulations of the minimum cost flow problem and another important network problem, the maximum flow problem, follow.

Minimum Cost Flow Problem

Consider a set of nodes $N_1, \ldots, N_k$ and a set of routes $R_{i,j}$ from N_i to N_j. Let $x_{i,j}$ denote the rate of flow from N_i to N_j, and let $c_{i,j}$ denote the

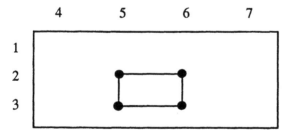

Figure 5   Representation of cycle on allocation tableau.

cost of a unit flow from N_i to N_j. Finally, let the total rate of flow F_i out of each node N_i be specified. Then the LP

$$\min \sum_{i,j} c_{i,j} x_{i,j}$$

$$\text{s.t.} \sum_m x_{i,m} - \sum_n x_{n,i} = F_i, \quad i \le k \quad (x_{i,j} \ge 0)$$

models the problem of maintaining specified rates of flow at minimum cost.

A related problem, to maximize the flow from N_1 to N_k when the rates of flow are limited by route capacities $M_{i,j}$, follows.

Maximum Flow Problem

The simplest version of this problem is the case where $F_i = 0$, $1 < i < k$, with the following LP model:

$$\max F$$

$$\text{s.t.} \sum_m x_{i,m} - \sum_n x_{n,i} = 0, \quad 1 < i < k$$

$$\sum_m x_{1,m} - \sum_n x_{n,1} = F$$

$$\sum_m x_{k,m} - \sum_n x_{n,k} = -F$$

$$0 \le x_{i,j} \le M_{i,j}.$$

Both a solution and its value F are called a flow. The largest value of F is called the maximum flow; several different flows on the network may be maximum flows; their values are all the maximum flow.

If I were to write a dual for this LP formulation and interpret it carefully, I could come up with a famous relationship, called the maximum flow–minimum cut theorem. Rather than plow through the details of formulating and interpreting an appropriate dual, I will simply state the theorem and then give a direct proof of it. We know that maximum problems have minimum dual problems. In this case the dual has an especially nice interpretation, which is explained below.

Maximum Flow–Minimum Cut Theorem

We can assume that the set of nodes and the set of links associated with the routes constitute a connected graph. Suppose that we cut the network into two pieces, without cutting through any nodes, so that node 1 and node k are in opposite pieces of the network. Let X denote the set of nodes in the piece that contains node 1 and let Y denote the set of nodes in the piece containing node k. (X,Y) is a pair of disjoint, nonempty sets of nodes; every node is in one of the sets. Node 1 is in X and node k is in Y; such a pair is called a *cut*. Look at the route capacities $M_{i,j}$ of the routes $R_{i,j}$ which go from a node i in X to a node j in Y. No flow can exceed the sum of these route capacities, which is called the *capacity* $C(X,Y)$ of the cut (X,Y). Consequently, the maximum flow is $\leq$ the minimum value of the cuts. The maximum flow–minimum cut theorem asserts that these values are equal; it is established by exhibiting a cut (X,Y) for which $C(X,Y)$ is equal to a flow from node 1 to node k.

Before exhibiting a maximal flow and corresponding minimal cut in the general case, I will show you an example. Both "path" and "minimal path" mean simple path, a path without cycles, in the example and ensuing discussion. The example uses links; links permit flow in both directions whereas a route $[i,j]$ only permits flow from node i to node j. You can think of a link (i,j) in a flow context as corresponding to two routes, $[i,j]$ and $[j,i]$.

Example 5.10.

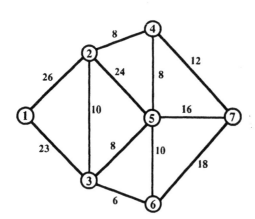

Augment the initial zero flow with the following sequence of maximal simple path flows

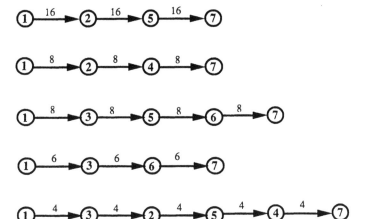

to get the flow

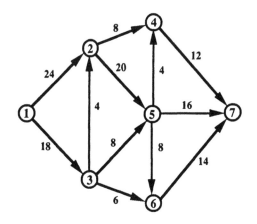

Observe that the path flow

can be augmented by 2 units to produce the flow

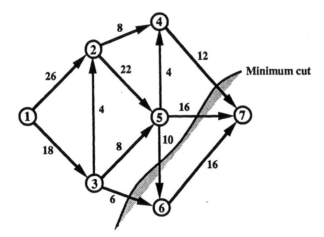

This flow produces a flow of 44 units from node 1 to node 7. The capacity of the cut $\{1,2,3,4,5\}$, $\{6,7\}$ is also 44. Thus we have a maximum flow and a minimum cut for this example and we return to the general case.

Our links are the links (i,j) corresponding to routes $R_{i,j}$. Consider a minimum path $\mathscr{P}$ from 1 to k. Links in the path are undirected, but their corresponding routes are directed; a link (t_i, t_{i+1}) in the path corresponds to either a *forward route* $R_{t_i,t_{i+1}}$ or a *backward route* R_{t_{i+1},t_i}. Now consider a flow $\mathscr{F} = \{x_{i,j}\}$ in the network. The path $\mathscr{P}$ is *flow augmenting* for $\mathscr{F}$ if $x_{i,j} \leq M_{i,j}$ for every forward arc in the path and $x_{i,j} > 0$ for every backward arc in the path. If $\mathscr{P}$ is a flow-augmenting path, we can increase the flow from node 1 to node k by modifying $\mathscr{F}$ on the routes in the minimum path $\mathscr{P}$ as follows. Because $x_{i,j} < M_{i,j}$ on every forward arc in $\mathscr{P}$ and $x_{i,j} > 0$ on every backward arc in $\mathscr{P}$, there is a largest positive number δ such that $x_{i,j} + \delta$ is $\leq M_{i,j}$ on every forward arc in $\mathscr{P}$ and $x_{i,j} - \delta \geq 0$ on every backward arc in $\mathscr{P}$. Thus we modify $\mathscr{F}$ by increasing $x_{i,j}$ to $x_{i,j} + \delta$ on forward arcs in $\mathscr{P}$ and decreasing $x_{i,j}$ to $x_{i,j} - \delta$ on backward arcs in $\mathscr{P}$. These modifications increase the flow from 1 to k by δ. We modify the flow on flow-augmenting paths until there is no flow-augmenting path from 1 to k. We will verify that this final flow $\mathscr{F}$ is an optimal flow by defining an appropriate cut as follows. Put node 1 in X and put node i in X if there is a flow-augmenting path for $\mathscr{F}$ from node 1 to node i. Let Y denote the set of nodes not in X. Then k is in Y and (X,Y) is a cut. Now suppose that i is in X and j is in Y; if $R_{i,j}$ is a route, $x_{i,j} = M_{i,j}$, and if $R_{j,i}$ is a route, $x_{i,j} = 0$, because otherwise j would be in X. Finally, we look at the LP formulation of the maximum flow problem; adding the constraint equations corresponding to the nodes in X, we get the cut capacity

$$C(X,Y) = \sum_{\substack{i \text{ in } X \\ j \text{ in } Y}} M_{i,j} = \sum_{\substack{i \text{ in } X \\ j \text{ in } Y}} x_{i,j} - \sum_{\substack{i \text{ in } Y \\ j \text{ in } X}} x_{i,j} = \bar{F},$$

where $\bar{F}$ is the flow from 1 to k corresponding to $\mathscr{F} = \{x_{i,j}\}$. Thus we have exhibited a maximum flow $\mathscr{F}$ and a corresponding minimum cut (X,Y).

In Example 5.10 flow can be in either direction through links and all the flow capacities are positive integers. When all the flow capacities are integers, every flow augmentation increases the flow by an integral amount. Consequently, we can be sure of arriving at a maximum flow after a finite number of augmentations. Example 5.10 merely illustrates flow augmentation, it does not present an algorithm for generating flow-augmenting paths. We return to Example 5.10 in Section 9.7, where an algorithm for generating flow-augmenting paths is discussed.

Some network flow algorithms are developed in Chapter 9. A basic idea in these algorithms is to use successive partial solutions to a problem to eliminate many possible solutions from consideration. This idea is utilized in Section 5.8, which considers a class of problems that can be modeled as a special kind of problem on a particular type of network (a NOcycle, connected network). This section concludes by defining two useful functions, *arg min* and *arg max*, which are used in Section 5.8 and Chapter 9.

Arg Min and Arg Max

Suppose that S is a finite set of positive integers and f is a function defined on S. Then

$arg \, min\{f(j); \, j \text{ in } S\}$ is the smallest integer in S for which f attains its minimum on S. Similarly,

$arg \, max\{f(j); \, j \text{ in } S\}$ denotes the smallest integer in S for which f attains its maximum on S.

For example, if $S = \{1,3,4,6,9\}$, $f(1) = 5$, $f(3) = 4.2$, $f(4) = 6.4$, $f(6) = 4.2$, and $f(9) = 6.4$, then $\arg \min\{f(j); j \text{ in } S\} = 3$ because 3 is the smallest integer in S for which f attains its minimum value, 4.2, on S. For the preceding example, $\arg \max\{f(j); j \text{ in } S\} = 4$ because 4 is the smallest integer in S at which f attains its maximum value, 6.4, on S.

5.8 INTRODUCTION TO DYNAMIC PROGRAMMING

Like linear programming, dynamic programming can be an effective way to model situations that have an appropriate structure. This section begins by introducing three characteristics of problems that fit into a (backward) dynamic programming format and outlining the steps involved in formulating a dynamic programming (DP) model. Then several examples are used to explain the terminology (stages, states, etc., used below) of DP, illustrate some types of problems that fit into a dynamic programming format, and show you how DP can be used to model these problems.

Three Characteristics Necessary for a (Backward) DP Model

1. Such problems can be organized into a finite sequence of stages, with a (policy) decision at each stage. A set of (initial) states enters a stage. A decision transforms each initial state to a terminal state; the terminal state leaves the stage and becomes an initial state for the next stage. Suppose that there are n stages for a problem. Then a solution to the problem amounts to making a sequence of n decisions, one for each stage. This sequence of decisions is called an optimal policy or strategy for the problem.

2. A key characteristic of problems that can be modeled effectively by dynamic programming is the *Markov property* or *principle of optimality*: Given any initial state at any stage, an optimal policy for the successive stages does not depend on the previous stages (i.e., an optimal policy for the future does not depend on the past).

3. The policies at the final state are determined (known in advance).

When these three properties are satisfied, the problem can be solved (dynamically) by moving backward stage by stage, making an optimal decision at each stage. Examples 5.11 and 5.15 include computations comparing the computational requirements of the DP model with a "look at all the cases" model.

Formulating a dynamic programming model involves the following steps:

1. Specifying a sequence of stages.
2. Noting that the final stage policies are determined.
3. Checking that the Markov property is satisfied by the stages.
4. Formatting each stage. This involves
    a. Specifying initial states
    b. Specifying how initial states are transformed
    c. Specifying terminal states
    d. Specifying a decision rule that determines the optimal terminal state (or states) corresponding to an initial state.

Example 5.11. Starting at $(1,0)$ in the (x_1,x_2)-plane, we will take n one-unit steps to the right, ending where $x_1 = n + 1$. At each of the first $n - 1$

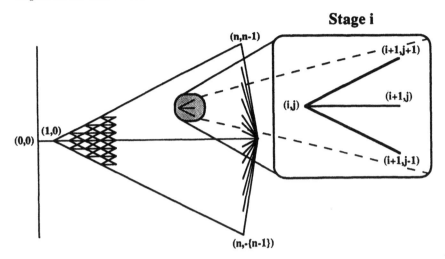

Figure 6 Path from $(1,0)$ to $(n + 1, 0)$ for which the sum of the costs is minimal.

steps, x_2 can remain constant or increase or decrease 1 unit; we can go from (i,j) to $(i + 1, j + 1)$, $(i + 1, j)$ or $(i + 1, j - 1)$. At step n we go from (n,j) to $(n + 1, 0)$. There is a specified cost $c(i,j,k)$ to move from (i,j) to $(i + 1, k)$. We wish to specify a path from $(1,0)$ to $(n + 1, 0)$ for which the sum of the costs is minimal. A diagram is shown in Fig. 6.

The transition from (i,j) to $(i + 1, k)$ occurs at stage i. There are 3^{n-1} paths. If we try to compute the cost of each path and then compare these costs, we are faced with 3^{n-1} multiplications of n numbers and then $3^{n-1} - 1$ comparisons. That is too much, so we try dynamic programming. We have already specified the stages. All the final stage policies are determined; at stage n, we go from (n,j) to $(n + 1, 0)$ at cost $c(n,j,0)$. The Markov property is satisfied; if an optimal path goes through (i,j), the part of that path from (i,j) to $(n + 1, 0)$ is an optimal path from (i,j) to $(n + 1, 0)$. Consequently, if we have an optimal path from (i,j) to $(n + 1, 0)$, we do not need to check all paths through (i,j) because we can use the optimal path that we have from (i,j) to $(n + 1, 0)$. We have optimal paths from (n,j) to $(n + 1, 0)$—there is only one path: it is optimal. We use these optimal paths to get optimal paths from $(n - 1, j)$ to $(n + 1, 0)$ as follows.

Stage n − 1

Choose an intermediate node (n,k_j) so that

$$k_j = \arg\min\{c(n - 1, j, k) + c(n,k,0) : k \text{ in } \{j - 1, j, j + 1\}\}$$

and put

$$f(n-1, j) = c(n-1, j, k_j) + c(n, k_j, 0).$$

Thus k_j tells you where to go at stage $n - 1$ and $f(n - 1, j)$ tells you the cost of this optimal path from $(n - 1, j)$ to $(n + 1, 0)$. Now we use this information to get optimal paths from $(n - 2, j)$ to $(n + 1, 0)$.

Stage n − 2

Given $(n - 2, j)$, choose an intermediate node $(n - 1, k_j)$ so that

$$k_j = \arg\min\{c(n-2, j, k) + f(n-1, k) : k \text{ in } \{j-1, j, j+1\}\}$$

and put

$$f(n-2, j) = c(n-2, j, k_j) + f(n-1, k_j).$$

Now we have optimal paths from $(n - 2, j)$ to $(n + 1, 0)$. We repeat this process for stage $n - 3$ down to stage 1 and get an optimal path from $(1, 0)$ to $(n + 1, 0)$. Let us look at the computations involved in this dynamic programing solution for the example.

At stage 1 there are 3 additions and 2 comparisons.
At stage 2 there are $3 \cdot 3$ additions and $3 \cdot 2$ comparisons.
At stage 3 there are $5 \cdot 3$ additions and $5 \cdot 2$ comparisons.
At stage k there are $(2k - 1)3$ additions and $(2k - 1)2$ comparisons.

Recall that

$$1 + 2 + \cdots + (2k - 1) = [1 + 2 + 3 + \cdots + (2k - 1)]$$
$$- [2 + 4 + 6 + \cdots + 2(k - 1)]$$
$$= \tfrac{1}{2}(2k - 1)(2k) - 2[1 + 2 + 3 + \cdots + (k - 1)]$$
$$= (2k - 1)k - 2[\tfrac{1}{2}(k - 1)k]$$
$$= k^2.$$

Adding the numbers of additions and comparisons required for stages 1 to $(n - 1)$ and applying the preceding formula with $k = n - 1$ shows that the dynamic programming solution requires $3(n - 1)^2$ additions and $2(n - 1)^2$ comparisons.

Example 5.12. Consider a list of 10 cities. Let the positive integer i denote the ith city in the list. The following table lists airline fares from city i to city j; a blank in the table indicates that a direct flight from city i to city j is not available. Let us use backward dynamic programming to find the least expensive way(s) to travel by air from city 1 to city 10.

	Destination									
Origin	1	2	3	4	5	6	7	8	9	10
1		5	4	6						
2					7	5				
3					5	6	4			
4						3	5			
5								5	3	
6								4	6	
7								4	2	
8										4
9										3
10										

Looking at the block structure of the table leads us to consider four stages:

Stage 1: Initial state: 1
         Terminal states: 2, 3, 4
Stage 2: Initial states: 2, 3, 4
         Terminal states: 5, 6, 7
Stage 3: Initial states: 5, 6, 7
         Terminal states: 8, 9
Stage 4: Initial states: 8, 9
         Terminal state: 10.

The stage 4 policies are determined: fly from cities 8 and 9 to city 10 at costs of 4 and 3 units, respectively. You should verify that the principle of optimality is valid for these stages; we will look at stage 3. Denote an initial state by s and a terminal state by t. For example, suppose that $s = 7$; then the optimality policy from 7 to 10 is obtained by comparing the cost of flying from 7 to 8 to 10 with the cost of flying from 7 to 9 to 10 and choosing a terminal state (either 8 or 9) for which the cost is minimal. Transition functions are used to formalize this procedure below. Put

$f_3(s,t)$ = cost of a flight from city s to city t
$f_4(t)$ = cost of an optimal policy from city t to city 10.

Choose $t = t(s)$ with $f_3(s,t(s)) + f_4(t(s)) = \min_t f_3(s,t) + f_4(t)$. Put $f_3(s) = f_3(s,t(s)) + f_4(t(s))$. For Example 5.12:

$$f_3(5,8) + f_4(8) = 9$$
$$f_3(5,9) + f_4(9) = 6, \qquad t(5) = 9, \qquad f_3(5) = 6;$$
$$f_3(6,8) + f_4(8) = 8$$

$f_3(6,9) + f_4(9) = 9,$ $t(6) = 8,$ $f_3(6) = 8;$

$f_3(7,8) + f_4(8) = 8$

$f_3(7,9) + f_4(9) = 5,$ $t(7) = 9,$ $f_3(7) = 5.$

Stage 2 is next; the initial states are 2, 3, and 4 and the terminal sates are 5, 6, and 7:

$f_2(2,5) + f_3(5) = 13$

$f_2(2,6) + f_3(6) = 13,$ $t_2(2) = 5$ or 6, $f_2(2) = 13;$

$f_2(3,5) + f_3(5) = 12$

$f_2(3,6) = f_3(6) = 14$

$f_2(3,7) + f_3(7) = 9,$ $t_2(3) = 7,$ $f_2(3) = 9;$

$f_2(4,6) + f_3(6) = 11$

$f_2(4,7) + f_3(7) = 10,$ $t_2(4) = 7,$ $f_2(4) = 10.$

Stage 1 has one initial state 1 and terminal states 2, 3, and 4:

$f_1(1,2) + f_2(2) = 18$

$f_1(1,3) + f_2(3) = 13$

$f_1(1,4) + f_2(4) = 16,$ $t_1(1) = 3,$ $f_1(1) = 13.$

The optimal policy is unique: fly from 1 to 3 to 7 to 10 at a total cost of 13 units.

Example 5.13.   You are to assign five medical teams to three countries to maximize total benefit, according to the following estimated benefit table:

Number of teams assigned	Assigned to country		
	1	2	3
0	0	0	0
1	45	20	50
2	70	45	70
3	90	75	80
4	105	110	100
5	120	150	130

Let us consider three stages. At stage 3 we will decide how many teams to send to country 3; at stage 2 we will decide how many teams to send to country 2; and at stage 1 we will decide how many teams to send to country 1. Beginning with stage 3, our initial states are the numbers of teams available to send to country 3. For these states the optimal decision is to send all available teams to country 3:

$$f_3(0) = \quad 0$$
$$f_3(1) = \quad 50$$
$$f_3(2) = \quad 70$$
$$f_3(3) = \quad 80$$
$$f_3(4) = 100$$
$$f_3(5) = 130.$$

Initial states for stage 2 are the numbers of teams available to send to countries 2 and 3; terminal states are the numbers of teams remaining to send to country 3; $f_2(s,t)$ = no. of units of benefit from sending $(s - t)$ teams to country 2:

$$f_2(0,0) + f_3(0) = 0, \quad t_2(0) = 0, \quad f_2(0) = 0;$$
$$f_2(1,0) + f_3(0) = 20 + 0 = 20$$
$$f_2(1,1) + f_3(1) = 0 + 50 = 50, \quad t_2(1) = 1, \quad f_2(1) = 50;$$
$$f_2(2,0) + f_3(0) = 45 + 0 = 45$$
$$f_2(2,1) + f_3(1) = 20 + 50 = 70$$
$$f_2(2,2) + f_3(2) = 0 + 70 = 70, \quad t_2(2) = 1 \text{ or } 2, \quad f_2(2) = 70;$$
$$f_2(3,0) + f_3(0) = 75 + 0 = 75$$
$$f_2(3,1) + f_3(1) = 45 + 50 = 95$$
$$f_2(3,2) + f_3(2) = 20 + 70 = 90$$
$$f_2(3,3) + f_3(3) = 0 + 80 = 80, \quad t_2(3) = 1, \quad f_2(3) = 95;$$
$$f_2(4,0) + f_3(0) = 110 + 0 = 110$$
$$f_2(4,1) + f_3(1) = 75 + 50 = 125$$
$$f_2(4,2) + f_3(2) = 45 + 70 = 115$$
$$f_2(4,3) + f_3(3) = 20 + 80 = 100$$
$$f_2(4,4) + f_3(4) = 0 + 100 = 100, \quad t_2(4) = 1, \quad f_2(f) = 125;$$
$$f_2(5,0) + f_3(0) = 150 + 0 = 150$$
$$f_2(5,1) + f_3(1) = 110 + 50 = 160$$
$$f_2(5,2) + f_3(2) = 75 + 70 = 145$$
$$f_2(5,3) + f_3(3) = 45 + 80 = 125$$

$f_2(5,4) + f_3(4) = 20 + 100 = 120$

$f_2(5,5) + f_3(5) = 0 + 130 = 130,$ $t_2(5) = 1,$ $f_2(5) = 160.$

There is one initial state for stage 1:

$f_1(5,0) + f_2(0) = 120 + 0 = 120$

$f_1(5,1) + f_2(1) = 105 + 50 = 155$

$f_1(5,2) + f_2(2) = 90 + 70 = 160$

$f_1(5,3) + f_2(3) = 70 + 95 = 165$

$f_1(5,4) + f_2(4) = 45 + 125 = 170$

$f_1(5,5) + f_2(5) = 0 + 160 = 160,$ $t_1(5) = 4,$ $f_1(5) = 170.$

The optimal policy is unique: send one team to country 1, three teams to country 2, and one team to country 3. (Check that the Markov property is satisfied!)

Let us focus on a stage, say stage k. We have initial states s. For each initial state s we have some terminal states t. For each initial/terminal pair (s,t) we have a kth-stage transition function $f_k(s,t)$ and a tail function $f_{k-1}(t)$, denoting the value of an optimal policy from t to the end of the problem through the stages following stage k. In Examples 5.12 and 5.13 the policy decision at stage k for initial state s involves choosing terminal states $t_k(s)$ so that

$$f_k(s,t_k(s)) + f_{k+1}(t_k(s)) = \min\{f_k(s,t) + f_{k+1}(t);$$

(s,t) is a stage k initial/terminal pair$\}$.

Example 5.14 illustrates another type of objective function that fits into a DP format.

Example 5.14. Three teams are trying to solve a problem. The probability of their solving the problem is estimated to be 0.6, 0.4, and 0.2. Two additional people become available to assign to the teams; the estimated probability of success with additional members is tabulated below.

	Team probability of success		
Additions	1	2	3
0	0.6	0.4	0.2
1	0.8	0.6	0.5
2	0.85	0.8	0.7

You are to assign the two additional people to the three teams to maximize the probability that at least one team solves the problem.

Let the number of people to be assigned to team k be decided on stage *k.* Maximizing the probability that at least one team succeeds is equivalent to minimizing the probability that all three teams fail. Focus on stage 3 first. Let *s* denote the number of people to assign to team 3 and let $f_3(s) =$ the probability that team 3 fails when *s* people are assigned to team 3: $f_3(0) = 0.8$, $f_3(1) = 0.5$, and $f_3(2) = 0.3$. Moving to stage 2, let *s* denote the number of people to assign to teams 2 and 3, let *t* denote the number of people to assign to team 3, and let $f_2(s,t) =$ the probability that team 2 fails when $s - t$ people are assigned to team 2. Then consider

$$f_2(s) = \min\{f_2(s,t)f_3(t); 0 \le t \le s\} = f_2(s,t_2(s))f_3(t_2(s)):$$

$$f_2(0) = (0.6)(0.8) = 0.48, \qquad t_2(0) = 0$$

$$f_2(1) = \min\{(0.4)(0.8),(0.6)(0.5)\} = 0.3, \qquad t_2(1) = 1$$

$$f_2(2) = \min\{(0.2)(0.8),(0.4)(0.5),(0.6)(0.3)\} = 0.16, \qquad t_2(2) = 0.$$

Continuing to stage 1, we obtain

$$f_1(2) = \min\{f_1(2,t)f_2(t); 0 \le t \le 2\}$$
$$= \min\{(0.15)(0.48),(0.2)(0.3),(0.4),(0.16)\} = 0.06, \qquad t_1(2) = 1.$$

Consequently, the maximum probability of success is 0.94; it occurs when one person is assigned to team 1 and one person is assigned to team 3. Verify that the principle of optimality is satisfied by this formulation.

Example 5.15. International Construction Corporation needs a piece of machinery for a 6-year construction project. They have purchased a 3-year-old machine to use during the first year. If they order now they can get a new machine delivered at the end of years 1 to 5 for $500,000. Estimated yearly operation costs, trade-in values, and salvage values (in thousands of dollars) are tabulated below.

Age at beginning of year	Cost to operate during year	Trade-in value at end of year	Salvage value at end of year
0	100	340	$250 = (0,C_0,T_0,S_0)$
1	130	220	$160 = (1,C_1,T_1,S_1)$
2	200	110	$70 = (2,C_2,T_2,S_2)$
3	400	60	$60 = (3,C_3,T_3,S_3)$
4	600	20	$0 = (4,C_4,T_4,S_4)$
5	700	0	$0 = (5,C_5,T_5,S_5)$
6	1000*	0	$0 = (6,C_6,T_6,S_6)$
7	1000*	0	$0 = (7,C_7,T_7,S_7)$
8	1000*	0	$0 = (8,C_8,T_8,S_8)$
9	1000*	0	$0 = (9,C_9,T_9,S_9)$

At the end of year 6, International plans to sell the machine for salvage and leave the construction site. The 1000* operating cost means that the machine is essentially useless and it costs $1,000,000 per year not to have such a machine available. International wishes to schedule purchases of machines to minimize the total cost associated with the machines. Let stage k correspond to year k. Let an initial state s represent the age of the machine at the beginning of year k and let a terminal state t denote the age of the machine at the beginning of year $k + 1$. Delivery of a new machine takes place just after the year has begun, so the pair $(2,1)$ denotes that a 2-year-old machine is replaced by a new machine at the beginning of the year. Let $f_k(s)$ = the minimal cost associated with the machine from the beginning of year k to the end of the project, and let $f_k(s,t)$ = the cost during year k associated with the pair (s,t). Thus $f_7(s) = -S_s$: at the end of year 6, the machine is sold for salvage. Focusing on stage k, $2 \le k \le 6$, if a new machine is delivered at the beginning of year k, then $t = 1$ and

$$f_k(s,1) = 100 + (500 - T_s) + f_{k+1}(1);$$

if the previous year's machine is retained for another year, then $t = s + 1$ and

$$f_k(s, s + 1) = C_s + f_{k+1}(s + 1).$$

You are asked to complete this example in Exercise 5.32a.

The following table lists the numbers of additions and comparisons done in the DP solution of Example 5.15.

Year	Additions	Comparisons
7		0
6	$6 \cdot 2$	6
5	$5 \cdot 2$	5
4	$4 \cdot 2$	4
3	$3 \cdot 2$	3
2	$2 \cdot 2$	2
1	$1 \cdot 2$	1
Total	$2(1 + \cdots + 6)$	$1 + \cdots + 6$

Using a DP model to look at an N-year time interval thus requires $3(1 + 2 + \cdots + N) = 1.5(N^2 + N)$ computations; in contrast, there are 2^{N-1} possible "buy" or "not-buy" sequences. When N is large, $1.5(N^2 + N)$ is much smaller than 2^{N-1}.

5.9 STABILITY AND SENSITIVITY

In this section we consider the response of the feasible set, the set of optimal points, and the value of the objective function with respect to changes in the parameters of an LP; to define the parameters, suppose that a "minimize" LP has m_1 " $\geq$ " constraints: $1, \ldots, m_1$, m_2 "=" constraints: $m_1 + 1, \ldots, m_1 + m_2$, m_3 " $\leq$ " constraints: $m_1 + m_2 + 1, \ldots$, $m_1 + m_2 + m_3 = m$, and n decision variables. The numbers m_1, m_2, m_3, m, and n, the $m \times n$ coefficient matrix A, the m-dimensional column vector b, and the n-dimensional row (cost) vector c are called parameters of the LP.

An LP will be called *stable* if small changes in the parameters A, b, or c result in small changes of the feasible set, the set of optimal points, and the optimal value of the objective function. *Sensitivity* relates to the rate of change of the optimal value of the objective function of a stable LP with respect to small changes in the parameters A, b, or c.

Two simple modifications of Example 3.1, Examples 5.16 and 5.17 below, illustrate lack of stability. Figure 6 in Chapter 3 provides a good visual reference for the following discussion. In the first example, the objective function and the feasible set are stable, but the set of optimal points is not stable under small changes in A or c.

Example 5.16. Add the objective $\min cx$, where $c = [-1, -\frac{2}{3}, -1]$ for now, to Example 3.1 to get the LP

$$\min \ cx$$

$$\text{s.t.} \quad Ax \leq b, \qquad x \geq 0,$$

where

$$A = \begin{bmatrix} 1 & \frac{2}{3} & 1 \\ \frac{1}{2} & 1 & 0 \end{bmatrix} \quad \text{and} \quad b = (1,1).$$

I have chosen c so that the planes on which cx is constant are parallel to the plane composed of the points $x = (x_1, x_2, x_3)$ satisfying the equation

$$x_1 + \tfrac{2}{3}x_2 + x_3 = 1.$$

Small changes in c will cause the set of optimal solutions to the LP to change drastically, depending on the direction in which c changes. With our current choice for c, the set of solution points for (5.16) is the convex set with the four corner points $(1,0,0)$, $(\frac{1}{2}, \frac{3}{4}, 0)$, $(0,1,\frac{1}{3})$ and $(0,0,1)$ corresponding to the inclined quadrilateral in the first octant, which forms the "top" of the feasible set in Fig. 6 of Chapter 3. The plane on which $cx = -1$ lies flat on this quadrilateral, but if we change c slightly, the solution set may change a lot. For instance, if we decrease the third component of c slightly (say that we put $c = [-1, -\frac{2}{3}, -1.00001]$), then $(0,1,1)$ is the only optimal solution to (1). If I leave c alone and change entries in the first row of A slightly, we can effect the same kind of behavior because we are changing the inclination of the top of the feasible set. Small changes in b do not change the set of optimal solutions much. The value of the objective function at optimal points does not change much in response to small changes in A, b, or c: it is stable with respect to the parameters of the LP.

Nothing is stable in Example 5.17 below, where an additional constraint causes the dimension of the feasible set to become lower than the dimension of the space in which the feasible set is located; when the latter situation occurs, a small change in B may cause the feasible set to disappear entirely.

Example 5.17. Add constraint 3: $\frac{3}{4}x_1 + \frac{5}{6}x_2 + \frac{1}{2}x_3 \geq 1$ to Example 5.16. That changes A to

$$\begin{bmatrix} 1 & \frac{2}{3} & 1 \\ \frac{1}{2} & 1 & 0 \\ -\frac{3}{4} & -\frac{5}{6} & -\frac{1}{2} \end{bmatrix}$$

and b to $(1,1,-1)$. The rank of A is two [row 1 + row 2 + 2(row 3) = 0] and the feasible set for (2) is the line interval between $(\frac{1}{2},\frac{3}{4},0)$ and $(0,1,\frac{1}{3})$ (see Fig. 6 in Chapter 3). All the instability of (5.16) remains. Moreover, if we decrease b_3 slightly, so that constraint 3 becomes $\frac{3}{4}x_1 + \frac{5}{6}x_2 + \frac{1}{2}x_3 \geq -b_3 > 1$, there are no feasible solutions to Example 5.17. These examples show clearly that stability involves a rather subtle study of the structure of an LP. Hence I will only discuss sensitivity for a stable situation.

Sensitivity

(This presentation does not assume that you have read Chapter 4.) Consider a standard-form LP that has a unique optimal point x. The fundamental theorem of LP assures us that x is a basic feasible solution. Fix the dimensions of the problem for discussion by supposing that A is an $m \times n$ matrix of rank m. Then x has $\leq m$ positive entries because the corresponding columns of A are linearly independent. Suppose that x has exactly m positive entries. Basic feasible solutions satisfying this condition are called nondegenerate. Let $x_{k_1}, \ldots, x_{k_m}$, where $1 \leq k_1 < k_2 < \cdots < k_m \leq n$, be the positive entries in x and put

$$B = [a_{k_1}, a_{k_2}, \ldots, a_{k_m}]$$

and

$$x_B = (x_{k_1}, x_{k_2}, \ldots, x_{k_m}).$$

Then

$$Ax = Bx_B = b.$$

Hence

$$x_B = B^{-1}b > 0.$$

Motivated by fact 2 in Section 2.5, put

$$\lambda_B = c_B B^{-1}.$$

Then

$$cx = c_B x_B = c_B(B^{-1}b) = (c_B B^{-1})b = \lambda_B b,$$

where

$$c_B = [c_{k_1}, c_{k_2}, \ldots, c_{k_m}].$$

The *reduced costs* $\tilde{c}_j$ *of the nonbasic variables* are given by the formula

$$\tilde{c}_j = c_j - \lambda_B a_j.$$

The reduced costs are all nonnegative when

$$c - \lambda_B A \geq 0,$$

or, equivalently, when λ_B is feasible for the dual problem.

Reduced costs are numbers associated with nonbasic variables. According to Section 4.9, all the reduced costs of the nonbasic variables with respect to a nondengerate basic feasible solution x are positive if, and only if, x is the unique optimal solution to the LP. Below, stability is verified and sensitivity is discussed for an LP satisfying this condition. You can use output from LP software to check for the possiblility of multiple optimal solutions because reduced costs often appear as output.

Example 5.18. min cx

$$\text{s.t. } Ax = b, \quad x \geq 0,$$
$$x_B > 0,$$
$$\text{all reduced costs} > 0.$$

Let δ denote a small number.

Change in c

Suppose that we change c_j by δ, so that c_j changes to $c_j + \delta$. Then $x_B = B^{-1}b$ does not change. If j does not correspond to a basic column, then $\lambda_B = c_B B^{-1}$ does not change and the reduced costs stay positive for small δ.

If $j = k_j$ corresponds to a basic column, let e_j denote the m-dimensional row vector with 1 in position j and zero elsewhere; $\lambda_B = c_B B^{-1}$ changes to $c_B B^{-1} + \delta e_j B^{-1}$. Hence the reduced costs change from $c_i - c_B B^{-1} a_i$ to $c_i - c_B B^{-1} a_i - \delta e_j B^{-1} a_i$. Thus, if δ is small, the reduced costs remain positive and we see that the unique solution to the problem stays constant. However, $cx = \sum_{i=1}^n c_i x_i$ changes to $\sum_{i=1}^n c_i x_i + \delta x_j$. Consequently, the ratio of the change δx_j in the objective to the change δ in c_j is equal to x_j:

x_j is the rate of change of the objective value with respect to a change in c_j.

Change in b

Suppose that we change b_i by δ so that b_i changes to $b_i + \delta$. Then $x_B = B^{-1} b$ changes to $x_B + \delta B^{-1} e_i'$, where the m-dimensional column vector e_i' has a 1 in position i and zero elsewhere. When δ is small, $x_B + \delta B^{-1} e_i'$ is close to x_B and has all positive entries. Moreover, $\lambda_B = c_B B^{-1}$ does not change. Consequently, the solution is stable (and unique). We will look at the objective value in the form $\lambda_B b$. Because $\lambda_B b$ changes to $\lambda_B b + (\lambda_B)_i \delta$:

$(\lambda_B)_i$ is the ratio of change of the objective value with respect to a change in b_i, where $(\lambda_B)_i$ denotes the ith entry in λ_B.

Change in A

Suppose that $a_{i,j}$ changes to $a_{i,j} + \delta$. We have two cases to consider.

a_j *not in* B. When a_j is not in B, $x_B = B^{-1} b$, $c_B x_B$ and $\lambda_B = c_B B^{-1}$ do not change. However, the reduced cost $\bar{c}_j$ change from $c_j - \lambda_B a_j$ to $c_j - \lambda_B a_j - \delta \lambda_B e_i'$. Thus the reduced costs remain positive, and consequently, uniqueness and stability are maintained under small changes in $a_{i,j}$.

a_j *in* B. In this case B^{-1} changes, so the situation is more complicated. However, B^{-1} changes only a little when $a_{i,j}$ changes by a small number δ, and we can show that the solution and the objective value are stable and uniquess is preserved. We estimate the rate of change of the objective value below.

Recall the formula

$$(1 + t)^{-1} = \frac{1}{1+t} = 1 - t + \frac{t^2}{1+t}, \qquad t \neq -1.$$

Consequently, $1 - t$ is a good approximation to $(1 + t)^{-1}$ when t is small. Now suppose that we change a_{ij} by an amount δ. Thus we change B by Δ_B, where Δ_B has δ in the (i, j) position and zero elsewhere. We put $t = \Delta_B B^{-1}$ and use the approximation $(1 + t)^{-1} \approx (1 - t)$:

$$f_{new} = c_B(B + \Delta_B)^{-1}b$$

$$= c_B([I + \Delta_B B^{-1}]B)^{-1}b$$

$$= c_B B^{-1}(I + \Delta_B B^{-1})^{-1}b$$

$$\approx c_B B^{-1}(I - \Delta_B B^{-1})b \quad \text{(approximation with } t = \Delta_B B^{-1})$$

$$= c_B B^{-1}b - c_B B^{-1}(\Delta_B B^{-1})b$$

$$= f_{old} - (c_B B^{-1})\Delta_B(B^{-1}b)$$

$$= f_{old} - \lambda_B \Delta_B x_B$$

$$= f_{old} - (\lambda_B)_i \delta x_{k_j},$$

so

$$\frac{f_{new} - f_{old}}{\delta} \approx -(\lambda_B)_i x_{k_j},$$

where $x_B = (x_{k_1}, \dots, x_{k_m})$. Consequently,

$-(\lambda_B)_i x_{k_j}$ is the rate of change of the objective value with respect to a change in a_{ij} if a_{ij} is in B.

Comment

If you are working on a small standard-form problem and your software gives you only an optimal basic feasible solution, you can use a matrix invertor to compute B^{-1}; then you can compute $\lambda_B = c_B B^{-1}$. Thus you can gain access to the information that you need to do a sensitivity analysis of your problem. After introducing some notation, Exercise 2.2 is used to illustrate.

Dual Variables or Dual Prices

The entries in λ_B are often called dual variables or dual prices in LP software output because they correspond to an optimal solution of the dual problem. The word "prices" is suggested by economic interpretation of a dual problem, as in Examples 2.1, 2.2, and 2.3.

Exercise 2.2 is an example of a symmetric primal LP:

$$\min cx$$

$$\text{s.t.} \quad Ax \geq b, \, x \geq 0.$$

The associated standard-form tableau for this problem includes the surplus variables; referring to Chapter 4 and applying the simplex method yields

$$\left[\begin{array}{c|c|c} A & -I & b \\ \hline c & 0 & 0 \end{array}\right] = \left[\begin{array}{c|c|c|c} B & V & -I & b \\ \hline c_B & c_V & 0 & 0 \end{array}\right]$$

$$\rightarrow \cdots \rightarrow \left[\begin{array}{c|c|c|c} I & B^{-1}V & -B^{-1} & B^{-1}b \\ \hline 0 & c_V - c_B B^{-1}V & c_B B^{-1} & -c_B B^{-1}b \end{array}\right];$$

the final simplex method tableau tells us that the reduced costs for the surplus variables are equal to the dual prices. Thus output from the following solution to Exercise 2.2 can be used to check whether there may be multiple optimal solutions.

MIN   25L + 17D + 22M
SUBJECT TO
    2)   .21L + .55D + .25M >= 3
    3)   .5L + .3D + .4M >= 7
    4)   .25L + .1D + .3M >= 5

OBJECTIVE FUNCTION VALUE

1)   380.000000

VARIABLE	VALUE	REDUCED COST
L	2.000000	.000000
D	.000000	3.000000
M	15.000000	.000000

ROW	SLACK OR SURPLUS	DUAL PRICES
2)	1.170000	0.000000
3)	.000000	−40.000000
4)	.000000	−20.000000

The basic variables in this optimal BFS are L, M, and the first surplus variable; all of these variables are positive (i.e., greater than zero) in the solution. Moreover, the reduced costs of the nonbasic variables (D and the other surplus variables) are all positive; consequently, the solution is nondegenerate and unique. Thus we have the stable situation described in Section 5.9. A discussion of sensitivity for this solution follows.

The rate of change of the optimal objective value with respect to a change in the right-hand side of constraint 3 should be equal to the value of the second dual variable; however, because of the manner in which the software package outputs values, the sign is opposite to what we expect it to be!

The right-hand side of constraint 3 is increased by 1 in the following LP.

MIN   25L + 17D + 22M
SUBJECT TO
2)   .21L + .55D + .25M >= 3
3)   .5L + .3D + .4M >= 8
4)   .25L + .1D + .3M >= 5

OBJECTIVE FUNCTION VALUE
1)   420.000000

VARIABLE	VALUE	REDUCED COST
L	8.000000	.000000
D	.000000	3.000000
M	10.000000	.000000

ROW	SLACK OR SURPLUS	DUAL PRICES
2)	1.180000	.000000
3)	.000000	−40.000000
4)	.000000	−20.000000

Note that the objective function value increased by the predicted amount, $40,000,000.

In each of the following three cases, values in the coefficient matrix are changed; then the predicted change and the actual change is compared.

Case 1

Change the coefficient of M in constraint 3 to 0.41. The linear estimate of the change is given by the following expression:

−(value of M) × (value of second dual variable) × (change in coefficient) = −15 × 40 × 0.01 = −6.

The actual change is computed below.

MIN   25L + 17D + 22M
SUBJECT TO
2)   .21L + .55D + .25M >= 3
3)   .5L + .3D + .41M >= 7
4)   .25L + .1D + .3M >= 5

OBJECTIVE FUNCTION VALUE
1)   373.684200

VARIABLE	VALUE	REDUCED COST
L	1.052634	.000000
D	.000000	2.789473
M	15.789470	.000000

ROW	SLACK OR SURPLUS	DUAL PRICES
2)	1.168421	.000000
3)	.000000	-42.105260
4)	.000000	-15.789470

Thus the actual change is about -6.32 in case 1.

Case 2

Increase the coefficient of L in constraint 4 to 0.26. In this case the predicted change is equal to

$-$(value of L) × (value of the third dual variable) × 0.01 = $- 2 \times 20 \times 0.01 = -0.4$.

According to the following solution, the actual change is 0.4348.

MIN   25L + 17D + 22M
SUBJECT TO
    2)   .21L + .55D + .25M >= 3
    3)   .5L + .3D + .4M >= 7
    4)   .26L + .1D + .3M >= 5

OBJECTIVE FUNCTION VALUE

1)   379.565200

VARIABLE	VALUE	REDUCED COST
L	2.173914	.000000
D	.000000	3.217391
̧M	14.782610	.000000

Case 3

In this final case increase the coefficient of L in constraint 2 to 0.23 and decrease the coefficient of M in constraint 4 to 0.29. The predicted change is $-[(2)(0)(0.02) + (15)(20)(-0.01)] = 3$, while the actual change is 10/3:

MIN   25L + 17D + 22M
SUBJECT TO
    2)   .23L + .55D + .25M >= 3
    3)   .5L + .3D + .4M >= 7
    4)   .25L + .1D + .29M >= 5

OBJECTIVE FUNCTION VALUE

1)   383.333300

VARIABLE	VALUE	REDUCED COST
L	.666665	.000000
D	.000000	3.111111
M	16.666670	.000000

You should explain why small changes in the coefficients of D in constraints 2, 3 or 4 would not cause any change in the value of the objective function.

EXERCISES

5.1 Referring to Exercise 2.6(a):
  (a) Find the optimal solution that comes closest to producing equal quantities of fir and pine lumber.
  (b) Find the optimal solution that comes closest to (1) producing equal quantities of fir and pine lumber *and* (2) maximizing plywood output.
5.2 Use your LP package to look for more solutions to Exercises 2.11 and 2.28.
5.3 Do Exercises 2.19 and 2.20 if you have not already done them.
5.4 Do Exercise 2.21(b).
5.5 (a) In Example 5.1 change the neighbor's per-barrel price for jet fuel from $100 to $80 and then to $70. At what price between $100 and $70 does the solution change? Why?
  (b) Suppose that the organization has only 10 million barrels of light crude available to process and the neighbor will sell fuel oil, gasoline, and jet fuel for the prices given in Example 5.1. Formulate an appropriate LP model and solve it.
5.6 (a) Do the exercise stated at the end of Section 5.2.
  (b) Using your answer to part (a), raise the maximum allowed levels of vitamins and fat in chicken feed to the minimum value of z from part (a) and solve the resulting LP to determine the composition of cattle feed and chicken feed. What are the vitamin and fat levels for your solution?
  (c) Returning to the original problem, find the minimum value of the sum of the excess levels of vitamins and fat in chicken feed.
  (d) Continuing with Example 5.3, suppose that the buyer is willing to buy chicken feed with surplus vitamins and fat at a discount of $10 per ton per unit of surplus vitamins and $20 per ton per unit of surplus fat. Formulate and solve an appropriate LP.
5.7 A company wishes to plan its production of two items with seasonal demands over a one-year period. The monthly demand of item 1 is

100,000 units during the months of October, November, and December; 10,000 units during the months of January, February, March, and April; and 30,000 units during the remaining months. The demand of item 2 is 50,000 during the months of October through February and 15,000 during the remaining months. Suppose that the unit product cost of items 1 and 2 is $5.00 and $8.00, respectively, provided that these were manufactured before June. Beginning June 1, the unit costs are reduced to $4.50 and $7.00 because of the installation of an improved manufacturing system. The total units of items 1 and 2 that can be manufactured during any particular month cannot exceed 120,000. Furthermore, each unit of item 1 occupies 2 cubic feet and each unit of item 2 occupies 4 cubic feet of inventory. Suppose that the maximum inventory space allocated to these items is 150,000 cubic feet and that the holding cost per cubic foot during any month is $0.10.
(a) Explain why there is no feasible solution.
(b) Suppose that storage space can be rented for $0.15 per cubic foot per month. Formulate and solve an appropriate LP.

5.8 A dairy wishes to produce milk containing exactly 4% butterfat, at least 5% protein, and at most 6% lactose. It can buy milk from three farms. Relevant information is tabulated below.

Farm	Fat (%)	Protein (%)	Lactose (%)	Cost/100 lb
1	3	5	8	$6.80
2	4	4	5	7.50
3	5	7	7	8.30

(a) Formulate an appropriate LP.
(b) Explain why there is no feasible solution composed only of milk from the three farms.
(c) Suppose that we can add water. What is the minimum percent of water that we can use to make an acceptable mix?
(d) Suppose that we can use all the water we wish. Formulate an appropriate LP.
(e) What is the minimal cost of 100 pounds of acceptable milk that can be mixed using the minimum possible amount of water?

5.9 The Exon Company blends gasoline from three components with the following characteristics:

Component	Vapor pressure	Octane number	Barrels available
1	8	83	2800
2	20	109	1400
3	4	74	4000

The company wishes to produce two types of lead-free gasoline, regular and premium, with the following characteristics.

Type	Maximum vapor pressure	Minimum octane number	Selling price per barrel
Regular	7	80	$ 9.80
Premium	6	100	12.00

(a) Assuming linear relationships, formulate an appropriate LP model.
(b) Explain why it is impossible to blend premium.
(c) Suppose that we need to keep vapor pressure ≤ 6. What is the highest octane number that we can achieve?
(d) How high must we allow vapor pressure to go in order to have an octane number ≥ 100?

5.10 Continuing Exercise 5.9, it is impossible to have the vapor pressure ≤ 6 and the octane number ≥ 100. Let

C_1 = units of component 1 in 1 unit of gasoline
C_2 = units of component 2 in 1 unit of gasoline
C_3 = units of component 3 in 1 unit of gasoline.

Put

$$8C_1 + 20C_2 + 4C_3 - 6 = P_1 - N_1$$
$$83C_1 + 109C_2 + 74C_3 - 100 = P_2 - N_2.$$

Formulate LPs to consider the following criteria of goodness:

(1) $[\max\{P_1, N_2\}]$ is small.
(2) $P_1 + N_2$ is small.
(3) A weighted sum of P_1 and N_2 is small; for example, a weighting like $16P_1 + N_2$ may be chosen to balance the relative sizes of P_1 and N_2.

5.11 Custom Lumber Company uses two types of wood, pine and walnut, to produce veneer lumber and paneling. Manufacturing 1000 feet of lumber requires 1000 feet of pine and 300 feet of walnut; producing 1000 feet of plywood requires 800 feet of pine and 120 feet of walnut. Lumber sells for $1.00 per foot and plywood sells for $0.75 per foot.

(a) Suppose that 40,000 feet of pine and 10,000 feet of walnut are in stock. Determine how much lumber and plywood to manufacture to maximize potential income from this stock.

(b) Suppose that Custom gets an order for 30,000 feet of lumber and 40,000 feet of plywood from Best Furniture Corporation. Show that the order cannot be filled using stock in hand.

(c) Suppose Best offers to permit Custom to deliver up to 20,000 feet of plywood and up to 15,000 feet of lumber 30 days later provided that they sell the late lumber for $0.85 a foot and the late plywood for $0.60 per foot. Custom can mill enough pine and walnut in 30 days to complete the order. How much lumber and plywood should they ship late?

5.12 A company produces refrigerators, stoves, and dishwashers. During the coming year, sales are expected to be as follows:

| | Quarter | | | |
Product	1	2	3	4
Refrigerators	1500	1000	2000	1200
Stoves	1500	1500	1200	1500
Dishwashers	1000	2000	1500	2500

Management has decided that the inventory level for each product should be at least 150 units at the end of the fourth quarter. There is no inventory of any product at the start of the first quarter.

During a quarter only 15,000 hours of production time are available. A refrigerator requires 2 hours, a stove 4 hours, and a dishwasher 3 hours of production time. Refrigerators cannot be manufactured in the second quarter because the company plans to modify tooling for a new product line.

Assume that refrigerators and stoves can be backlogged at a loss of $20 per item per quarter for the late manufacture, while dishwashers incur a backlog penalty of only $10 per item per quarter. Assume that each item left in inventory at the end of a quarter incurs a holding cost of $5. The company wants to plan its production schedule over the year in a way that minimizes the total backlogging and inventory cost. Formulate an appropriate LP.

5.13 In this exercise a cheese company is sailing along smoothly. The opportunity arises to double its sales. The company produces two types of cheese: cheddar cheese and Swiss cheese. Fifty experienced production workers have been producing 10,000 and 6000 pounds of cheddar and Swiss per week. It takes 1 worker-hour to produce 10 pounds of cheddar and 1 worker-hour to produce 6 pounds of Swiss. A workweek is 40 hours. Management has decided to double production by putting on a second shift over an 8-week period. The weekly demands (in thousands of pounds) during the 8-week period are tabulated below.

Type	1	2	3	4	5	6	7	8
Cheddar	10	10	12	12	16	16	20	20
Swiss	6	7.2	8.4	10.8	10.8	12	12	12

An experienced worker can train three new employees in a 2-week training period during which all involved contribute nothing to production; nevertheless, each trainee receives full salary as an experienced worker during this period. One hundred experienced workers are required by the end of week 8. Experienced workers are willing to work overtime at time and a half during the 8-week period. Experienced workers earn $360 per week. Management would consider the possibility of backlogging orders during the 8-week transition period at a discounted price of $0.50 per pound for cheddar and $0.60 per pound for Swiss for each week that shipment is delayed. All back orders must be filled by the end of week 8.

5.14 The Westslope Jelly Company manufactures apple jelly, cherry jelly, and A-C, an apple–cherry jam. Westslope uses apples with a 70% juice yield: 6/7 kilogram of apples and 0.4 kilogram of sugar produce 1 kilogram of apple jelly. Westslope's cherries are sour and have a 50% juice yield: 0.8 kilogram of cherries and 0.6 kilogram of sugar produces 1 kilogram of cherry jelly. After apples and cherries are processed for juice, the residue can be processed for jam mash; each kilogram of apple and cherry residue yields 0.5 and 0.2 kilogram of jam mash during the second process. Each kilogram of A-C jam consists of 0.2 kilogram of jam mash, 0.3 kilogram of fruit juice, and 0.5 kilogram of corn sweetener. Equal weights of apple and cherry products are used in A-C jam. Apples, cherries, sugar, and corn sweetener cost $0.6, $1.10, $0.43, and $0.32 per kilogram; 4000 kilograms of apples and 1200 kilograms of cherries are available. Apple jelly, cherry jelly, and A-C jam sell for $1.20, $1.40, and $0.95 per 500-gram jar. Additional production costs are $0.20 per jar.

(a) Formulate an appropriate LP model.

(b) Now suppose that the second process costs $50 extra to set up for each type of fruit processed. Formulate LP models to handle this additional factor.

5.15 Referring to Exercise 2.23, suppose that if river water is processed by either process, then at least 40,000 gallons must be processed by the process.

5.16 Referring to Exercise 2.25, the programmer decides not to have a feedlot. She also decides to have at most 10 horses and at most 5 cows. Moreover, she decides that if she will have feeder calves, she will have at least 25 feeder calves. Incorporate these additional constraints into your model for Exercise 2.25.

5.17 Incorporate an integer variable into your LP model for Exercise 2.13 to handle the additional requirement that at most one of foods 2 and 4 can be used in the diet.

5.18 (a) Use two integer variables to act as switches in Exercise 5.14(b).

(b) Use integer variables in Exercise 5.15.

(c) Use an integer variable in Exercise 5.16.

5.19 Do Exercise 2.21(a).

5.20 Formulate a model that will incorporate all four possibilities (a) to (d) in Example 5.6.

5.21 A paper recycling machine can produce toilet paper, writing pads, and paper towels, which sell for 18, 29, and 25 cents and consume 0.5, 0.22, and 0.85 kilograms of newspaper and 0.2, 0.4, and 0.22 minute. Each day 10 hours and 1500 kilograms of newspaper are available, and at least 1000 rolls of toilet paper and 400 rolls of paper towels are required. If any writing pads are manufactured, then at least 500 must be made; moreover, the government will pay a bonus of $20 if at least 1200 rolls of toilet paper are produced. Formulate a model to solve this problem.

5.22 Three sources of supply are available for a product that is needed during two consecutive periods of time: 3000 units of the product are needed during period 1 and 4000 units are needed during period 2. It is necessary to consider three types of costs:

Type 1: cost to start a source
Type 2: cost to operate a source for a period
Type 3: cost per unit produced by a source

Relevant information is tabulated below.

Source	Type 1	Type 2	Type 3	Capacity/period
1	3100	750	5.4	2200
2	2300	900	4.2	1700
3	1700	1000	6.8	3100

Formulate a model to solve this problem.

5.23 A product is assembled from three parts that can be manufactured on two types of machines, A and B. Neither machine can process different parts on the same day. The number of parts processed by each machine per hour is summarized below.

Part	Machine A	Machine B
1	12	6
2	15	12
3	–	25

Each assembly uses one part 1, two part 2, and one part 3. Management seeks a daily schedule of the machines so that the number of assemblies is maximized. Currently, the company has three machines of type A and five machines of type B. Assume that each machine is available 16 hours per day. Formulate a model to solve this problem.

5.24 You may use some zero–one integer variables to formulate a model to maximize the number of plastic rolling pins that can be manufactured in $24 \times 6 = 144$ hours. Each pin requires two ends and one cylinder. There are six machines on which to make these parts. The first four machines cn be set up to make either part at the beginning of the week, but they cannot be changed to the other part during the week. The last two machines are programmed to make both kinds of parts and they can be switched during the week. Production rates are tabulated below.

Machine	Parts produced per 8-hour shift	
	Cylinders	Ends
1	30	50
2	40	70
3	50	100
4	60	130
5	165	300
6	235	480

5.25 A power authority operates four power plants located near Denver, Fort Collins, Pueblo, and Salt Lake City. It can purchase coal from three suppliers located in Colorado, Utah, and Wyoming. It must have at least 400, 80, 120, and 560 boxcars of coal per month to keep the plants operating. It would like to have 1000, 230, 430, and 1400 boxcars per month at the power plants. The suppliers can provide 800, 600, and 1000 boxcars per month. The following table lists costs (including shipping) for a boxcar of coal delivered from a supplier to a power plant.

	Denver	Fort Collins	Pueblo	Salt Lake City
Colorado	403	441	458	430
Utah	530	510	550	350
Wyoming	360	340	380	410

Formulate this problem as a transportation problem in tableau format.

5.26 Four research centers are requesting research grants from four government agencies. The centers need 20, 30, 40, and 50 million dollars next year to keep operating. They would each like an infinite amount of money. The agencies have 40, 30, 30, and 80 million dollars available for research grants to the four centers next year. The government has decided to keep the centers operating and has prepared the following table of estimated benefit per million dollars granted.

		Center			
		1	2	3	4
	1	2	5	7	6
Agency	2	6	3	3	6
	3	8	9	6	4
	4	5	7	4	10

Formulate this problem as a transportation problem in tableau format.

5.27 A relief organization wishes to ship 1000 tons of supplies per week to a city (at minimum cost). The organization has five supply sites S_i, $i \leq 5$. It can ship supplies from these sites to three ports, P_1, P_2, and P_3, at the following costs in dollars per ton:

| | Port | | |
	1	2	3
Site 1	100	170	220
2	140	110	180
3	190	160	70

It can transport up to 250 and 300 tons of supplies per week from P_1 and P_2 to the city at a cost of $70 and $40 per ton. It can transport up to 300 tons per week from each of P_3 and S_4 to an intermediate city T at costs of 30 and 50 dollars per ton. It can transport up to 400 tons per week from T to the city at a cost of $20 per ton. It can ship up to 500 tons per week by train from S_5 to an intermediate border B at a cost of $150 per ton. At the border, it can ship up to 100 tons per week from the border B to the city at a cost of $60 per ton. It can ship supplies free on a train from B to the city, but only 60% of these supplies arrive at the city. Assume that plenty of supplies are available at S_1, S_2, S_3, and S_5 but only 200 tons are available per week at S_4. What percent of the supplies shipped to the city on the free train from the border would have to arrive at the city before the relief organization would consider using that train to transport supplies to the city?

5.28 A city has eight main stormwater flow junctions, $N_1, \ldots, N_8$. We will denote a route from N_i to N_j by (i,j). We will consider 16 routes, listed below with cost per unit flow and capacity in units.

Route	Cost/unit	Capacity	Route	Cost/unit	Capacity
(1,6)	0	20	(4,2)	0	10
(1,7)	12	20	(4,3)	0	10
(2,1)	0	10	(4,5)	0	10
(2,3)	0	10	(5,1)	0	10
(2,8)	0	10	(5,6)	9	10
(3,1)	0	10	(5,7)	17	100
(3,6)	14	10	(6,7)	5	100
(4,1)	0	10	(7,8)	4	100

A storm occurs. The following table lists holding capacity units at each junction and water units generated at each junction.

Node	Capacity	Water	Node	Capacity	Water
1	10	23	5	0	17
2	0	11	6	12	8
3	7	16	7	0	6
4	0	14	8	Infinite	—

Formulate a model to solve this problem.

5.29 Consider the diagram

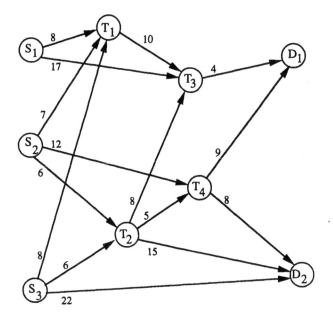

(a) Suppose that the numbers represent costs along the routes indicated by the arrows. Suppose that S_1, S_2, and S_3 are supply points with supplies 20, 10, and 30, D_1 and D_2 are demand points with demands 25 and 15, and T_1, T_2, T_3, and T_4 are transshipment points, each with demand 5. Formulate an appropriate model.

(b) Suppose that the numbers represent route capacities. Formulate a model to maximize flow from the three supply points in the west to the two demand points in the east.

(c) Modify part (a) to include paying 5% of each flow that comes into a transshipment point as a toll instead of paying 5 units at each transshipment point.

5.30 Apply Example 5.12 to the following data.

	Destination									
Origin	1	2	3	4	5	6	7	8	9	10
1		2	4	3						
2					7	4	6			
3					3	2	4			
4					4	1	5			
5								1	4	
6								6	3	
7								3	3	
8										3
9										4
10										

5.31 Assign two medical and one dental teams to three countries according to the following benefit table.

	Country		
Team composition	1	2	3
1 dental	25	40	30
1 medical	40	20	50
2 medical	70	60	60
1 dental + 1 medical	70	65	70
1 dental + 2 medical	110	100	105

5.32 (a) Complete Example 5.15.

(b) Suppose that it was possible to lease a new machine on an annual basis. What is the maximum price that International would be willing to pay to lease a machine?

5.33 A college student has 2 weeks before final exams begin in her courses. She is taking four courses and has one project and two term papers to complete. She has decided to spend 3 days on the project and 2 days on each term paper, leaving her 7 days to review the four courses. She prefers to concentrate on one course during a day and has decided to spend at least one day on each course. She compiled the following benefit table:

Study days	Course (benefit units)			
	1	2	3	4
1	2	2	3	5
2	3	5	5	5
3	8	6	7	5
4	9	7	9	6

Use backward dynamic programming to decide how many days to spend reviewing each course.

5.34 Given the following table of expected profit, in hundreds of dollars, from shipping truckloads of apples to fruit stands. Use dynamic programming to decide where to ship five truckloads of apples.

Number of truckloads	Stand		
	1	2	3
1	3	5	4
2	7	10	6
3	9	11	11
4	12	11	12
5	13	11	12

5.35 Suppose that you have eight shiploads of bananas to ship to three ports. At least one shipload must go to each port. The following table lists expected profit from sending i shiploads to port j. Use backward dynamic programming to solve this problem.

Number of shiploads	Port		
	1	2	3
1	1	1	2
2	3	5	4
3	7	10	6
4	9	11	11
5	12	11	12
6	13	12	13

5.36 (a) A company wishes to plan next year's production of two products, A and B. Demands and labor costs for these products vary quarterly as indicated in the tables below. Total production capacity is at most 120,000 items per quarter. Inventory costs are $1.50 per quarter for each unit of product A and $2.50 per quarter for each unit of product B. The company has 100,000 cubic feet of inventory space available and they can rent additional space at a cost of $0.50 per unit per quarter. A unit of product A requires 2 cubic feet and a unit of product B requires 4 cubic feet.

Labor costs (in dollars per unit):

		Quarter		
Product	1	2	3	4
A	7	9	8	8
B	10	14	12	12

Demand (in thousands of units);

		Quarter		
Product	1	2	3	4
A	40	10	30	100
B	45	25	45	100

Profit (exclusive of inventory cost, in dollars per unit):

		Quarter		
Product	1	2	3	4
A	3	1	2	2
B	6	2	4	4

Formulate and solve an appropriate LP.

(b) The marketing department of the company estimates that additional third-quarter demand for the products could be generated, at a cost of $1 per unit for item A and $1.50 per unit for item B. Incorporate this additional consideration into your model.

5.37 Consider a transportation problem with m supply points and n demand points (Example 2.3 and Section 5.7). Show that a node-incidence matrix for the problem has rank $m + n - 1$.

5.38 Show that rows 2 through 7 in the node-incidence matrix for Example 5.8 are linearly independent.

5.39 In Section 5.7 we found an initial BFS for Example 5.8 by lexicographically assigning flows that are as large as possible.

    (a) Find another BFS by replacing one of the routes in the initial BFS by the route [3,4].

    (b) Draw both the network and allocation tableau representations of the cycle flow corresponding to the change that occurs in the solution when you bring a flow on route [3,4] into the solution.

5.40 Solve Example 5.7 for the numerical data given at the end of the example.

5.41 (a) Find all solutions to Exercise 2.22.

    (b) How many integer-valued solutions can you find?

SOLUTIONS TO EXERCISES

5.1 An LP model for Exercise 2.6(a) is formulated in the Chapter 2 exercise solutions. A printout of a solution follows.

MAX   500FL + 640PL + 700FP + 800PP
SUBJECT TO
   2)   140FP + 160PP > = 1000
   3)   1000FL + 800PL > = 5000
   4)   FL + FP < = 10
   5)   PL + PP < = 10

OBJECTIVE FUNCTION VALUE

1)   14000.0000

VARIABLE	VALUE	REDUCED COST
FL	.000000	.000000
PL	6.250000	.000000
FP	10.000000	.000000
PP	3.750000	.000000

From this printout we see that

$$X = (FL,PL,FP,PP) = (0,6.25,10,3.75)$$

is an optimal solution to the problem. However, because the reduced cost of FL is equal to zero, there may be another optimal solution with $FL > 0$. To test for such a solution, we will increase the coefficient of FL slightly, making it a bit more advantageous to use FL, and solve this new problem below.

MAX   500.001FL + 640PL + 700FP + 800PP
SUBJECT TO
   2)   140FP + 160PP > = 1000
   3)   1000FL + 800PL > = 5000
   4)   FL + FP < = 10
   5)   PL + PP < = 10

OBJECTIVE FUNCTION VALUE

1)   14000.0000

VARIABLE	VALUE	REDUCED COST
FL	5.000000	.000000
PL	.000000	.000793
FP	5.000000	.000000
PP	10.000000	.000000

Thus we have found another optimal BFS,

$$Y = (FL,PL,FP,PP) = (5,0,5,10),$$

to Exercise 2.6(a); X and Y are the corner points of the set of feasible solutions to Exercise 2.6(a). The points on the interval from X to Y are parameterized below.
   For $0 \le T \le 1$, let

$$XT = X + T(Y - X)$$
$$= (5T, 6.25(1 - T), 10 - 5T, 3.75 + 6.25T)$$
$$= (X_1T, X_2T, X_3T, X_4T).$$

Then $\{XT; 0 \le T \le 1\}$ is the set of optimal solutions to the LP; $X_0 = X$ and $X_1 = Y$ are the corner points of the set of optimal solutions.
   The output vectors for Y and X are (5000,0,700,1600) and (0,5000,1400,600), respectively; X produces 2000 sheets of plywood and Y produces 2300 sheets of plywood.
(a) To use LP to solve this problem, put

$$1000FL - 800PL = P_1 - N_1$$

and solve the following LP:

$$\min \ P_1 + N_1$$

subject to the constraints in Exercise 2.6(a) and the additional constraints

$$500FL + 640PL + 700FP + 800PP = 14{,}000$$

and

$$1000FL - 800PL - P_1 + N_1 = 0.$$

You could also find the solution that you are looking for by solving the equation

$$1000X_1T = 800X_2T$$
$$1000(5T) = 800(6.25 - 6.25T)$$

for T:

$$5000T = 5000 - 5000T$$
$$T = \tfrac{1}{2}.$$

Thus $XT = (2.5, 3.125, 7.5, 6.875)$ is the solution we are looking for.

(b) Three measures of closeness, d_1, d_2, and d_m, have been considered; LP is used below to solve Exercise 5.1(b) for d_1 and d_m.

d_1: Add the new constraint

$$2300 - 140FP - 160PP = P_2$$

to the constraints in part (a), and minimize $P_1 + N_1 + P_2$.

d_m: In addition to the constraints for d_1, introduce two new constraints, $P_1 + N_1 \leq M$ and $P_2 \leq M$; then minimize M.

d_2: To find the solution minimizing d_2, minimize the quadratic function

$$(1000X_1T - 800X_2T)^2 + (2300 - 140X_3T - 160X_4T)^2$$

subject to $0 \leq T \leq 1$.

5.5 (a) We begin with $B_J = 80$.

MIN $25L + 17D - 10F - 10G - 10J + 32BF + 50BG + 80BJ$
SUBJECT TO
     2) $.21L + .55D - F + BF = 3$

3) .5L + .3D − G + BG = 7
4) .25L + .1D − J + BJ = 5

OBJECTIVE FUNCTION VALUE

1)   458.000000

VARIABLE	VALUE	REDUCED COST
L	20.000000	.000000
D	.000000	1.340000
F	1.200000	.000000
G	3.000000	.000000
J	.000000	61.600000
BF	.000000	22.000000
BG	.000000	40.000000
BJ	.000000	8.400002

Because the reduced cost of B_J is 8.4, we can expect the solution to change when the neighbor offers to sell jet fuel for less than $71.4 per barrel. Let us change the price to $70 per barrel and run the problem again.

MIN   25L + 17D − 10F − 10G − 10J + 32BF + 50BG + 70BJ
SUBJECT TO
2) .21L + .55D − F + BF = 3
3) .5L + .3D − G + BG = 7
4) .25L + .1D − J + BJ = 5

OBJECTIVE FUNCTION VALUE

1)   455.714300

VARIABLE	VALUE	REDUCED COST
L	14.285720	.000000
D	.000000	.452381
F	.000000	1.904762
G	.142857	.000000
J	.000000	60.000000
BF	.000000	20.095240
BG	.000000	40.000000
BJ	1.428571	.000000

To see why the solution has changed, let us analyze the organizations marginal cost of producing jet fuel. Producing 1 barrel of jet fuel requires processing 4 barrels of light crude at a cost of $100. However, processing 4 barrels of light crude also produces

.84 barrel of fuel oil and 2 barrels of gasoline. These products can be sold for $28.40. Hence it costs the organization $71.60 to produce a barrel of jet fuel. Consequently, the organization will consider buying jet fuel at a price that is less than $71.60 per barrel.

(b) MIN   25L + 17D − 10F − 10G − 10J + 32BF + 50BG + 100BJ
SUBJECT TO
2)   .21L + .55D − F + BF = 3
3)   .5L + .3D − G + BG = 7
4)   .25L + .1D − J + BJ = 5
5)   L < = 10

OBJECTIVE FUNCTION VALUE

1)   491.500000

VARIABLE	VALUE	REDUCED COST
L	10.000000	.000000
D	25.000000	.000000
F	12.850000	.000000
G	5.500000	.000000
J	.000000	75.000000
BF	.000000	22.000000
BG	.000000	40.000000
BJ	.000000	15.000000

5.6 To begin, notice that the minimum vitamin, calcium, and fat requirements for chicken feed are satisfied automatically by any combination of available ingredients, so we will not consider these constraints; similarly, we will omit the minimum fat requirement for cattle feed.

Now let us compute the mix of available feeds that minimizes z in Example 5.3. The decision variables used in the following model are defined in the text.

MIN   Z
SUBJECT TO
2)   C + L + S + F = 1
3)   − Z + 8C + 6L + 10S + 4F < = 0
4)   10C + 5L + 12S + 8F > = 6
5)   − Z + 8C + 6L + 6S + 9F < = 0

OBJECTIVE FUNCTION VALUE

1)   6.25531900

VARIABLE	VALUE	REDUCED COST
Z	6.255319	.000000
C	.000000	.723404
L	.808511	.000000
S	.106383	.000000
F	.085106	.000000

This result will be compared later with the solution to Exercise 5.6(c). With that comparison in mind, notice that the slack in rows 2 and 5 is equal to zero; consequently, the sum of the numbers of units of vitamins and fat in this mix is equal to $2z = 12.5106+$. Exercise 5.6(a) will be solved next.

(a) Let C = metric tons of corn to put in the chicken feed used to fill the order; similarly for L, S, F

CC = metric tons of corn to put in the cattle feed used to fill the order; similarly for LC, SC, FC.

MIN   Z
SUBJECT TO

2) $C + L + S + F = 8$
3) $-8Z + 8C + 6L + 10S + 4F < = 0$
4) $10C + 5L + 12S + 8F > = 48$
5) $-8Z + 8C + 6L + 6S + 9F < = 0$
6) $CC + LC + SC + FC = 10$
7) $8CC + 6LC + 10SC + 4FC > = 60$
8) $10CC + 5LC + 12SC + 8FC > = 60$
9) $6CC + 10LC + 6SC + 6FC > = 70$
10) $8CC + 6LC + 6SC + 9FC < = 80$
11) $C + CC < = 5$
12) $L + LC < = 8$
13) $S + SC < =4$
14) $F + FC < = 3$

OBJECTIVE FUNCTION VALUE

1)   6.41666700

VARIABLE	VALUE	REDUCED COST
Z	6.416667	.000000
C	.000000	.083333
L	5.500000	.000000
S	1.388889	.000000
F	1.111111	.000000
CC	3.000000	.000000

LC	2.500000	.000000
SC	2.611111	.000000
FC	1.888889	.000000

(c) Let WP = percent above 6 of the vitamin level of chicken feed
ZP = percent above 6 of the fat level of chicken feed.

MIN   WP + ZP
SUBJECT TO
2)   C + L + S + F = 1
3)   8C + 6L + 10S + 4F − WP + WN = 6
4)   10C + 5L + 12S + 8F > = 6
5)   8C + 6L + 6S + 9F − ZP + ZN = 6

OBJECTIVE FUNCTION VALUE

1)   .461538500

VARIABLE	VALUE	REDUCED COST
C	.000000	1.307692
L	.769231	.000000
S	.076923	.000000
F	.153846	.000000
WP	.000000	.192308
ZP	.461538	.000000
WN	.000000	.807682
ZN	.000000	1.000000

Compare the levels of vitamins and fat in this solution with the corresponding levels in case where we minimized the maximum of WP and ZP.

(d) I will modify the formulation of Exercise 5.6(a) by incorporating WP and WN in constraint 3, ZP and ZN in constraint 5, and putting Z equal to the value of the objective function. I will also write the objective function in "tens of dollars."

MIN   Z
SUBJECT TO
2)   C + L + S + F = 8
3)   10C + 5L + 12S + 8F > = 48
4)   CC + LC + SC + FC = 10
5)   8CC + 6LC + 10SC + 4FC > = 60
6)   10CC + 5LC + 12SC + 8FC > = 60
7)   6CC + 10LC + 6SC + 6FC > = 70
8)   8CC + 6LC + 6SC + 9FC < = 80

9)	$C + CC < = 5$
10)	$L + LC < = 8$
11)	$S + SC < = 4$
12)	$F + FC < = 3$
13)	$8C + 6L + 10S + 4F - 8WP + 8WN = 48$
14)	$8C + 6L + 6S + 9F - 8ZP + 8ZN = 48$
15)	$- Z + 20C + 12L + 24S + 12F + 20CC + 12LC$
	$+ 24SC + 12FC + 8WP + 16ZP = 0$

OBJECTIVE FUNCTION VALUE

1)   290.000000

VARIABLE	VALUE	REDUCED COST
Z	290.000000	.000000
C	.000000	2.000000
L	5.500000	.000000
S	2.000000	.000000
F	.500000	.000000
CC	5.000000	.000000
LC	2.500000	.000000
SC	.000000	.000000
FC	2.500000	.000000
WP	.875000	.000000
WN	.000000	8.000000
ZP	.187500	.000000
ZN	.000000	16.000000

Notice that the reduced cost of SC is equal to zero; consequently, there may be another optimal BFS with $SC > 0$. To look for such a solution, decrease the coefficient of SC slightly in the objective function and solve the revised problem; I put the coefficient of SC equal to 23.999 below.

MIN   Z
SUBJECT TO

2)	$C + L + S + F = 8$
3)	$10C + 5L + 12S + 8F > = 48$
4)	$CC + LC + SC + FC = 10$
5)	$8CC + 6LC + 10SC + 4FC > = 60$
6)	$10CC + 5LC + 12SC + 8FC > = 60$
7)	$6CC + 10LC + 6SC + 6FC > = 70$
8)	$8CC + 6LC + 6SC + 9FC < = 80$
9)	$C + CC < = 5$

10)   L + LC < = 8
11)   S + SC < = 4
12)   F + FC < = 3
13)   8C + 6L + 10S + 4F − 8WP + 8WN = 48
14)   8C + 6L + 6S + 9F − 8ZP + 8ZN = 48
15)   − Z + 20C + 12L + 24S + 12F + 20CC + 12LC
        + 23.999SC + 12FC + 8WP + 16ZP = 0

OBJECTIVE FUNCTION VALUE

1)   289.998800

VARIABLE	VALUE	REDUCED COST
Z	289.998800	.000000
C	.000000	1.999336
L	5.500000	.000000
S	.833333	.000000
F	1.666667	.000000
CC	5.000000	.000000
LC	2.500000	.00000
SC	1.166667	.000000
FC	1.333333	.000000
WP	.000000	.001337
WN	.000000	7.998663
ZP	.625000	.000000
ZN	.000000	16.000000

5.17 Let X be a 0–1 integer variable that is equal to 1 when F_2 is considered for inclusion in the diet and equal to 0 when F_4 is considered for inclusion in the diet.
The following constraints have the desired effect. Put

$F_2 \le 10X$ or, equivalently, $-10X + F_2 \le 0$

$F_4 \le 10(1 - X)$ or, equvialently, $10X + F_4 \le 10$.

I used the constant 10 because it is clear from the constraints that we do not need to put more than 10 units of any food in the diet.

MIN .15F1 + .23F2 + .79F3 + .47F4 + .52F5
SUBJECT TO
2)   F2 + 5F3 + 2F4 > = 3
3)   7F1 + 2F4 + 3F5 > = 10
4)   F1 + 3F2 + 4F3 + F4 < = 2
5)   F1 + F2 + 3F4 + 2F5 < = 3
6)   4F2 + F3 + F5 > = 2

7) $-10X + F2 < = 0$
8) $10X + F4 < = 10$

OBJECTIVE FUNCTION VALUE

1)   1.66729700

VARIABLE	VALUE	REDUCED COST
X	.000000	−5.970269
F1	.297297	.000000
F2	.000000	.000000
F3	.135135	.000000
F4	1.162162	.000000
F5	1.864865	.000000

For comparison, a solution to Exercise 2.13 is included below.

MIN .15FL + .23F2 + .79F3 + .47F4 + .52F5
SUBJECT TO
2) $F2 + 5F3 + 2F4 > = 3$
3) $7F1 + 2F4 + 3F5 > = 10$
4) $F1 + 3F2 + 4F3 + F4 < = 2$
5) $F1 + F2 + 3F4 + 2F5 > = 3$
6) $4F2 + F3 + F5 > = 2$

OBJECTIVE FUNCTION VALUE

1)   1.61836100

VARIABLE	VALUE	REDUCED COST
F1	.295082	.000000
F2	.081967	.000000
F3	.000000	.362131
F4	1.459016	.000000
F5	1.672131	.000000

5.18 (b) Let us begin by recalling the solution to Exercise 2.23.

MIN R1 + 3R2 + 4WI + 5.5W2
SUBJECT TO
2) $W1 + W2 < = 4$
3) $R1 + R2 + W1 + W2 = 10$
4) $15R1 + 7.5R2 + 5WI + W2 < = 100$

OBJECTIVE FUNCTION VALUE
1)   23.333300

VARIABLE	VALUE	REDUCED COST
R1	3.333333	.000000
R2	6.666667	.000000
W1	.000000	.333333
W2	.000000	.766667

Now I will use two zero-one valued integer variables, X and Y, to act as switches on the processing of river water as follows:

If $X = 0$, then $R_1 = 0$; if $X = 1$, then $R_1 \geq 4$.
If $Y = 0$, then $R_2 = 0$; if $Y = 1$, then $R_2 \geq 4$.

This is accomplished through constraints 5 to 8 in the following model.

MIN   R1 + 3R2 + 4WI + 5.5W2
SUBJECT TO
   2)   W1 + W2 < = 4
   3)   R1 + R2 + W1 + W2 = 10
   4)   15R1 + 7.5R2 + 5W1 + W2 < = 100
   5)   - 4X + R1 > = 0
   6)   - 10X + R1 < = 0
   7)   - 4Y + R2 > = 0
   8)   - 10Y + R2 < = 0

Before we look at a solution, student questions have indicated that I should comment on constraints 5 to 8.

Constraint 5: $R_1 \geq 4X$ says that if $X = 1$, then $R_1 \geq 4$.
Constraint 6: $R_1 \leq 10X$ says that if $X = 0$, then $R_1 \leq 0$. But we have $R_1 \geq 0$ automatically; so if $X = 0$, then $R_1 = 0$. It is reasonable to ask where I got the number 10 for constraint 6. I got it from constraint 3. Constraint 3 tells us that if R_1 is feasible, then $R_1 \leq 10$; thus we can put $R_1 \leq 10$ without changing the set of feasible solutions. Of course, any number ≥ 10 would work just fine!

Constraints 7 and 8 use the 0–1 integer variable Y to switch R_2 between the cases $R_2 \geq 4$ and $R_2 = 0$, just as constraints 5 and 6 use X to switch R_1. A solution to the problem follows.

OBJECTIVE FUNCTION VALUE
1)   23.9230800

VARIABLE	VALUE	REDUCED COST
X	1.000000	3.538462
Y	1.000000	.000000
R1	4.000000	.000000
R2	5.230769	.000000
W1	.000000	.038462
W2	.769231	.000000

Now suppose that well water is not available; a model that includes this additional supposition and a solution appear below.

MIN   R1 + 3R2 + 4WI + 5.5W2
SUBJECT TO
   2)   W1 + W2 < = 4
   3)   R1 + R2 + W1 + W2 = 10
   4)   15R1 + 7.5R2 + 5W1 + W2 < = 100
   5)   − 4X + R1 > = 0
   6)   − 10X + R1 < = 0
   7)   − 4Y + R2 > =0
   8)   − 10Y + R2 < = 0
   9)   W1 = 0
   10)   W2 = 0

OBJECTIVE FUNCTION VALUE

1)   30.0000000

VARIABLE	VALUE	REDUCED COST
X	.000000	− 20.000000
Y	1.000000	.000000
R1	.000000	.000000
R2	10.000000	.000000
W1	.000000	.000000
W2	.000000	.000000

The preceding model requires that exactly 100,000 gallons of water be processed. The following model incorporates the option of being able to process surplus river water.

MIN   R1 + 3R2 + 4WI + 5.5W2
SUBJECT TO
   2)   W1 + W2 < = 4
   3)   R1 + R2 + W1 + W2 > = 10
   4)   5R1 − 2.5R2 − 5W1 − 9W2 < = 0
   5)   − 4X + R1 > = 0

```
6)   - 10X + R1 < = 0
7)   - 4Y + R2 > = 0
8)   - 10Y + R2 < = 0
9)      W1 = 0
10)     W2 = 0
```

OBJECTIVE FUNCTION VALUE

1)   28.0000000

VARIABLE	VALUE	REDUCED COST
X	1.000000	28.000000
Y	1.000000	.000000
R1	4.000000	.000000
R2	8.000000	.000000
W1	.000000	.000000
W2	.000000	.000000

(c) As in Exercise 5.16, let

H = no. of horses to buy
C = no. of cows to buy
G = no.of goats to buy
F = no. of calves to buy.

In Exercise 5.16 I wrote and solved two LPs to find that it was preferable to have 25 calves. Here I will use a zero–one integer-valued variable Z to act as a switch, switching between the cases $F = 0$ and $F \geq 25$: $Z = 0$ will imply that $F = 0$ and $Z = 1$ will imply that $F \geq 25$.

MAX   320H + 140C + 20G + 210F
SUBJECT TO

```
2)    1.5H + .8C + .25G + F < = 35
3)    H > = 3
4)    C > = 1
5)    G > = 4
6)    H < = 10
7)    C < = 5
8)    - 25Z + F > = 0
9)    - 35Z + F < = 0
```

Before looking at a solution, let me explain how constraints 8 and 9 work. Constraint 8 can be rewritten in the form $F \geq 25Z$; thus we see that if $Z = 1$, then $F \geq 25$. Hence $F \geq 25$ when $Z = 1$. Constraint 9 can be rewritten in the form $F \leq 35Z$; thus we see that if $Z = 0$, then $F \leq 0$. However, we automatically have $F \geq 0$; hence $F = 0$ when $Z = 0$.

Because of the fact that either $Z = 0$ or $Z = 1$ (i.e., because exactly one of these two possibilities always occurs), we can conclude that either $F = 0$ or $F \geq 25$ (i.e., exactly one of these latter two possibilities always occurs); a solution follows.

OBJECTIVE FUNCTION VALUE

1)   7219.33300

VARIABLE	VALUE	REDUCED COST
Z	1.000000	83.333210
H	5.466667	.000000
C	1.000000	.000000
G	4.000000	.000000
F	25.000000	.000000

For comparison, I will delete the constraints involving F and solve the resulting LP.

MAX   320H + 140C + 20G + 210F
SUBJECT TO
    2)   1.5H + 1.8C + .25G + F < = 35
    3)   H > = 3
    4)   C > = 1
    5)   G > = 4
    6)   H < = 10
    7)   C < = 5

OBJECTIVE FUNCTION VALUE

1)   7242.00000

VARIABLE	VALUE	REDUCED COST
H	10.000000	.000000
C	1.000000	.000000
G	4.000000	.000000
F	18.200000	.000000

At this point, what would you advise the programmer to do? Let us look next at a model with a few more integer variables.

Comparing the $F = 0$ and $F \geq 25$ cases from Exercise 5.16, we know that $F \geq 25$; we also know that $H \geq 3$, $C \geq 1$, and $G \geq 4$. Thus I decided to delete all the requirements for 3 horses, 1 cow, 4 goats, and 25 calves from the available resources to get the model max $32H + 14C + 2G + 21F$

s.t. (2) $1.5H + .8C + .25G + F \leq 3.7$.

From the preceding constraint, we see that $H \leq 2$, $C \leq 4$, $G \leq 14$, and $F \leq 3$. I will use binary expansions of the integers H, C, G, and F to write them in terms of 0–1 integer variables as follows. Put

$H = H_1 + 2H_2$, where H_1 and H_2 are 0–1 integer variables.

$C = C_1 + 2C_2 + 4C_3$, where C_1, C_2, and C_3 are 0–1 integer variables

$G = G_1 + 2G_2 + 4G_3 + 8G_4$, where G_1, G_2, G_3, and G_4 are 0–1 integer variables

$F = F_1 + 2F_2$, where F_1 and F_2 are 0–1 integer variables.

Now we have the model

MAX   320H + 140C + 20G + 210F
SUBJECT TO
        2)   1.5H + .8C + .25G + F < = 3.7
        3)   − H1 − 2H2 + H = 0
        4)   − C1 − 2C2 − 4C3 + C = 0
        5)   − G1 − 2G2 − 4G3 − 8G4 + G = 0
        6)   −F1 − 2F2 + F = 0
END
INTEGER-VARIABLES =
        11: C1,C2,C3,F1,F2,G1,G2,G3,G4,H1,H2.

Two solutions follow, first without and then with the integer requirements present.

LP solution (no integer restrictions active)

OBJECTIVE FUNCTION VALUE

1)   789.333400

VARIABLE	VALUE	REDUCED COST
H1	1.000000	.000000
H2	.733333	.000000
C1	.000000	.000000
C2	.000000	.000000
C3	.000000	.000000
G1	.000000	33.333330
G2	.000000	66.666660
G3	.000000	133.333300
G4	.000000	266.666700

F1	.000000	3.333328
F2	.000000	6.666656
H	2.466667	.000000
C	.000000	30.666670
G	.000000	.000000
F	.000000	.000000

IP solution

OBJECTIVE FUNCTION VALUE

1)   740.000000

VARIABLE	VALUE	REDUCED COST
H1	1.000000	−320.000000
H2	.000000	−640.000000
C1	.000000	−140.000000
C2	.000000	−280.000000
C3	.000000	−560.000000
G1	.000000	−20.000000
G2	.000000	−40.000000
G3	.000000	−80.000000
G4	.000000	−160.000000
F1	.000000	−210.000000
F2	1.000000	−420.000000
H	1.000000	.000000
C	.000000	.000000
G	.000000	.000000
F	2.000000	.000000

5.19 Suppose that we have 0–1 integer-valued variables available to use to model this situation. Our decisions in this problem are of two types: (1) whether or not to ship from warehouse I to lumberyard J, and (2) if we decide to send a truck from warehouse I to lumberyard J, we need to decide how many bags of mix to put on the truck. For $1 \le I, J \le 3$, let T_{IJ} be a 0–1 integer-valued variable which is equal to 1 if mix is delivered from warehouse I to lumberyard J, and equal to zero otherwise; let Q_{IJ} = number of bags of mix shipped from warehouse I to lumberyard J.

MIN   45 T11 + 55 T12 + 103 T13 + 80 T21 + 112 T22 + 96 T23
+ 128 T31 + 54 T32 + 70 T33
QIJ < = 5000TIJ, 1 < =I,J < = 3
Q11 + Q21 + Q31 = 5000
Q12 + Q22 + Q32 = 3500

$$Q13 + Q23 + Q33 = 2500$$
$$Q11 + Q12 + Q13 < = 3300$$
$$Q21 + Q22 + Q23 < = 6700$$
$$Q31 + Q32 + Q33 < = 4600.$$

5.21 Let

T = no. of rolls of toilet paper to produce in a 10-hour day
W = no. of writing pads to produce in a 10-hour day
P = no. of rolls of paper towels to produce in a 10-hour day.

Let Y be a 0–1 integer variable that is equal to 1 when writing pads are produced, and let Z be a 0–1 integer variable that is equal to 1 if at least 1200 rolls of toilet paper are produced. The following comments explain how the integer variables work in the model below.

Constraints 6 and 7 take care of writing pads. I chose to use the number 3000 in constaint 7 because it is clear from constraint 3 that $W \le 3000$ whenever W is feasible; 1500 or any larger number would work equally well. Constraint 8, $T \ge 1200Z$, has the desired effect because the presence of the term $20Z$ in the objective function causes the value of Z to pop up to 1 as soon as the value of T reaches 1200.

MAX $20Z + .18T + .29W + .25P$
SUBJECT TO

2)	$.5T + .22W + .85P < = 1500$	
3)	$.2T + .4W + .22P < = 600$	
4)	$T > = 1000$	
5)	$P > = 400$	
6)	$-500Y + W > = 0$	
7)	$-3000Y + W < = 0$	
8)	$-1200Z + T > = 0$	

OBJECTIVE FUNCTION VALUE

1)   562.818200

VARIABLE	VALUE	REDUCED COST
Y	1.000000	82.272740
Z	1.000000	36.727270
T	1200.000000	.000000
W	500.000000	.000000
P	727.272700	.000000

5.22 Let U_I be a 0–1 integer variable that is equal to 1 if source I can be used, $I \le 3$, and V_{IJ} be a 0–1 integer variable that is equal to 1 if source I can be used in period J, $I \leq 3$, $J \leq 2$. Then let

X_{IJ} = hundreds of units produced by source I in period J, $I \leq 3$, J
   $< = 2$

X_I = hundreds of units produced by source I, $I \leq 3$.

  MIN 31U1 + 23U2 + 17U3
      + 7.5V11 + 7.5V12 + 9V21 + 9V22 + 10V31 + 10V32
      + 5.4X1 + 4.2X2 + 6.8X3
  SUBJECT TO
     2) − X1 + X11 + X12 = 0
     3) − X2 + X21 + X22 = 0
     4) − X3 + X31 + X32 = 0
     5)   X11 + X21 + X31 > = 30
     6)   X12 + X22 + X32 > = 40
     7) − 22V11 + X11 < = 0
     8) − 22V12 + X12 < = 0
     9) − 17V21 + X21 < = 0
    10) − 17V22 + X22 < = 0
    11) − 31V31 + X31 < = 0
    12) − 31V32 + X32 < = 0
    13) − 2U1 + V11 + V12 < = 0
    14) − 2U2 + V21 + V22 < = 0
    15) − 2U3 + V31 + V32 < =0

OBJECTIVE FUNCTION VALUE

1)   452.600000

VARIABLE	VALUE	REDUCED COST
U1	1.000000	.000000
U2	1.000000	.000000
U3	1.000000	17.000000
V11	1.000000	23.000000
V12	1.000000	−7.800003
V21	1.000000	.099995
V22	1.000000	−23.700000
V31	.000000	10.000000
V32	1.000000	10.000000
X1	35.000000	.000000
X2	34.000000	.000000
X3	1.000000	.000000
X11	13.000000	.000000
X12	22.000000	.000000
X21	17.000000	.000000

X22	17.000000	.000000
X31	.000000	1.400000
X32	1.000000	.000000

5.25

		Power Plants								
		MD	OD	MF	OF	MP	OP	MS	OS	Supply
Suppliers	C	403	403	441	441	458	458	430	430	800
	U	530	530	510	510	550	550	350	350	600
	W	360	360	340	340	380	380	410	410	1000
	Dummy	M	O	M	O	M	O	M	O	660
	Demand	400	600	80	150	120	310	560	840	3060

5.27 I will begin by numbering the nodes, starting a list of the routes, and drawing a sketch of the situation described in the problem (Fig. 7). The routes will be listed lexicographically: denote a route from node i to node j by $[i,j]$, then $[p,q]$ denotes a route from node p to node q; $[i,j]$ precedes $[p,q]$ in the list of routes if (i) $i < p$ or (ii) $i = p$ and $j < q$.

Node	Correspondence	Beginning of a list of the routes	
1	Site 1	1.	[1,4]
2	Site 2	2.	[1,5]
3	Site 3	3.	[1,6]
4	Port 1	4.	[2,4]
5	Port 2	.	.
6	Port 3	.	.
7	Site 4	.	.
8	Intermediate city		
9	Site 5		
10	Border		
11	Destination city		

Let X_I = tens of tons shipped via route I. Since only 60% of the supplies shipped by "free" train from the border to the city arrive at the city, we have

X_{17} = 0.6 of the quantity shipped from border

or the quantity shipped from border = 1/0.6 of X_{17}; thus

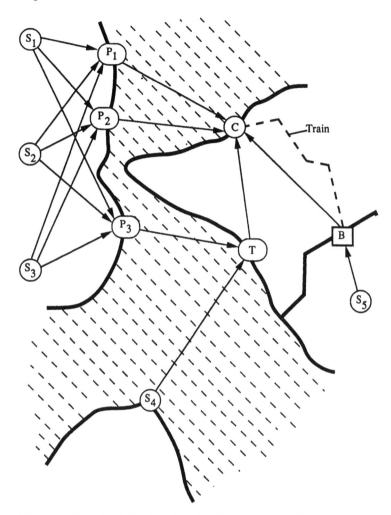

Figure 7 Sketch of the situation described in the problem.

$$X_{16} + 1.667X_{17} = X_{15}.$$

An LP model for the problem follows.

MIN   10X1 + 17X2 + 22X3 + 14X4 + 11X5 + 18X6 + 19X7
    + 16X8 + 7X9 + 7X10 + 4X11 + 3X12 + 5X13
    + 2X14 + 15X15 + 6X16
SUBJECT TO
    2)   X1 + X4 + X7 − X10 = 0

3) $X2 + X5 + X8 - X11 = 0$
4) $X3 + X6 + X9 - X12 = 0$
5) $X12 + X13 - X14 = 0$
6) $X10 < = 25$
7) $X11 < = 30$
8) $X12 < = 30$
9) $X13 < = 30$
10) $X14 < = 40$
11) $X10 + X11 + X14 + X16 + X17 = 100$
12) $X15 < = 50$
13) $X16 < = 10$
14) $X17 < = 24$
15) $X13 < = 20$
16) $X15 - X16 - 1.667X17 = 0$

OBJECTIVE FUNCTION VALUE

1)   1360.00000

VARIABLE	VALUE	REDUCED COST
X1	25.000000	.000000
X2	.000000	6.000000
X3	.000000	15.000000
X4	.000000	4.000000
X5	30.000000	.000000
X6	.000000	11.000000
X7	.000000	9.000000
X8	.000000	5.000000
X9	20.000000	.000000
X10	25.000000	.000000
X11	30.000000	.000000
X12	20.000000	.000000
X13	20.000000	.000000
X14	40.000000	.000000
X15	5.000000	.000000
X16	5.000000	.000000
X17	.000000	4.005001

Because $0 < X_{16} < 10$, the effective cost of shipping from S_5 to the city via the free train must decrease to \$210 per ton before we will consider using that train to transport supplies. To see what percentage of those supplies must be delivered, I examine the routes from S_5 to the city from a different perspective below; look at the paths from S_5 to the city without explicitly considering the border.

Let

X_{15} = tens of tons shipped from S_5 to the city without using the free train from the border to the city

X_{16} = tens of tons shipped from S_5 to the city via the free train from the border of the city.

Now the cost contribution of X_{15} is $(150 + 60)X_{15}$, and the cost contribution of X_{16} is $(1.667 \times 150)X_1$. Let f denote the fraction of the supplies shipped on the free train that will be delivered. For the free train to be competitive we must have $150/f \leq 210$ or, equivalently, $f \geq 150/210$.

5.29 Begin by ordering the nodes as follows:

$$S_1, S_2, S_3, T_1, T_2, T_3, T_4, D_1, D_2.$$

Then list the routes lexicographically according to this ordering of the nodes; thus X_1 denotes the flow along route (1,4), which goes from S_1 to T_1, and so on.

(a) The following LP models the minimum cost flow problem with a 5-unit fee at each transshipment point.

MIN   C
SUBJECT TO
    2)   X1 + X2 < = 20
    3)   X3 + X4 + X5 < = 10
    4)   X6 + X7 + X8 < = 30
    5)   X13 + X14 − I1 = 0
    6)   X8 + X12 + X15 − I2 = 0
    7)   − C + 8X1 + 17X2 + 7X3 + 6X4 + 12X5 + 8X6
        + 6X7 + 22X8 + 10X9 + 8X10 + 5X11
        + 15X12 + 4X13 + 9X14 + 8X15 = 0
    8)   I1 = 25
    9)   I2 = 15
   10)   X1 + X3 + X6 − X9 = 5
   11)   X4 + X7 − X10 − X11 − X12 = 5
   12)   X2 + X9 + X10 − X13 = 5
   13)   X5 + X11 − X14 − X15 = 5

OBJECTIVE FUNCTION VALUE

1)   975.000000

VARIABLE	VALUE	REDUCED COST
C	975.000000	.000000

X1	5.000000	.000000
X2	15.000000	.000000
X3	.000000	2.000000
X4	10.000000	.000000
X5	.000000	1.000000
X6	.000000	3.000000
X7	30.000000	.000000
X8	.000000	3.000000
X9	.000000	1.000000
X10	15.000000	.000000
X11	20.000000	.000000
X12	.000000	2.000000
X13	25.000000	.000000
X14	.000000	2.000000
X15	15.000000	.000000
I1	25.000000	.000000
I2	15.000000	.000000

5.35 Stage 3

Initial state	$f_3(i)$	(i denotes initial state)
1	2	
2	4	
3	6	
4	11	
5	12	
6	13	

Stage 2

Initial state	$f_2(i)$	Optimal terminal state(s)
2	3	1
3	7	1
4	12	1
5	14	2
6	16	3,4
7	21	4

Stage 1

Initial state	$f_1(8)$	Optimal terminal state
8	22	7

The optimal solution is unique: one ship to port 1, three ships to port 2, and four ships to port 3.

6
Duality

In this short chapter we present the duality theorem of linear programming, introduce the concept of complementary slackness, and present the Kuhn–Tucker conditions for a symmetric primal LP. Exposure to these concepts can be quite helpful when utilizing output from LP packages and also when analyzing nonlinear programs. In Chapter 7 we show you how an appropriate version of the Kuhn–Tucker conditions permits you to write a quadratic program in a form called a linear complementarity problem.

The duality theorem of linear programming is an expanded statement of the second fundamental fact about standard primal–dual pairs that was discussed in Section 2.5, Exercise 2.33, and its solution. Suppose that you have a standard or symmetric primal–dual pair. The first fundamental fact about the pair says that if x is feasible for the primal and y is feasible for the dual, then $yb \leq cx$, and the second fundamental fact adds that if $yb = cx$, then x and y are optimal for the primal and dual.

6.1 THE DUALITY THEOREM OF LINEAR PROGRAMMING

Suppose that you have a standard primal–dual pair, if either the primal or the dual has a finite optimal solution, so does the other and the

202

values are equal. If either the primal is unbounded below or the dual is unbounded above, the other has no feasible solution.

Proof: Consider the case where the primal has a finite optimal solution. According to the fundamental theorem of linear programming, there is an optimal basic feasible solution $x = (x_B, 0)$. According to Chapter 4, we can require that all the reduced costs are ≥ 0. Put $y = c_B B^{-1}$. Then $yA = c_B B^{-1}[B \mid V] = [c_B \mid c_B B^{-1} V] \leq [c_B, c_V] = c$ because $0 \leq \bar{c}_V = c_V - c_B B^{-1} V$ is the reduced cost vector for x_B. Hence y is feasible for the dual and $cx = c_B x_B = c_B(B^{-1}b) = (c_B B^{-1})b = yb$, so y is optimal for the dual. Consequently, if the primal has a finite optimal solution, so does the dual, and the values are equal. If either is unbounded in the indicated fashion, the first fundamental fact asserts that the other can have no feasible solution to block the values of the unbounded objective function. To verify that when the dual has a finite optimal solution, so does the primal, and the optimal values are equal, recall that the second example in Appendix 1 shows that the dual of the dual is a standard primal. Consequently, the proof for the primal applies to the dual when the dual is written as a standard primal.

Remark

Referring again to Appendix 1, notice that the duality theorem of linear programming carries over to other primal–dual pairs.

6.2 COMPLEMENTARY SLACKNESS

Complementary slackness is a rephrasing of the two fundamental facts about primal–dual pairs that were stated in Section 2.5. For easy reference the tableau in Section 2.4 and the two fundamental facts are repeated below.

Pairs		
Primal	Dual	
min cx s.t. $Ax \geq b, x \geq 0$	max yb s.t. $yA \leq c, y \geq 0$	Symmetric pair
min cx s.t. $Ax = b, x \geq 0$	max yb s.t. $yA \leq c$	Standard pair

Two Fundamental Facts About Standard and Symmetric Primal–Dual Pairs

1. If x is feasible for the primal and y is feasible for the dual, then

$$yb \leq y(Ax) = (yA)x \leq cx.$$

2. If x is feasible for the primal and y is feasible for the dual and $yb = cx$, then x and y are optimal solutions to the primal and dual (i.e., both are optimal solutions).

The standard pair will be considered first; suppose that x and y are feasible. Observe that

$$cx - yb = cx - y(Ax) = (c - yA)x.$$

Consequently, the condition $cx = yb$ is equivalent to $(c - yA)x = 0$ or, equivalently, $(c_i - ya_i)x_i = 0$, $1 \leq i \leq n$. The latter form simply says that for each i, at least one of $c_i - ya_i$ and x_i must be zero. These conditions are called complementary slackness. They may be rewritten as follows:

(i) if $x_i > 0$, then $ya_i = c_i$, and
(ii) if $ya_i < c_i$, then $x_i = 0$.

Theorem 6.1 summarizes these observations below.

Theorem 6.1. Let x and y be feasible for a standard primal–dual pair. Then x and y are optimal if and only if the following complementary slackness conditions are satisfied:

(i) if $x_i > 0$, then $ya_i = c_i$, and
(ii) if $ya_i < c_i$, then $x_i = 0$.

A symmetric set of complementary slackness conditions can be found for the symmetric primal–dual pair by supposing that x and y are feasible and examining $cx - yb$ again. Since

$$cx \geq (yA)x = y(Ax) \geq yb,$$

$cx = yb$ if and only if $cx = (yA)x$ and $y(Ax) = yb$.
Or, equivalently, $cx = yb$ if and only if

$$(c - yA)x = 0 \quad \text{and} \quad y(Ax - b) = 0.$$

We have seen that the first of the latter two conditions is equivalent to conditions (i) and (ii) above. In like manner, the second condition, $y(Ax - b) = 0$, is equivalent to conditions

(iii) if $y_j > 0$, then $a^j x = b_j$, and

(iv)   if $a^j x > b_j$, then $y_j = 0$, where a^j denotes the jth row of A.

Theorem 6.2 summarizes the preceding discussion.

Theorem 6.2.   Let x and y be feasible for a symmetric primal–dual pair. Then both are optimal if and only if the complementary slackness conditions (i) to (iv) are satisfied.

6.3 KUHN–TUCKER CONDITIONS

The Kuhn–Tucker conditions for a LP pertain to (1) primal feasibility, (2) dual feasibility, and (3) complementary slackness. These conditions are written below in the form of a theorem for a symmetric primal LP; $c - yA$ is identified with a (row) vector named z in condition 2.

Theorem 6.3 (Kuhn–Tucker Conditions).   A vector x is an optimal solution to the symmetric primal LP

$$\min \ cx$$

$$\text{s.t.} \ \ Ax \geq b, \quad x \geq 0$$

if, and only if, the following three conditions are satisfied:

1. $Ax \geq b, x \geq 0$.
2. There exist nonnegative y and z with $c - yA - z = 0$.
3. $y(Ax - b) = 0$ and $zx = 0$.

Conditions 1 to 3 are called Kuhn–Tucker conditions for a symmetric primal LP. They are necessary and sufficient conditions in order that x be an optimal solution to a symmetric primal LP. Condition 1 states that x is feasible for the primal, condition 2 states that y is feasible for the corresponding dual, and condition 3 is a statement of the appropriate complementary slackness conditions, where $z = c - yA$. Condition 3 indicates when there might be multiple optimal solutions; if both terms in one of the entries are zero (i.e., if, for some i, $\{y_i = 0$ and $a^i x - b_i = 0\}$ or if, for some j, $\{c_j - ya_j = 0$ and $x_j = 0\}$), there might be multiple optimal solutions.

These conditions are considered from a different point of view in Chapter 7. This section concludes by using Exercise 4.1(e) for a numerical example.

Example 6.1.   The standard-form primal simplex tableau for Exercise 4.1(e) is

$$\begin{bmatrix} 1 & 4 & 2 & | & 1 & 0 & | & 10 \\ 2 & 3 & 4 & | & 0 & 1 & | & 8 \\ -2 & -1 & -1 & | & 0 & 0 & | & 0 \end{bmatrix} \xrightarrow[\substack{1/2(2) \\ -(2) \text{ to } (1) \\ 2(2) \text{ to } (3)}]{} \begin{bmatrix} 0 & \frac{5}{2} & 0 & | & 1 & -\frac{1}{2} & | & 6 \\ 1 & \frac{3}{2} & 2 & | & 0 & \frac{1}{2} & | & 4 \\ 0 & 2 & 3 & | & 0 & 1 & | & 8 \end{bmatrix}:$$

$$\begin{bmatrix} A & | & I & | & b \\ -c & | & 0 & | & 0 \end{bmatrix} \rightarrow \begin{bmatrix} B^{-1}A & | & B^{-1} & | & B^{-1}b \\ -c + c_B B^{-1}A & | & c_B B^{-1} & | & c_B B^{-1}b \end{bmatrix}.$$

Consequently, the optimal value of 8 occurs at $x = (4,0,0)$ and $y = c_B B^{-1} = [0,1]$ is an optimal solution of the dual problem

$$\min \; 10y_1 + 8y_2$$
$$\text{s.t.} \quad y_1 + 2y_2 \geq 2$$
$$4y_1 + 3y_2 \geq 1$$
$$2y_1 + 4y_2 \geq 1, \quad y \geq 0.$$

To find the Kuhn–Tucker conditions, let's rewrite the problem as a symmetric primal:

$$\min \; [-2,-1,-1]x$$
$$\text{s.t.} \quad \begin{bmatrix} -1 & -4 & -2 \\ -2 & -3 & -4 \end{bmatrix} x \geq \begin{bmatrix} -10 \\ -8 \end{bmatrix}, \quad x \geq 0.$$

In this form the Kuhn–Tucker condition are

(i) $\quad -\begin{bmatrix} 1 & 4 & 2 \\ 2 & 3 & 4 \end{bmatrix} x \geq -\begin{bmatrix} 10 \\ 8 \end{bmatrix}, \quad x \geq 0,$

(ii) $\quad [-2,-1,-1] - y\left\{ -\begin{bmatrix} 1 & 4 & 2 \\ 2 & 3 & 4 \end{bmatrix} \right\} - z = 0, \quad y,z \geq 0,$

(iii) $\quad y\left\{ -\begin{bmatrix} 1 & 4 & 2 \\ 2 & 3 & 4 \end{bmatrix} x - \begin{bmatrix} -10 \\ -8 \end{bmatrix} \right\} = 0 \quad \text{and} \quad zx = 0,$

They can be rewritten as follows:

(i) $\quad \begin{bmatrix} 1 & 4 & 2 \\ 2 & 3 & 4 \end{bmatrix} x \leq \begin{bmatrix} 10 \\ 8 \end{bmatrix}, \quad x \geq 0,$

(ii) $\quad -[2,1,1] + y\begin{bmatrix} 1 & 4 & 2 \\ 2 & 3 & 4 \end{bmatrix} - z = 0, \quad y,z \geq 0,$

(iii) $\quad y\left\{ \begin{bmatrix} 1 & 4 & 2 \\ 2 & 3 & 4 \end{bmatrix} x - \begin{bmatrix} 10 \\ 8 \end{bmatrix} \right\} = 0 \quad \text{and} \quad zx = 0.$

We can use the fact that $x = (4,0,0)$ is an optimal solution to the primal and Kuhn–Tucker conditions in the latter form to find an optimal solution to the dual as follows: (iii) implies that

$$y\left\{\begin{bmatrix} 1 & 4 & 2 \\ 2 & 3 & 4 \end{bmatrix}\begin{bmatrix} 4 \\ 0 \\ 0 \end{bmatrix} - \begin{bmatrix} 10 \\ 8 \end{bmatrix}\right\} = y\left\{\begin{bmatrix} 4 \\ 8 \end{bmatrix} - \begin{bmatrix} 10 \\ 8 \end{bmatrix}\right\} = -6y_1 = 0 \quad \text{and} \quad z_1 = 0.$$

Then (ii) becomes

$$-[2,1,1] + [0,y_2]\begin{bmatrix} 1 & 4 & 2 \\ 2 & 3 & 4 \end{bmatrix} - [0,z_2,z_3] = 0, \qquad \text{so } y_2 = 1.$$

Of course, we can also find λ_B via the formula

$$\lambda_B = c_B B^{-1} = [0,-2]\begin{bmatrix} -1 & \frac{1}{2} \\ 0 & -\frac{1}{2} \end{bmatrix} = [0,1];$$

here B^{-1} is the negative of the B^{-1} in the tableau because of the sign changes that were made to put the problem in symmetric primal form.

EXERCISES

6.1 Referring to Exercise 4.1, write the dual problem, solve the dual problem and write the Kuhn–Tucker conditions for the problem [parts (a) to (j)].

6.2 Consider the LP

$$\max 3x_1 \quad + 4x_3$$
$$\text{s.t.} \quad x_1 + x_2 + 2x_3 \le 8$$
$$3x_1 + x_2 - 2x_3 \le 4, \qquad x > 0.$$

(a) Use the simplex method to solve the LP.
(b) Use the revised simplex method to check your answer.
(c) Write the dual problem.
(d) Write a solution to the dual problem.
(e) Write the Kuhn–Tucker conditions for the LP.

6.3 Write Kuhn–Tucker conditions for a standard-form primal LP. (Here is a good opportunity to look at Appendix 1.)

7

Quadratic Programming

A linear program has a linear objective function and linear constraints; a quadratic program has a quadratic objective function and linear constraints. The primary goal of Chapter 7 is to explain how a quadratic program with a convex objective function can be solved by solving an associated linear complementarity problem; a linear complementarity problem is a linear system with an additional complementary slackness condition on the variables. On the way to this goal, several important ideas are discussed in Sections 7.1, 7.2, and 7.3.

In Section 7.1 we introduce quadratic functions and gradients and discuss minimum and maximum values. Conditions for existence and uniqueness of a minimum correspond to properties named positive semi-definite and positive definite. In Section 7.2 we define convexity of a quadratic function and show that a quadratic function is convex if, and only if, it is positive semidefinite. Section 7.2 also relates convexity to a relationship between the function and its gradient which is used in Section 7.3 to establish the Kuhn–Tucker conditions for a convex quadratic program. In Section 7.4 we show how to rewrite the Kuhn–Tucker conditions as a linear complementarity problem. Algorithms are available for linear complementarity problems, therefore we show you how to put a convex quadratic program into linear complementarity format. Nonnegative quadratic functions are convex, so the entire chapter applies to the investment application, which is presented in Section 7.5.

208

7.1 QUADRATIC FUNCTIONS

The discussion of quadratic functions on R^n to R begins with a review of the case $n = 1$.

$n = 1$

A quadratic function g on R to R is defined for all t in R by the formula

$$g(t) = \alpha_0 + \alpha_1 t + \alpha_2 t^2, \quad \alpha_0, \alpha_1, \alpha_2 \text{ in } R.$$

If $\alpha_2 = 0$, g is also a linear function on R to R.

The value $g(z)$ of g at z in R can be compared with the value $g(x)$ of g at x in R as follows. Notice that because $z = x + [z - x]$.

$$
\begin{aligned}
g(z) &= g(x + [z - x]) \\
&= \alpha_0 + \alpha_1(x + [z - x]) + \alpha_2(x + [z - x])^2 \\
&= \alpha_0 + \alpha_1 x + \alpha_1(z - x) + \alpha_2 x^2 + 2\alpha_2 x(z - x) + \alpha_2(z - x)^2 \\
&= \alpha_0 + \alpha_1 x + \alpha_2 x^2 + (\alpha_1 + 2\alpha_2 x)(z - x) + \alpha_2(z - x)^2 \\
&= g(x) + (\alpha_1 + 2\alpha_2 x)(z - x) + \alpha_2(z - x)^2.
\end{aligned}
$$

Thus

$$g(z) - g(x) = (\alpha_1 + 2\alpha_2 x)(z - x) + \alpha_2(z - x)^2.$$

The *tangent line* $\{(z, T(z)) : z \text{ in } R\}$ to the graph of the function g at the point $(x, g(x))$ in R^2 is defined for all z in R by the formula

$$T(z) = g(x) + (\alpha_1 + 2\alpha_2 x)(z - x).$$

The *gradient* ∇g of g at x in R is defined by the formula

$$\nabla g(x) = \alpha_1 + 2\alpha_2 x, \tag{7.1}$$

and the formulas

$$T(z) - T(x) = \nabla g(x)(z - x) \tag{7.2}$$

and

$$g(z) - g(x) = \nabla g(x)(z - x) + \alpha_2(z - x)^2 \tag{7.3}$$

apply.

Max and Min

Formulas (7.1)–(7.3) lead to the following facts about maxima and minima: (1) If the function g has a minimum or a maximum at x, then

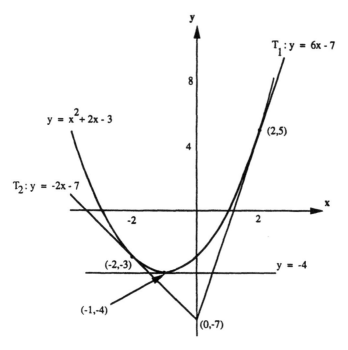

Figure 1

$\nabla g(x) = 0$ because $\nabla g(x)$ is the slope of the tangent line $T(x)$ to the graph of g at x. (2) If $\nabla g(x) = 0$ and $\alpha_2 > 0$, then g has a unique minimum at x. (3) If $\nabla g(x) = 0$ and $\alpha_2 < 0$, then g has a unique maximum at x. (4) If $\nabla g(x) = 0$ and $\alpha_2 = 0$, then $g(x) = \alpha_0$ for all x in R. Therefore, a discussion of maximum points for a quadratic function is symmetric to a discussion of minimum points; consequently, attention is henceforth restricted to minimum points.

Before going on to the case $n > 1$, a numerical example is given below for the case $n = 1$. Figure 1 displays the graph and tangent lines.

Example 7.1. Suppose that g is defined on R by the formula

$$g(x) = y = x^2 + 2x - 3.$$

Then

$$\nabla g(x) = 2 + 2x;$$

$\nabla g(x) = 0$ if $x = -1$ and $\alpha_2 = 1$. Thus g has a unique minimum point at -1. The tangent line to the graph of g at the point $(-1, -4)$ is given by the horizontal line

$$y = T_0(z) = -4.$$

At $x = 2$, the tangent line has slope 6,

$$y = T_1(z) = 6z - 7,$$

and at $x = -2$, the tangent line has slope -2,

$$y = T_2(z) = -2z - 7.$$

$n \geq 1$

The following discussion parallels the discussion given above for the case $n = 1$. A quadratic function f defined on R^n has the form

$$f(x) = c_0 + \sum_{i=1}^{n} c_i x_i + \sum_{i,j=1,1}^{n,n} q_{i,j} x_i x_j = c_0 + cx + x^t Qx, \qquad (7.4)$$

where Q is the $n \times n$ matrix with entries $q_{i,j}$, $c = [c_1, \ldots, c_n]$ and $x = (x_1, \ldots, x_n)$.

Notice that

$$x^t Qx = \sum_{i,j=1,1}^{n,n} q_{i,j} x_i x_j = \sum_{i,j=1,1}^{n,n} \tfrac{1}{2}(q_{i,j} + q_{j,i}) x_i x_j = x^t Sx,$$

where S has entries $s_{i,j} = \tfrac{1}{2}(q_{i,j} + q_{j,i})$; S is a symmetric matrix $s_{i,j} = s_{j,i}$; $S^t = S$.

Thus *we will suppose that Q is symmetric*; if we are given a Q that is not symmetric, we will replace it by a symmetric matrix as above.

The computation to compare $f(z)$ with $f(x)$ follows [because Q is symmetric, $(z - x)^t Qx = x^t Q(z - x)$].

$$f(z) = c_0 + c(x + [z - x]) + (x + [z - x])^t Q(x + [z - x])$$
$$= c_0 + cx + c(z - x) + x^t Qx + x^t Q(z - x) + (z - x)^t Qx$$
$$\quad + (z - x)^t Q(z - x)$$
$$= c_0 + cx + x^t Qx + [c + 2x^t Q](z - x) + (z - x)^t Q(z - x)$$
$$= f(x) + [c + 2x^t Q](z - x) + (z - x)^t Q(z - x).$$

The *gradient* $\nabla f(x)$ of f at x in R^n is defined by the formula

$$\nabla f(x) = c + 2x^t Q; \qquad (7.5)$$

thus we have the relationship

$$f(z) - f(x) = \nabla f(x)(z - x) + (z - x)^t Q(z - x). \qquad (7.6)$$

The function, defined for all z in R^n by the formula

$$T(z) = f(x) + \nabla f(x)(z - x),$$

gives us the tangent hyperplane $\{(z,T(z)) : z$ in $R^n\}$ to the graph of f at the point $(x,f(x))$ in R^{n+1}. The tangent hyperplane will not be explained further because it is not needed here.

Minimization Theorem

The following theorem contains four facts about minimization of f.

Theorem 7.1.

1. If f has a minimum at x, then $\nabla f(x) = 0$.
2. If there exists w in R^n with $w^t Q w < 0$, then f has no minimum and f is unbounded below.
3. Suppose that $\nabla f(x) = 0$. Then f has a minimum at x if and only if Q is *positive semidefinite*: $w^t Q w \geq 0$ for all w in R^n.
4. Suppose that $\nabla f(x) = 0$ and Q is *positive definite*: $w^t Q w > 0$ whenever $w \neq 0$; then f has a unique minimum at x.

Proof: Parts 1 and 2 will be verified; parts 3 and 4 are left for you to do. To verify part 1, put $v = [v_1, \ldots, v_n] = \nabla f(x)$ and suppose that $v_i \neq 0$. Let e_i denote the unit vector in R^n with ith component equal to 1 and all other components equal to zero. Look at the function g defined on R by putting $z = x + te_i$ in the formula for $f(z) - f(x)$. This reduction to a one-dimensional function puts us back in the setting of high school algebra, looking at a linear or quadratic function of the variable t. Put

$$g(t) = f(x + te_i) - f(x) = tv_i + t^2 q_{ii}.$$

Thus we see that f has no minimum at x because as we go from x in the direction e_i, f increases if $v_i > 0$ and decreases if $v_i < 0$; the slope of the tangent line is $v_i \neq 0$. To verify part 2, suppose that w is in R^n with $w^t Q w < 0$. Let x be in R^n; for t in R, put

$$g(t) = f(x + tw) - f(x) = t(\nabla f(x)w) + t^2(w^t Q w).$$

Again we have a quadratic function g on R to R. As in part 1, conclude from g that f does not have a minimum at x. Moreover, notice that $f(x + tw)$ goes to $-\infty$ as t gets large, positively or negatively.

Corollary. If f is bounded below, then Q is positive semidefinite.

Test for Positive Definiteness

Another useful fact from algebra is that one can check for positive definiteness by checking the principal minors of Q: It can be shown that Q is positive definite if and only if the determinants of the following matrices:

$$[q_{11}], \begin{bmatrix} q_{11} & q_{12} \\ q_{21} & q_{22} \end{bmatrix}, \ldots, Q,$$

are all positive (i.e., greater than zero). The corresponding test for positive semidefiniteness does not work; for instance,

$$Q = \begin{bmatrix} 0 & 0 & 0 \\ 0 & 1 & 0 \\ 0 & 0 & -1 \end{bmatrix}$$

has all nonnegative principal minors, but it is not positive semidefinite. We consider the case $n = 2$ with $q_{11} \neq 0$ below:

$$x^t Q x = [x_1, x_2] \begin{bmatrix} q_{11} & q_{12} \\ q_{12} & q_{22} \end{bmatrix} \begin{bmatrix} x_1 \\ x_2 \end{bmatrix}$$

$$= q_{11} x_1^2 + 2q_{12} x_1 x_2 + q_{22} x_2^2$$

$$= \frac{1}{q_{11}} \{ (q_{11} x_1 + q_{12} x_2)^2 + (q_{11} q_{22} - q_{12}^2) x_2^2 \}.$$

Convince yourself that Q is positive definite if, and only if, $q_{11} > 0$ and $|Q| = q_{11} q_{22} - q_{12}^2 > 0$. The determinant $|q_{11}|$ is defined to be equal to q_{11}.

Minimization Procedure

When Q is positive definite, $|Q| \neq 0$ and Q^{-1} exists; in this case f attains its minimum at $x = -\frac{1}{2} Q^{-1} c^t$ because the gradient of f is zero at this point. We can compute x by solving the linear system defined by $\nabla f(x) = 0$. When Q is positive semidefinite but not positive definite, Q^{-1} does not exist; however, we can still look for minimum points by trying to solve the linear system defined by $\nabla f(x) = 0$.

Examples for $n = 2$

Let f be defined on R^2 by

$$f(x) = (x_1 - 2)^2 + (x_1 - 2x_2)^2 - 4$$

$$= 2x_1^2 - 4x_1 x_2 + 4x_2^2 - 4x_1$$

$$= x^t \begin{bmatrix} 2 & -2 \\ -2 & 4 \end{bmatrix} x - [4, 0] x. \tag{7.7}$$

Computing $|q_{11}| = 2 > 0$ and $|Q| = 8 - 4 > 0$, we see that f is positive definite. To find the unique minimum point x by solving the vector equation $\nabla f(x) = c + 2x^tQ = 0$, compute

$$\nabla f(x) = [-4,0] + 2[x_1,x_2]\begin{bmatrix} 2 & -2 \\ -2 & 4 \end{bmatrix}$$

$$= [-4,0] + 2[2x_1 - 2x_2, -2x_1 + 4x_2]$$

$$= [4x_1 - 4x_2 - 4, -4x_1 + 8x_2]$$

$$= 4[x_1 - x_2 - 1, -x_1 + 2x_2]$$

and solve the linear system

$$x_1 - x_2 = 1$$
$$-x_1 + 2x_2 = 0.$$

The minimum value of -4 occurs at $x = (2,1)$.

$$y = f(x_1,x_2) = x_1^2 + x_2^2$$
$$= x^tQx;$$

$$Q = \begin{bmatrix} 1 & 0 \\ 0 & 1 \end{bmatrix}. \tag{7.8}$$

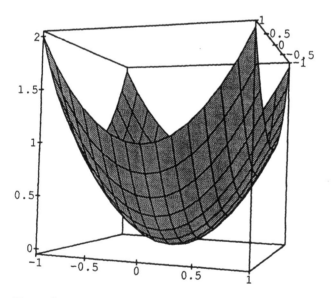

Figure 2

$$y = f(x_1, x_2) = x_1^2$$
$$= x^t Q x;$$
$$Q = \begin{bmatrix} 1 & 0 \\ 0 & 0 \end{bmatrix}. \tag{7.9}$$

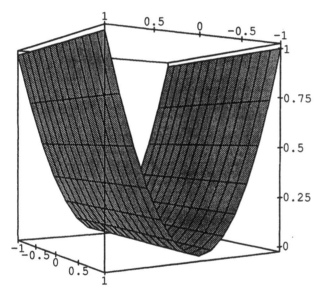

Figure 3

$$y = f(x_1, x_2) = x_1^2 - x_2^2$$
$$= x^t Q x;$$
$$Q = \begin{bmatrix} 1 & 0 \\ 0 & -1 \end{bmatrix}. \tag{7.10}$$

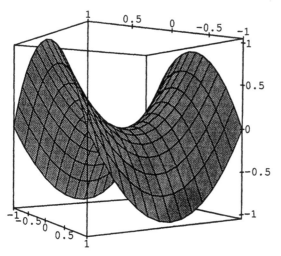

Figure 4

$$y = f(x_1, x_2) = x_1^2 + x_2$$
$$= cx + x^t Q x;$$

$$c = [0,1], \quad Q = \begin{bmatrix} 1 & 0 \\ 0 & 0 \end{bmatrix}. \tag{7.11}$$

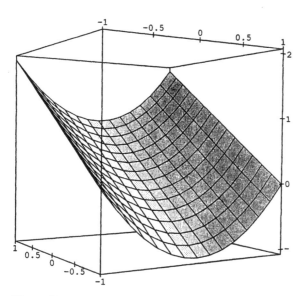

Figure 5

7.2 CONVEX QUADRATIC FUNCTIONS

Convex functions on R^n will be defined and two necessary and sufficient conditions for a quadratic function on R^n to be convex will be established.

Definition. A function f on R^n to R is *convex* if whenever x and z are in R^n and $0 \leq t \leq 1$,

$$f(tx + (1 - t)z) \leq tf(x) + (1 - t)f(z).$$

Convexity says that chords between points on the graph of f lie above the graph of f.

The following theorem gives two characterizations of convex quadratric functions. The first characterization says that a quadratic function f is convex if and only if Q is positive semidefinite. The second characterization says that f is convex if and only if all the tangent hyperplanes to the graph of f lie below the graph of f.

Theorem 7.2. Suppose that f is a quadratic function on R^n to R. Then the following three statements are equivalent:

1. f is convex on R^n.
2. Q is positive semidefinite.
3. For all x and z in R^n, $f(z) - f(x) \geq \nabla f(x)(z - x)$.

Proof: Put

$$A(x,z,u) = f(uz + [1 - u]x)$$
$$= f(x + u[z - x]) = f(x) + \nabla f(x)(u[z - x]) + u^2(z - x)'Q(z - x)$$
$$= f(x) + u\{\nabla f(x)(z - x) + u(z - x)'Q(z - x)\}$$

and

$$B(x,z,u) = uf(z) + [1 - u]f(x)$$
$$= f(x) + u\{f(z) - f(x)\}$$
$$= f(x) + u\{\nabla f(x)(z - x) + (z - x)'Q(z - x)\}.$$

Then f is convex $\Leftrightarrow$ for all x and z in R^n and $0 < u \leq 1$,

$$A(x,z,u) \leq B(x,z,u)$$

$\Leftrightarrow$ for all x and z in R^n and $0 < u \leq 1$,

$$u(z - x)'Q(z - x) \leq (z - x)'Q(z - x)$$

$\Leftrightarrow$ for all w in R^n, $w'Qw \geq 0$.

Thus 1 and 2 are equivalent; 2 and 3 are equivalent because

$$f(z) - f(x) = \nabla f(x)(z - x) + (z - x)'Q(z - x).$$

7.3 KUHN-TUCKER CONDITIONS FOR CONVEX QUADRATIC PROGRAMS

A *convex quadratic program* is a quadratic program with a convex objective function. A linear program is a special case because $Q = 0$ is positive semidefinite. The stage will be set to consider a quadratic objective by referring to the Kuhn–Tucker conditions for a symmetric primal LP in Theorem 6.3. The first condition, $Ax \geq b$, $x \geq 0$, can be rewritten $b - Ax \leq 0, -x \leq 0$ or, equivalently, $g(x) \leq 0$, where

$$g(x) = (g_1(x), \ldots, g_m(x), g_{m+1}(x), \ldots, g_{m+n}(x))$$
$$g_i(x) = b_i - a^i x, \qquad 1 \leq i \leq m$$
$$g_{m+i}(x) = -x_i, \qquad 1 \leq i \leq n;$$

g is a function from R^n to R^{m+n} and each g_j is a function from R^n to R, $1 \leq j \leq m + n$. Define the *gradient* ∇g of g to be the $(m + n) \times n$ matrix whose jth row is

$$\nabla g_j : \nabla g = \begin{bmatrix} -A \\ -I \end{bmatrix};$$

for all x and z in R^n, $g(z) - g(x) = \nabla g(z - x)$. Put $f(x) = cx$, and consider f to be a quadratic function with $Q = 0$. Then $\nabla f = c$. The second Kuhn–Tucker condition $c - yA - z = 0$, $y \geq 0$, $z \geq 0$ becomes

$$\nabla f + [y,z]\begin{bmatrix} -A \\ -I \end{bmatrix} = \nabla f + [y,z]\nabla g = 0 \qquad \text{or}$$
$$\nabla f + \lambda \nabla g = 0, \qquad \text{where } \lambda = [y,z] \geq 0.$$

The third Kuhn–Tucker condition becomes $\lambda g = 0$. Thus Theorem 6.3 can be restated as follows.

Kuhn–Tucker Conditions for a LP

Given $f(x) = cx$ and

$$g(x) = \begin{bmatrix} b - Ax \\ -x \end{bmatrix}.$$

$$\min f(x)$$
$$\text{s.t.} \quad g(x) \leq 0.$$

A feasible point x is optimal if and only if it satisfies the Kuhn–Tucker conditions:

$$\nabla f(x) + \lambda \nabla g(x) = 0 \quad \text{for some } \lambda \geq 0 \text{ for which } \lambda g(x) = 0.$$

It will be verified below that the corresponding Kuhn–Tucker conditions for a convex quadratic program are sufficient for a feasible point x to be optimal.

Theorem 7.3 (Kuhn–Tucker Conditions). Given a convex quadratic function $f(x) = c_0 + cx + x^t Q x$, $Q = Q^t$, and

$$g(x) = \begin{bmatrix} b - Ax \\ -x \end{bmatrix}.$$

Consider the convex quadratic program

$$\min \ f(x)$$
$$\text{s.t.} \ \ g(x) \leq 0.$$

A sufficient condition for a feasible point x to be optimal is that

$$\nabla f(x) + \lambda \nabla g(x) = 0 \quad \text{for some } \lambda \geq 0 \text{ for which } \lambda g(x) = 0.$$

Proof. Suppose that the conditions of the theorem are satisfied at a feasible point x. To show that x is optimal, suppose that z is a feasible point; Theorem 7.2 and the fact that f is convex will be used to show that $f(z) \geq f(x)$ as follows. If $g_i(x) < 0$, then $\lambda_i = 0$, so $\lambda_i \nabla g_i(x)(z - x) = 0$. On the other hand, if $g_i(x) = 0$, then $0 \geq g_i(z) = g_i(z) - g_i(x) = \nabla g_i(x)(z - x)$; so $\lambda_i \nabla g_i(x)(z - x) \leq 0$ because $\lambda_i \geq 0$. Hence $\lambda \nabla g(x)(z - x) \leq 0$. Because $[\nabla f(x) + \lambda \nabla g(x)] = 0$.

$$0 = [\nabla f(x) + \lambda \nabla g(x)](z - x)$$
$$= \nabla f(x)(z - x) + \lambda \nabla g(x)(z - x)$$

or, equivalently,

$$\nabla f(x)(z - x) = -\lambda \nabla g(x)(z - x) \geq 0.$$

Theorem 7.2 applies because f is convex, so

$$f(z) - f(x) \geq \nabla f(x)(z - x) \geq 0$$

and Theorem 7.3 is established.

7.4 LINEAR COMPLEMENTARITY FORMULATION OF KUHN–TUCKER CONDITIONS

Theorem 7.3 will be rewritten below. The condition $Ax \geq b$ can be rewritten $Ax - w = b$. Also, putting $\lambda = [y,z]$,

$$0 = \nabla f(x) + \lambda \nabla g(x)$$

$$= c + 2(Qx)^t + [y,z]\begin{bmatrix} -A \\ -I \end{bmatrix}$$

$$= c + 2(Qx)^t - yA - z,$$

or (transposing and rearranging)

$$2Qx - A^t y^t - z^t = -c^t.$$

Putting everything together, a solution $(x,y^t,z^t,w) \geq 0$ to the system

$$2Qx - A^t y^t - z^t = -c^t$$
$$Ax - w = b, \qquad \text{with}$$
$$zx = yw = 0$$

produces a solution x to the quadratic program

$$\min\ f(x) = c_0 + cx + x^t Qx$$
$$\text{s.t.}\quad Ax \geq b, \qquad x \geq 0,$$

where f is a convex function.

Definition. The conditions $zx = yw$ are called *complementarity conditions*. A *linear complementarity problem* is a linear system with complementarity conditions. A linear complementarity formulation of a quadratic program is a linear complementarity problem, whose solution produces a solution to the quadratic program.

Other forms for the linear constraints in a convex quadratic program lead to similar linear complementarity formulations, for instance to

$$\min\ f(x) = c_0 + cx + x^t Qx$$
$$\text{s.t.}\quad Ax \leq b, \qquad x \geq 0,$$

solve the system

$$2Qx + A^t y^t - z^t = -c^t$$
$$Ax + w = b,$$

with $zx = yw = 0, \qquad x,y,w,z \geq 0$.

This system can be rewritten in matrix format as follows:

$$\left[\begin{array}{c|c|c|c} -2Q & -A^t & I & 0 \\ \hline A & 0 & 0 & I \end{array}\right]\begin{bmatrix} x \\ y^t \\ z^t \\ w \end{bmatrix} = \begin{bmatrix} c^t \\ b \end{bmatrix},$$

$$zx = yw = 0, \qquad x,y,z,w \geq 0$$

or, modifying the form,

$$\begin{bmatrix} z^t \\ w \end{bmatrix} - \begin{bmatrix} 2Q & A^t \\ \hline -A & 0 \end{bmatrix} \begin{bmatrix} x \\ y^t \end{bmatrix} = \begin{bmatrix} c^t \\ b \end{bmatrix},$$

$$zx = yw = 0, \qquad x,y,z,w \geq 0.$$

Putting

$$u = \begin{bmatrix} z^t \\ w \end{bmatrix}, \quad M = \begin{bmatrix} 2Q & A^t \\ \hline -A & 0 \end{bmatrix}, \quad v = \begin{bmatrix} x \\ y^t \end{bmatrix}, \quad \text{and} \quad d = \begin{bmatrix} c^t \\ b \end{bmatrix},$$

we get the system in *linear complementarity problem format*:

$$u - Mv = d, \qquad u,v \geq 0, \qquad u^t v = 0.$$

In closing we observe that

$$v^t M v = [x^t, y] \begin{bmatrix} 2Q & A^t \\ \hline -A & 0 \end{bmatrix} \begin{bmatrix} x \\ y^t \end{bmatrix} = [x^t, y] \begin{bmatrix} 2Qx + A^t y^t \\ -Ax \end{bmatrix}$$

$$= 2x^t Q x + x^t A^t y^t - yAx$$

$$= 2x^t Q x;$$

hence M is positive semidefinite if and only if Q is positive semidefinite.

7.5 INVESTMENT APPLICATION

Suppose that you have a fixed number M of money units to invest for k time periods and there are n investment opportunities available. You believe that opportunity i returns r_{ij} units per unit invested at the end of period j, $j = 1,2, \ldots ,k$. You require an average total profit of at least $p\%$ on the amount M by the end of the kth time period; subject to that requirement, you wish to minimize the variance V of the returns at the ends of the periods.

To determine how many money units to invest in each opportunity, let x_i = number of money units to invest in opportunity i. Denote by T_j the total return at the end of period j,

$$T_j = \sum_{i=1}^{n} r_{ij} x_i.$$

Let T denote the total return,

$$T = \sum_{j=1}^{k} T_j;$$

the average total return per period is T/k. The variance V to be minimized is given by the formula

$$V = \sum_{j=1}^{k} \left(T_j - \frac{T}{k} \right)^2,$$

and the profit requirement takes the form

$$T \ge (1 + p)M.$$

According to the way the problem is worded, it is not required that all the money be invested. Thus the additional constraint $\sum_{i=1}^{n} x_i \le M$ must be included. Consequently, V may attain its minimum when $\sum_{i=1}^{n} x_i < M$; if all the returns r_{ij} were positive, we might wish to require that $\sum_{i=1}^{n} x_i = M$, even if doing so would cause the variance to increase.

Some computation is done below to put V in a nice form; after that, a numerical example is given. The total per-unit return R_i for opportunity i over the k periods is given by the formula

$$R_i = \sum_{j=1}^{k} r_{ij}.$$

Thus

$$T = \sum_{j=1}^{k} T_j = \sum_{j=1}^{k} \left(\sum_{i=1}^{n} r_{ij} x_i \right) = \sum_{i=1}^{n} \left(\sum_{j=1}^{k} r_{ij} \right) x_i = \sum_{i=1}^{n} R_i x_i.$$

Hence

$$V = \sum_{j=1}^{k} \left(T_j - \frac{T}{k} \right)^2$$

$$= \sum_{j=1}^{k} \left(\sum_{i=1}^{n} r_{ij} x_i - \frac{1}{k} \sum_{i=1}^{n} R_i x_i \right)^2$$

$$= \sum_{j=1}^{k} \left[\sum_{i=1}^{n} \left(r_{ij} - \frac{R_i}{k} \right) x_i \right]^2$$

$$= \sum_{j=1}^{k} \left[\sum_{u=1}^{n} \sum_{v=1}^{n} \left(r_{uj} - \frac{R_u}{k} \right) \left(r_{vj} - \frac{R_v}{k} \right) x_u x_v \right]$$

$$= \sum_{u=1}^{n} \sum_{v=1}^{n} \left[\sum_{j=1}^{k} \left(r_{uj} - \frac{R_u}{k} \right) \left(r_{vj} - \frac{R_v}{k} \right) \right] x_u x_v.$$

But

$$\sum_{j=1}^{k} \left(r_{uj} - \frac{R_u}{k} \right)\left(r_{vj} - \frac{R_v}{k} \right)$$

$$= \sum_{j=1}^{k} \left(r_{uj}r_{vj} - r_{uj}\frac{R_v}{k} - r_{vj}\frac{R_u}{k} + \frac{R_uR_v}{k^2} \right)$$

$$= \left(\sum_{j=1}^{k} r_{uj}r_{vj} \right) - \left(\sum_{j=1}^{k} r_{uj} \right)\frac{R_v}{k} - \left(\sum_{j=1}^{k} r_{vj} \right)\frac{R_u}{k} + k\frac{R_uR_v}{k^2}$$

$$= \left(\sum_{j=1}^{k} r_{uj}r_{vj} \right) - \frac{1}{k}R_uR_v.$$

Thus, putting

$$q_{uv} = \left(\sum_{j=1}^{k} r_{uj}r_{vj} \right) - \frac{1}{k}R_uR_v,$$

$$r = [r_{ij}],$$

$R = (R_1, \ldots, R_n)$, a n-dimensional column vector, and

$$Q = [q_{uv}] = rr^t - \frac{1}{k}RR^t,$$

we get

$$V = x^t Q x.$$

Example 7.2. Let

$$Q = \begin{bmatrix} 3 & 1 & -0.5 \\ 1 & 2 & -0.4 \\ -0.5 & -0.4 & 1 \end{bmatrix}, \quad R = (1.3, 1.2, 1.08), \quad \text{and } p = 0.12;$$

not all the money is required to be invested, and no more than 75% of the available funds can be invested any single opportunity. Normalize by putting $M = 1$; then the x_i's represent fractions. After multiplying the first constraint $1.3x_1 + 1.2x_2 + 1.08x_3 \geq 1.12$ by -1 to put it in $\leq$ form, the QP model is

$$\min \; x'Qx$$

$$\text{s.t.} \quad Ax \le b, \; x \ge 0$$

$$A = \begin{bmatrix} -1.3 & -1.2 & -1.08 \\ 1 & 1 & 1 \\ 1 & 0 & 0 \\ 0 & 1 & 0 \\ 0 & 0 & 1 \end{bmatrix} \quad \text{and} \quad b = \begin{bmatrix} -1.12 \\ 1 \\ 0.75 \\ 0.75 \\ 0.75 \end{bmatrix}.$$

Since $c_0 = 0$ and $c = 0$ in this example, the linear complementarity formulation

$$2Qx + A'y' - z' = -c'$$

$$Ax + w = b \quad (zx = wy = 0; \; x, y, w, z \ge 0)$$

can be written in inequality form

$$2Qx + A'y' \ge 0$$

$$Ax \le b.$$

EXERCISES

7.1 Verify parts 3 and 4 of Theorem 7.1.

7.2 Show that the function f defined for $x = (x_1, x_2)$ in R^2 by $f(x) = x_1^2$ has a line of minimum points.

7.3 Show that the function f defined for $x = (x_1, x_2)$ in R^2 by $f(x) = x_1^2 + x_2$ has no minimum.

7.4 Check the following functions for definiteness and find minimum points when they exist.
  (a) $f(x_1, x_2) = x_1^2 + 4x_1x_2 + 5x_2^2 - 2x_2$.
  (b) $f(x_1, x_2) = 2x_1^2 - 6x_1x_2 + 5x_2^2 + 4x_1 - 4x_2$.
  (c) $f(x_1, x_2) = 9x_1^2 - 6x_1x_2 + 2x_2^2 - 2x_2$.
  (d) $f(x_1, x_2) = 8x_1^2 - 4x_1x_2 + x_2^2 - 2x_2$.

7.5 Find w with $w'Qw < 0$ for the functions f defined on R^2 below.
  (a) $f(x) = x_1^2 - x_2^2$.
  (b) $f(x) = x_1x_2$.

7.6 Referring to Exercise 2.28, Traction needs to borrow the money to pay for the new presses. Three different sources have agreed to supply all or part of the money. However, they have different repayment schedules.

Source	Percent of principal to be repaid at end of year					
	1	2	3	4	5	6
1	40	40	20	15	15	15
2	5	15	25	35	40	45
3	0	0	30	40	50	55

Traction requires that no more than 40% of the total amount repaid be for interest. Subject to that constraint, Traction wishes to borrow so that the total amounts to be repaid each year are as nearly equal as possible. Formulate an appropriate QP.

7.7 An engineer who plans to retire in 2 years wishes to sell some mutual fund shares and put $60,000 in more stable investments. He considers three. The first pays nothing for 3 years and then returns 16% per year thereafter; the second returns 21% every second year; and the third returns 32% every third year. His goal is to receive equal amounts of money from these investments each year during his first 15 years of retirement. Formulate an appropriate QP.

7.8 Give an example of a nontrivial ($Q \neq 0$), negative semidefinite 2×2 matrix Q for which the determinants of both principal minors are nonnegative.

7.9 Show that the convex quadratic program (Q is positive semidefinite)

$$\min \ f(x) = c_0 + cx + x^t Q x$$
$$\text{s.t.} \quad Ax \leq b$$
$$Bx = d$$
$$x \geq 0$$

can be rewritten in the form

$$\min \ f(x)$$
$$\text{s.t.} \quad g(x) \leq 0$$
$$h(x) = 0$$
$$x \geq 0$$

Show that a sufficient condition for a feasible point x to be optimal is that $\nabla f(x) + \lambda \nabla g(x) + \mu \nabla h(x) = 0$ for some $\lambda \geq 0$ for which $\lambda g(x) = 0$ and some μ (μ may be positive, negative, or zero).

7.10 (a) If you have access to QP software, run Example 7.2 for the given numerical data. Then run the problem again with the requirement that all the money is to be invested and compare the answers.

(b) You can use 0–1 integer variables to implement the complementary slackness conditions in a small linear complementarity problem. Write the linear complementarity formulation of Example 7.2 as an LP with some 0–1 integer variables used to model the conditions $zx = yw = 0$ and solve it.

8
Minimizing a Quadratic Function

Quadratic functions were introduced and discussed in Chapter 7. In this chapter we discuss three methods for minimizing a quadratic function

$$f(x) = c_0 + cx + x'Qx,$$

where x is in R^n, $c_0 = f(0)$ is a number, c is in R_n, and Q is a positive definite symmetric matrix; fully written out, the function becomes

$$f(x) = c_0 + \sum_{j=1}^{n} c_j x_j + \sum_{i,j=1}^{n} q_{ij} x_i x_j.$$

The three methods are called Newton's method, steepest descent, and the conjugate gradient method; they are basic ingredients in many nonlinear programming algorithms. The problem of minimizing a quadratic function provides a learner-friendly environment in which to introduce them to you. Section 8.1 contains eigenvalue conditions for positive definiteness and positive semidefiniteness. Section 8.2 discusses Newton's method, which entails computing Q^{-1} or solving the linear system corresponding to the equation $\nabla f(x) = 0$; these two computations were presented in Section 7.1, where Newton's method was discussed briefly without naming it. In Section 8.3 we discuss steepest descent steps, which are very simple computationally; Section 8.3 also contains a formula that you can use to estimate the rate at which the values of f at the steps converge to the minimum value $f(x^*)$ of f. Steps in the conjugate gradient method are more complicated computationally than steepest descent steps; but you do not need to compute Q^{-1} or solve the linear system $\nabla f(x) = 0$, and this method gets to the minimum in no more than n steps. The conjugate gradient method combines ideas about conjugate directions and the gradient. Section 8.4 presents some important properties of conjugate directions. The conjugate gradient algorithm is developed in Section 8.5; its final form and an example appear in Section 8.6.

8.1 EIGENVALUE CONDITIONS FOR POSITIVE (SEMI)DEFINITENESS

Begin with the following useful fact from linear algebra. A symmetric $n \times n$ matrix Q has *eigenvalues* $\lambda_1 \leq \lambda_2 \leq \cdots \leq \lambda_n$ and corresponding *orthogonal unit eigenvectors* $v_1, v_2, \ldots, v_n$ satisfying

$$v_i^t v_j = 0 \qquad i \neq j \qquad \text{(orthogonal vectors)}$$
$$v_i^t v_i = 1 \qquad 1 \leq i \leq n \quad \text{(unit vectors)}$$
$$Q v_i = \lambda_i v_i, \qquad 1 \leq i \leq n \quad \text{(eigenvectors)}.$$

Put $V = [v_1, v_2, \ldots, v_n]$ and let D denote the diagonal matrix whose i,i entry is λ_i (off-diagonal entries are zero). Then (Exercise 8.1)

$$V^t = V^{-1} \qquad \text{and} \qquad QV = VD,$$

so

$$V^t Q V = D \qquad \text{and} \qquad Q = VDV^t.$$

The preceding equations imply (Exercise 8.2) that Q is positive definite if, and only if, $0 < \lambda_1$ and Q is positive semidefinite if, and only if, $0 \leq \lambda_1$. The case $n = 2$ with $q_{11} > 0$ is used below to illustrate.

If λ is an eigenvalue with unit eigenvector $v = (v_1, v_2)$, then $v_1^2 + v_2^2 = 1$ and $Qv = \lambda v$ or, equivalently, $(Q - \lambda I)v = 0$. Linear algebra tells us that λ is an eigenvalue when the determinant $|Q - \lambda I|$ of $Q - \lambda I$ equals 0. Accordingly, to compute the eigenvalues, consider the determinant

$$|Q - \lambda I| = \begin{vmatrix} q_{11} - \lambda & q_{12} \\ q_{12} & q_{22} - \lambda \end{vmatrix} = (q_{11} - \lambda)(q_{22} - \lambda) - (q_{12})^2$$
$$= (\lambda - \lambda_1)(\lambda - \lambda_2),$$

where

$$\lambda_1 = \tfrac{1}{2}[(q_{11} + q_{22}) - [(q_{11} - q_{22})^2 + 4(q_{12})^2]^{(1/2)}]$$

and

$$\lambda_2 = \tfrac{1}{2}[(q_{11} + q_{22}) + [(q_{11} - q_{22})^2 + 4(q_{12})^2]^{(1/2)}].$$

Thus $0 < \lambda_1$ if, and only if, $[(q_{11} - q_{22})^2 + 4(q_{12})^2]^{(1/2)} < (q_{11} + q_{22})$. Squaring both sides of the preceding inequality and simplifying, we find that if $0 < \lambda_1$, then $q_{12}^2 < q_{11}q_{22}$, which implies that q_{11} and q_{22} have the same sign; q_{11} and q_{22} must both be positive. Conversely, if q_{11} is positive and $q_{11}q_{22} > q_{12}^2$, we can reverse the preceding computation and get $\lambda_1 > 0$. For instance, in formula (7.8) because $f(x) = x_1^2 + x_2^2$ and

$$Q = \begin{bmatrix} 1 & 0 \\ 0 & 1 \end{bmatrix},$$

$\lambda_1 = \lambda_2 = 1$; similarly, in formula (7.10), $f(x) = x_1^2 - x_2^2$,

$$Q = \begin{bmatrix} 1 & 0 \\ 0 & -1 \end{bmatrix},$$

$\lambda_1 = -1$, and $\lambda_2 = 1$.

8.2 NEWTON'S METHOD

Q^{-1} exists because Q is positive definite; consequently, you can solve the linear system

$$c + 2(Qx)^t = 0.$$

Equivalently,

$$c^t + 2Qx = 0$$

or

$$Qx = -\tfrac{1}{2}c^t,$$

and find that the unique minimum of f occurs at

$$x^* = -\tfrac{1}{2}Q^{-1}c^t.$$

You can also compute x^* by computing Q^{-1} and plugging it into the preceding formula. This is Newton's method; it computes the minimum point in one big step. Before considering steepest descent and the conjugate gradient method, let us look at an example.

Example 8.1.

$$f(x) = (x_1 + 2x_2 - 5)^2 + 2(-2x_1 + x_2)^2$$
$$= 9x_1^2 - 4x_1x_2 + 6x_2^2 - 10x_1 - 20x_2 + 25$$
$$= 25 + [-10,-20]x + x^t \begin{bmatrix} 9 & -2 \\ -2 & 6 \end{bmatrix} x.$$

In this example, $c = [-10,-20]$ and

$$Q = \begin{bmatrix} 9 & -2 \\ -2 & 6 \end{bmatrix}.$$

Thus

$$Q^{-1} = \frac{1}{50} \begin{bmatrix} 6 & 2 \\ 2 & 9 \end{bmatrix}$$

and, according to Newton's method, the minimum occurs at

$$\begin{bmatrix} -\frac{1}{2} \end{bmatrix} \frac{1}{50} \begin{bmatrix} 6 & 2 \\ 2 & 9 \end{bmatrix} \begin{bmatrix} -10 \\ -20 \end{bmatrix} = \frac{1}{100} \begin{bmatrix} 100 \\ 200 \end{bmatrix} = \begin{bmatrix} 1 \\ 2 \end{bmatrix}.$$

To find the eigenvalues of

$$Q = \begin{bmatrix} 9 & -2 \\ -2 & 6 \end{bmatrix},$$

consider the determinant

$$|Q - \lambda I| = \begin{vmatrix} 9 - \lambda & -2 \\ -2 & 6 - \lambda \end{vmatrix}$$
$$= (9 - \lambda)(6 - \lambda) - 4$$
$$= 50 - 15\lambda + \lambda^2$$
$$= (\lambda - 5)(\lambda - 10).$$

For $\lambda = \lambda_1 = 5$,

$$(Q - 5I)v_1 = \begin{bmatrix} 4 & -2 \\ -2 & 1 \end{bmatrix} v_1,$$

which is zero if

$$v_1 = \frac{1}{\sqrt{5}} \begin{bmatrix} 1 \\ 2 \end{bmatrix}.$$

For $\lambda = \lambda_2 = 10$,

$$(Q - \lambda I)v_2 = \begin{bmatrix} -1 & -2 \\ -2 & -4 \end{bmatrix} v_2,$$

which is zero if

$$v_2 = \frac{1}{\sqrt{5}} \begin{bmatrix} -2 \\ 1 \end{bmatrix}.$$

8.3 STEEPEST DESCENT

Referring to formula (7.6), when $\nabla f(x) \neq 0$ and $z - x$ is small, the term $\nabla f(x)(z - x)$ (which is linear in $z - x$) generally has a much greater effect on $f(z) - f(x)$ than $(z - x)'Q(z - x)$ (which is a pure quadratic in $z - x$ without any linear terms in $z - x$). Thus $\nabla f(x)(z - x)$ is examined below.

To simplify the formulas, put

$$u = [u_1, u_2, \ldots, u_n] = \nabla f(x)$$

and

$$v = (v_1, \ldots, v_n) = z - x$$

to get

$$u_1 v_1 + \cdots + u_n v_n = \nabla f(x)(z - x).$$

The *length* of a vector u is denoted by $\|u\|$ and defined by the formula

$$\|u\| = [(u_1)^2 + (u_2)^2 + \cdots + (u_n)^2]^{1/2} = \|\nabla f(x)\|.$$

Accordingly,

$$\|v\| = [(v_1)^2 + \cdots + (v_n)^2]^{1/2} = \|z - x\|, \quad \text{the length of } z - x.$$

One more computational simplification will be made by using unit vectors (vectors of legnth 1) w and y parallel to u and v: put

$$w = \|u\|^{-1} u \quad \text{and} \quad y = \|v\|^{-1} v.$$

Now observe that

$$0 \leq \|w - y\|^2 = [(w_1 - y_1)^2 + \cdots + (w_n - y_n)^2]$$
$$= (w_1^2 - 2w_1 y_1 + y_1^2) + \cdots + (w_n^2 - 2w_n y_n + y_n^2)$$
$$= (w_1^2 + \cdots + w_n^2) + (y_1^2 + \cdots + y_n^2) - 2(w_1 y_1 + \cdots + w_n y_n)$$
$$= 1 + 1 - 2(w_1 y_1 + \cdots + w_n y_n)$$
$$= 2[1 - (w_1 y_1 + \cdots + w_n y_n)];$$

thus

$$w_1y_1 + \cdots + w_ny_n \leq 1.$$

If you replace $w - y$ by $w + y$ and repeat the preceding computation, you will find that $w_1y_1 + \cdots + w_ny_n \geq -1$. Taken together, these last two inequalities are equivalent to the inequality

$$|w_1y_1 + \cdots + w_ny_n| \leq 1.$$

Recalling that $w_i = \|u\|^{-1}u_i$ and $y_i = \|v\|^{-1}v_i$, establishes the formula

$$|u_1v_1 + \cdots + u_nv_n| \leq \|u\| \cdot \|v\|.$$

If $v = z - x$ is chosen to be in the direction of $u = \nabla f(x)$: $v = tu$ for some positive number t, then

$$u_1v_1 + \cdots + u_nv_n = t(u_1^2 + \cdots + u_n^2)$$
$$= (u_1^2 + \cdots + u_n^2)^{1/2}[(tu_1)^2 + \cdots + (tu_n)^2]^{1/2}$$
$$= \|u\| \cdot \|v\|.$$

The latter inequality tells us that near x (when $z - x$ is a small vector), f increases most rapidly when $z - x$ is a vector in the direction of $\nabla f(x)$. A similar computation shows that if $u = t(-\nabla f(x))$, then $u_1v_1 + \cdots + u_nv_n = -\|u\| \cdot \|v\|$, so f decreases most rapidly when $z - x$ is a positive multiple of the negative gradient $-\nabla f(x)$ of f at x. Consequently, $-\nabla f(x)$ is called the *direction of steepest descent* of f at x.

Steepest descent for a function defined on R^2 can be visualized because the graph of the function corresponds to a surface in R^3. The figure accompanying formula (7.8) is an illustration. In (7.8) the paraboloid in R^3 is defined by the equation

$$x_3 = f(x_1,x_2) = x_1^2 + x_2^2.$$

This parabolic surface in R^3 is the graph of f; f is defined on R^2. Suppose that you start from a point $p = (p_1,p_2,p_3) > 0$ on the surface and ski down to $0 = (0,0,0)$ in such a way that at each point in time you are skiing in the direction of steepest descent. Since $-\nabla f(x) = -2[x_1,x_2]$ always points to $(0,0)$, your descent path will be part of a parabola lying in a vertical plane containing the point p and the x_3-axis. The points (x_1,x_2,x_3) on the descent path project on points (x_1,x_2) on the line inteval in R^2 from (p_1,p_2) to $(0,0)$. That is not surprising. But suppose that f is changed to $f(x_1,x_2) = x_1^2 + 2x_2^2$. Now the graph of f is an elliptic paraboloid in R^3. Using differential equations, it can be shown that the descent path on this paraboloid projects on a curve in R^2. This curve is on the parabola in R^2 defined by the equation $x_2 = (p_2/p_1^2)x_1^2$ (see Fig. 1). If f is changed to $f(x_1,x_2) =$

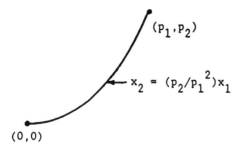

$(0,0)$

Figure 1

$x_1^2 + 3x_2^2$, the points on the projection of the descent path are defined by the cubic $x_2 = (p_2/p_1^3)x_1^3$.

The method of steepest descent does not try to follow projections of descent paths. Given a starting reference point p in the domain of the function f, a steepest descent step minimizes the function on the half-line emanating from p in the direction of the negative gradient $-\nabla f(x)$ of f at x to obtain a new reference point from which to repeat the process. Restricting $z - x$ to be a positive multiple of the negative gradient results in a one-dimensional function $g(t)$ to be minimized for $t \geq 0$: put

$$z(t) = x + t(-\nabla f(x)) \qquad \text{and} \qquad g(t) = f(z(t)).$$

To visualize this process, suppose that you scoop out half a watermelon and brace the shell open side up on a table. Choose a reference point somewhere on the side of the convex bowl that forms its inside surface and estimate the direction of steepest descent. Slice the shell vertically in that direction and look at the edge of the bigger piece. You see a convex curve. Estimate the minimum point on that curve. This point becomes your new reference point; estimate the direction of steepest descent, slice the shell, estimate the minimum point on the new curve, and obtain a new reference point at which to repeat the process. An algorithm for the descent step follows.

Steepest Descent Algorithm

Choose a point x^0 at which to begin.

(1)   Compute $g_0' = -\nabla f(x^0)$. If $g_0 = 0$, stop; otherwise,
(2)   minimize f on the half-line emanating from x^0 in the direction g_0 to find a new point x^1 at which to apply (1).

Example 8.1 Continued.   To illustrate the method of steepest descent, it is applied to Example 8.1 with $x^0 = 0$ below.

Put $x^0 = 0$. Then $\nabla f(0) = [-10,-20] = -10[1,2]$. Drop the factor of 10 to simplify the computation. Put

$$h(u) = f\left(u\begin{bmatrix}1\\2\end{bmatrix}\right)$$

$$= f(u[1,2]^t)$$

$$= 9u^2 - 4(u)(2u) + 6(2u)^2 - 10(u) - 20(2u) + 25$$

$$= 25u^2 - 50u + 25$$

$$= 25(u^2 - 2u + 1)$$

$$= 25(u - 1)^2, \quad u \geq 0.$$

By inspection, we see that h is minimized when $u = 1$, so $x^1 = u\begin{bmatrix}1\\2\end{bmatrix} = \begin{bmatrix}1\\2\end{bmatrix}$. Put $x^1 = \begin{bmatrix}1\\2\end{bmatrix}$. Compute $\nabla f(x^1)$; $\nabla f\begin{bmatrix}1\\2\end{bmatrix} = 0$, so we stop with the knowledge that the unique minimum occurs at $x = \begin{bmatrix}1\\2\end{bmatrix}$ because we already know that Q is positive definite.

Steepest descent finds the minimum for Example 8.1 in one step. In Example 8.2 below, steepest descent does not terminate in a finite number of steps. Do a steepest descent step for Example 8.1 starting at $(1,1)$.

New Form for the Function f

Requiring that $c_0 = 0$ does not affect the location of the minimum and writing f in the form

$$f(x) = \tfrac{1}{2}x^t Qx - b^t x \tag{8.1}$$

makes

$$[\nabla f(x)]^t = Qx - b, \tag{8.2}$$

so the unique minimum occurs at x when $Qx = b$ and Q is positive definite. *This form will be used throughout the rest of this chapter.*

Example 8.2. Consider (7.7):

$$f(x) = 2x_1^2 - 4x_1 x_2 + 4x_2^2 - 4x_1$$

$$= (x_1 - 2)^2 + (x_1 - 2x_2)^2 - 4$$

$$= \tfrac{1}{2}x^t \begin{bmatrix} 4 & -4 \\ -4 & 8 \end{bmatrix} x - [4,0]x;$$

$$\nabla f(x) = [4x_1 - 4x_2 - 4, -4x_1 + 8x_2] = 4[x_1 - x_2 - 1, -x_1 + 2x_2].$$

To apply Newton's method, solve the linear system

$$4x_1 - 4x_2 - 4 = 0$$
$$-4x_1 + 8x_2 \quad\quad = 0$$

and find $x = \begin{bmatrix} 2 \\ 1 \end{bmatrix} = (2,1)$; notice that $|4| = 4$ and $|Q| = 16$, so Q is positive definite and $(2,1)$ is the unique point at which f attains its minimum value. The method of steepest descent follows.
Put $x^0 = 0$. Then $g_0 = -[\nabla f(x^0)]^t = b = (4,0)$. Put

$$h(u) = f(x^0 + ug_0) = f\left(\begin{bmatrix} 4u \\ 0 \end{bmatrix} \right) = 32u^2 - 16u.$$

The *slope of a quadratic function* $h(u) = \alpha + \beta u + \gamma u^2$ is given by the formula $\nabla h(u) = \beta + 2\gamma u$. Here the slope $\nabla h(u) = 64u - 16 = 0$ when $u = \frac{1}{4}$, so $u = \frac{1}{4}$ and

$$x^1 = \begin{bmatrix} 1 \\ 0 \end{bmatrix} = (1,0).$$

Then $\nabla f(x^1) = [0,-4] = -(g_1)^t$.
Put

$$h(u) = f(x^1 + ug_1) = f\left(\begin{bmatrix} 1 \\ 4u \end{bmatrix} \right) = -2 - 16u + 64u^2.$$

Then the slope $\nabla h(u) = 128u - 16 = 0$ when $u = \frac{1}{8}$, so $x^2 = x^1 + \frac{1}{8}(0,4) = (1,\frac{1}{2})$. You can do the next two iterations for homework (Exercise 8.4).

Alternative Computation

There is another way to minimize $f(x) = \frac{1}{2}x^tQx - b^tx$ along the half-line emanating from x in the direction $-\nabla f(x)$ which is instructive; it is presented below. Let $g(x)$ denote $-(\nabla f(x))^t = -Qx + b$. Put $h(u) = f(x + ug)$, $u \geq 0$. Then the slope $\nabla h(u)$ of $h(u)$ is given by the formula

$$\nabla h(u) = [\nabla f(x + ug)]g$$
$$= [-(Q(x + ug))^t + b^t]g$$
$$= [-(Qx)^t - ug^tQ + b^t]g$$
$$= [\{-(Qx)^t + b^t\} - ug^tQ]g$$
$$= [g^t - ug^tQ]g$$
$$= g^tg - ug^tQg,$$

so $\nabla h(u) = 0$ when $ug^tQg = g^tg$, or

$$u = \frac{g^t g}{g^t Q g}.$$

Consequently, the descent points can be computed recursively as follows.

Recursion Formula for Steepest Descent Points

Choose x^0. For $k \geq 0$ put

$$g_k = -(\nabla f(x^k))^t$$

and put

$$x^{k+1} = x^k + \frac{g_k^t g_k}{g_k^t Q g_k} g_k.$$

(For $g \neq 0$, $g^t Q g > 0$ because Q is positive definite.) Application of this method to Example 8.2 follows.

Put $x^0 = 0$. Then

$$g_0 = (4,0)$$

$$g_0^t g_0 = 16$$

$$g_0^t Q g_0 = [4,0] \begin{bmatrix} 4 & -4 \\ -4 & 8 \end{bmatrix} - \begin{bmatrix} 4 \\ 0 \end{bmatrix} = [4,0](16,-16) = 64$$

$$x^1 = x^0 + \tfrac{16}{64} g_0 = (1,0)$$

$$g_1 = (0,4)$$

$$g_1^t g_1 = 16$$

$$g_1^t Q g_1 = [0,4] \begin{bmatrix} 4 & -4 \\ -4 & 8 \end{bmatrix} (0,4) = [0,4] \begin{pmatrix} 16 \\ -32 \end{pmatrix} = 128$$

$$x^2 = (1,0) + \tfrac{16}{128}(0,4) = (1,\tfrac{1}{2}).$$

Two methods to minimize $f(x) = \frac{1}{2} x^t Q x - b^t x$ have been discussed. Newton's method finds the point x^* at which f attains its minimum in one step, and steepest descent generates a sequence $x_1, x_2, \ldots$ of approximations to x^* at which the function has progressively lower values. An estimate of the rate at which the values $f(x_k)$ of f at x_k converge to the minimum value $f(x^*)$ follows.

Estimate for the Rate of Convergence of Steepest Descent Steps

For notational simplicity, put $m = \lambda_1$ and $M = \lambda_n$. Observe that

$$f(x) = \tfrac{1}{2} x^t Q x - b^t x$$
$$= \tfrac{1}{2} x^t Q x - x^t Q x^*,$$

where the minimum of f occurs at x^* with $Q x^* = b$.

Applying this formula to $f(x^*)$ and subtracting, we have

$$f(x) - f(x^*) = \tfrac{1}{2}x'Qx - x'Qx^* - \tfrac{1}{2}x^{*'}Qx^* + x^{*'}Qx^*$$
$$= \tfrac{1}{2}x'Qx - x'Qx^* + \tfrac{1}{2}x^{*'}Qx^*$$
$$= \tfrac{1}{2}(x - x^*)'Q(x - x^*).$$

Put

$$E(x) = \tfrac{1}{2}(x - x^*)'Q(x - x^*) = f(x) - f(x^*);$$

$E(x)$ is positive unless $x = x^*$. After a nontrivial discussion which is omitted, we come to the following formula:

$$E(x_{k+1}) \le \frac{(M - m)^2}{(M + m)^2} E(x_k),$$

which is valid for $k = 1, 2, \ldots$.

8.4 CONJUGATE DIRECTIONS

To motivate the concept of conjugate directions, recall that a set $\{d_0, \ldots, d_k\}$ of nonzero vectors in R^n is said to be orthogonal if $d_i'd_j = 0$, $0 \le i \ne j \le k$. By analogy, the set is said to be *Q-orthogonal* or *Q-conjugate* or simply *conjugate* (with a symmetric, positive definite matrix Q being associated implicitly) if $d_i'Qd_j = 0$, $0 \le i \ne j \le k$.

In this section we assume that Q-orthogonal sets are available to use; in the next section gradients are used to generate them. Many useful properties of orthogonal sets carry over to Q-orthogonal sets. For instance, Proposition 8.1 below shows that a Q-orthogonal set of vectors is a linearly independent set. Consequently, a Q-orthogonal set of n vectors in R^n is a basis for R^n. After Proposition 8.1 is established, you will learn how to write the optimal solution x^* as a linear combination of vectors in a Q-conjugate basis. Proposition 8.2 verifies that the partial sums in this representation are the best approximations to x^* in terms of the corresponding Q-conjugate vectors.

Proposition 8.1. Given a symmetric, positive definite, $n \times n$ matrix Q, suppose that $\{d_i\}_{i=0}^k$ is a Q-orthogonal set in R^n. Then $\{d_i\}_{i=0}^k$ is a linearly independent set in R^n.

Proof: Suppose that $x = \sum_{i=0}^k x_i d_i = 0$. Then

$$0 = x^t Q x = \left(\sum_{i=0}^{k} x_i d_i \right)^t Q \left(\sum_{j=0}^{k} x_j d_j \right)$$

$$= \sum_{j=0}^{k} \sum_{i=0}^{k} x_i x_j d_i^t Q d_j = \sum_{i=0}^{k} x_i^2 d_i^t Q d_i;$$

hence $x_i = 0$, $0 \le i \le k$, which implies linear independence.

Formula for x* in Terms of a Given Q-orthogonal Basis

Suppose that $\{d_0, d_1, \ldots, d_{n-1}\}$ is a Q-orthogonal set in R^n. Then this set is a basis for R^n according to Proposition 8.1. Recall that $f(x) = \frac{1}{2} x^t Q x - b^t x$ attains its minimum at x^* where $Qx^* = b$. Because $\{d_0, \ldots, d_{n-1}\}$ is a basis for R^n, there are numbers $\alpha_0, \alpha_1, \ldots, \alpha_{n-1}$ such that $x^* = \sum_{i=0}^{n-1} \alpha_i d_i$. The coefficients $\{\alpha_i\}_{i=0}^{n-1}$ of the basis vectors $\{d_i\}_{i=0}^{n-1}$ specify x^*, so computing the α_i's will specify x^*, which minimizes $f(x) = \frac{1}{2} x^t Q x - bx$. This can be done as follows. We know that

$$Qx^* = Q \left(\sum_{j=0}^{n-1} \alpha_j d_j \right) = \sum_{j=0}^{n-1} \alpha_j Q d_j = b.$$

so

$$d_i^t b = d_i^t Q x^* = d_i^t \left(\sum_{j=0}^{n-1} \alpha_j Q d_j \right) = \sum_{j=0}^{n-1} \alpha_j d_i^t Q d_j = \alpha_i d_i^t Q d_i.$$

Consequently,

$$\alpha_i = \frac{d_i^t b}{d_i^t Q d_i}, \qquad 0 \le i < n$$

and

$$x^* = \sum_{i=0}^{n-1} \frac{d_i^t b}{d_i^t Q d_i} d_i.$$

Partial Sums of the Formula for x*

The partial sums x^k of x^* are defined for $1 \le k \le n$ by the formula

$$x^k = \sum_{i=0}^{k-1} \frac{d_i^t b}{d_i^t Q d_i} d_i.$$

A sequence of subspaces $\{V_k\}_{k=1}^{n}$ associated with the Q-orthogonal basis $\{d_k\}_{k=0}^{n-1}$ is defined below; the vector x^k is in V_k for $1 \le k \le n$, and Proposition 8.2 shows that x^k is the unique point in V_k at which f attains its minimum value.

Definition. Denote by $V_k = L\{d_0,d_2, \ldots ,d_{k-1}\}$ the linear subspace of R^n spanned by $\{d_i\}_{i=0}^{k-1}$.

Proposition 8.2. f attains its minimum value on V_k at x^k.

Proof: Consider the function h defined on R^n by

$$h(u) = f\left(\sum_{i=0}^{n-1} u_i d_i\right), \quad \text{where } u = (u_0, u_1, \ldots ,u_{n-1}),$$

$$= \tfrac{1}{2}\left(\sum_{i=0}^{n-1} u_i d_i\right)^t Q \left(\sum_{i=0}^{n-1} u_i d_i\right) - b^t \left(\sum_{i=0}^{n-1} u_i d_i\right)$$

$$= \tfrac{1}{2} \sum_{i=0}^{n-1} u_i^2 d_i^t Q d_i - \sum_{i=0}^{n-1} u_i b^t d_i.$$

The slope $\nabla_i h$ of h considered as a quadratic function in u_i is given by the formula

$$\nabla_i h = (d_i^t Q d_i) u_i - b^t d_i.$$

The minimum value of h on the subspace

$$S_k = \{u \text{ in } R^n;\, u = (u_0, u_1, \ldots , u_{k-1}, 0, \ldots , 0)\}$$

$$= \{u = (u_0, \ldots , u_{n-1}) \text{ in } R^n;\, u_i = 0 \text{ for } i \ge k\}$$

occurs when $\nabla_i h = 0$ for $0 \le i \le k$, that is, when

$$u_i = \frac{b^t d_i}{d_i^t Q d_i} = \frac{d_i^t b}{d_i^t Q d_i}, \quad 0 \le i < k.$$

Hence the minimum value of f on V_k occurs at x^k.

Comparison Between $f(x^k)$ and $f(x)$, where x Is in V_k

The value of f at an arbitrary point $x = \sum_{i=0}^{k-1} u_i d_i$ in V_k can be compared with $f(x^k)$ as follows:

$$f\left(\sum_{i=0}^{k-1} u_i d_i\right) = \frac{1}{2} \sum_{i=0}^{k-1} u_i^2 d_i^t Q d_i - \sum_{i=0}^{k-1} u_i b^t d_i$$

$$= \frac{1}{2} \sum_{i=0}^{k-1} [\alpha_i + (u_i - \alpha_i)]^2 d_i^t Q d_i - \sum_{i=0}^{k-1} [\alpha_i + (u_i - \alpha_i)] b^t d_i$$

$$= \frac{1}{2} \sum_{i=0}^{k-1} [\alpha_i^2 + 2(u_i - \alpha_i)\alpha_i + (u_i - \alpha_i)^2] d_i^t Q d_i$$

$$- \sum_{i=0}^{k-1} \alpha_i b^t d_i - \sum_{i=0}^{k-1} (u_i - \alpha_i) b^t d_i$$

$$= f(x^k) + \frac{1}{2} \sum_{i=0}^{k-1} (u_i - \alpha_i)^2 d_i^t Q d_i$$

because $\alpha_i d_i^t Q d_i = b^t d_i, 0 \le i < k$.

Generating Conjugate Directions in Two Dimensions

When $n = 2$, pairs of conjugate directions can be found by inspection. Example 8.2 will be used again to illustrate; for that Q, a pair x,y of vectors in R^2 is Q-conjugate if

$$x^t Q y = [x_1, x_2] \begin{bmatrix} 4 & -4 \\ -4 & 8 \end{bmatrix} \begin{bmatrix} y_1 \\ y_2 \end{bmatrix} = 4[x_1, x_2] \begin{bmatrix} y_1 - y_2 \\ -y_1 + 2y_2 \end{bmatrix}$$

$$= 4[x_1(y_1 - y_2) + x_2(-y_1 + 2y_2)] = 0.$$

For instance, we can put $d_0 = (1,0) = x$ and $d_1 = (1,1) = y$. Then

$$d_0^t Q d_0 = [1,0](4,-4) = 4$$
$$d_0^t b = [1,0](4,0) = 4$$
$$d_1^t Q d_1 = [1,1](0,4) = 4$$
$$d_1^t b = [1,1](4,0) = 4.$$

Thus $\alpha_0 = 1$, $\alpha_1 = 1$, $x^0 = d_0 = (1,0)$, and $x^1 = d_0 + d_1 = (2,1)$. In the next section we demonstrate how to generate conjugate directions.

8.5 CONJUGATE GRADIENT METHOD

One would like to generate conjugate directions so that $f(x^1) > f(x^2) > \cdots > f(x^k) = f(x^*)$; Exercise 8.6 illustrates that this need not occur unless we exercise some care. We can ensure that improvement occurs at each step by using the gradient $\nabla f(x)$ to generate conjugate directions — hence the name "conjugate gradient method."

When Q is the identity, the Gram–Schmidt orthogonalization process generates orthogonal vectors from linearly independent vectors. The Q-orthogonal analog of that process is used to generate Q-orthogonal vectors from gradients below. After the basic method is presented, it is simplified significantly.

Basic Method

Starting at $x^0 = 0$, compute $g_0 = -[\nabla f(x^0)]^t$. If $g_0 = 0$, then the minimum occurs at 0; stop. Otherwise, put $d_0 = g_0$. Then f attains its minimum on $L\{d_0\}$ at $x^1 = \alpha_0 d_0$ with $\alpha_0 \neq 0$ and $f(x^1) < f(0)$.

Compute $g_1 = -[\nabla f(x^1)]^t$. If $g_1 = 0$, then the minimum of f on R^n occurs at x^1; stop. Otherwise, because x^1 is the point at which f attains its minimum on $L\{d_0\}$, g_1 is perpendicular to $L\{d_0\}$ and hence linearly independent to $\{d_0\}$.
Put

$$d_1 = g_1 - \frac{d_0^t Q g_1}{d_0^t Q d_0} d_0.$$

Then $\{d_0, d_1\}$ is a Q-orthogonal set, $L\{d_0, d_1\} = L\{g_0, g_1\}$, and the minimum of f on $L\{d_0, d_1\}$ occurs at $x^2 = \alpha_0 d_0 + \alpha_1 d_1$ with $\alpha_1 \neq 0$ and $f(x^2) < f(x^1)$.

Compute $g_2 = \nabla f(x^2)$. If $g_2 = 0$, then the minimum of f on R^n occurs at x^2; stop. Otherwise, because x^2 is the point at which f attains its minimum on $L\{d_0, d_1\}$, g_2 is perpendicular to $L\{d_0, d_1\}$ and hence linearly independent to $\{d_0, d_1\}$.
Put

$$d_2 = g_2 - \frac{d_0^t Q g_2}{d_0^t Q d_0} d_0 - \frac{d_1^t Q g_2}{d_1^t Q d_1} d_1.$$

Then $\{d_0, d_1, d_2\}$ is a Q-orthogonal set, $L\{d_0, d_1, d_2\} = L\{g_0, g_1, g_2\}$, and the minimum of f on $L\{d_0, d_1, d_2\}$ occurs at $x^3 = \alpha_0 d_0 + \alpha_1 d_1 + \alpha_2 d_2$ with $\alpha_2 \neq 0$ and $f(x^3) < f(x^2)$. Iterate this process; after k steps,

$$x^k = \sum_{i=0}^{k-1} \alpha_i d_i,$$

where

$$\alpha_i = \frac{d_i^t b}{d_i^t Q d_i}.$$

Compute $g_k = -[\nabla f(x^k)]^t$. If $g_k = 0$, then the global minimum of f occurs at x^k; stop. Otherwise, because x^k is the point at which f attains its minimum on $L\{d_0, \ldots, d_{k-1}\}$, g_k is perpendicular to $L\{d_0, \ldots, d_{k-1}\}$.
Put

$$d_k = g_k - \sum_{i=0}^{k-1} \frac{d_i^t Q g_k}{d_i^t Q d_i} d_i.$$

Then $\{d_0, \ldots, d_k\}$ is a Q-orthogonal set, $L\{d_0, \ldots, d_k\} = L\{g_0, \ldots, g_k\}$ and the minimum of f on $L\{d_0, \ldots d_k\}$ occurs at $x^{k+1} = \sum_{i=0}^{k} \alpha_i d_i = x^k + \alpha_k d_k$ with $a_k \neq 0$ and $f(x^{k+1}) < f(x^k)$. Continue until $g_k = 0$, which occurs after at most n steps.

Simplification of the Formula for d_k

Mathematical induction will be used to simplify the formula for d_k significantly. To begin, suppose that $g_0 \neq 0$. Then $d_0 = g_0 = b \neq 0$, $x^1 = \alpha_0 d_0$, $\alpha_0 \neq 0$, $g_1 = -[\nabla f(x^1)]^t = -\alpha_0 Q d_0 + b$, which implies that

$$Qd_0 = \frac{-1}{\alpha_0} g_1 + \frac{1}{\alpha_0} b$$

is in $L\{d_0, g_1\} = L\{d_0, d_1\}$:

(S₁) Qd_0 is in $L\{d_0, d_1\}$.

The corresponding jth statement follows:

(Sⱼ) Qd_{j-1} is in $L\{d_0, \ldots, d_j\}$.

The truth of (S₁) has been verified above. Suppose that (Sᵢ) is true for $i < j$; showing below that this supposition implies that (Sⱼ) is true verifies, according to the principle of induction, that (Sⱼ) is true for all (relevant) positive integers.

$$g_j = -[\nabla f(x_j)]^t = -Q(x^j) + b = -\sum_{i=0}^{j-1} \alpha_i Q d_i + b, \qquad \alpha_{j-1} \neq 0,$$

so applying the inductive supposition to $Q d_i$, $i < j - 1$,

$$Qd_{j-1} = \frac{-1}{\alpha_{j-1}} g_j - \sum_{i=0}^{j-2} \frac{\alpha_i}{\alpha_{j-1}} Q d_i + \frac{1}{\alpha_{j-1}} b$$

is in $L\{d_0, \ldots, d_{j-1}, g_j\} = L\{d_0, \ldots, d_{j-1}, d_j\}$. Consequently, (Sⱼ) is true whenever $g_j \neq 0$. Looking at the formula for d_k, when $i < k - 1$, Qd_i is in $L\{d_0, \ldots, d_{i+1}) \subset L\{d_0, \ldots, d_{k-1}\}$ and g_k is orthogonal to this latter subspace. Hence $(Qd_i)^t g_k = d_i^t Q g_k = 0$ for $i < k - 1$ and the formula for d_k becomes simply

$$d_k = g_k - \beta_{k-1} d_{k-1},$$

where

$$\beta_{k-1} = \frac{d_{k-1}^t Q g_k}{d_{k-1}^t Q d_{k-1}}.$$

Alternative Formula for α_k

The following remarks lead to an alternative formula for α_k. First, since

$$x^{k+1} = x^k + \alpha_k d_k$$

$$\begin{aligned}
g_{k+1} &= -Q(x^k + \alpha_k d_k) + b \\
&= -Qx_k - Q(\alpha_k d_k) + b \\
&= (-Qx^k + b) - \alpha_k Q d_k \\
&= g_k - \alpha_k Q d_k,
\end{aligned}$$

thus

$$0 = d_k^t g_{k+1} = d_k^t g_k - \alpha_k d_k^t Q d_k,$$

which implies that

$$\alpha_k = \frac{d_k^t g_k}{d_k^t Q d_k}.$$

Moreover, since

$$d_k = g_k - \beta_{k-1} d_{k-1}$$

$$g_k^t d_k = g_k^t g_k - \beta_{k-1} g_k^t d_{k-1} = g_k^t g_k,$$

thus

$$\alpha_k = \frac{g_g^t g_k}{d_k^t Q d_k}.$$

Alternative Formula for β_{k-1}

Now the formula for β_{k-1} can be rewritten in a form that does not involve Q explicitly. Algorithms to minimize nonquadratic functions use various combinations and variations of the methods that have been discussed. Some of them use the form of β_{k-1} which is presented below. Since

$$g_k = g_{k-1} - \alpha_{k-1} Q d_{k-1}$$

and

$$g_k^t g_{k-1} = 0,$$

$$g_k^t g_k = g_k^t g_{k-1} - \alpha_{k-1} g_k^t Q d_{k-1} = -\alpha_{k-1} g_k^t Q d_{k-1};$$

thus

$$\beta_{k-1} = \frac{g_k^t Q d_{k-1}}{d_{k-1}^t Q d_{k-1}} = \frac{-g_k^t g_k}{\alpha_{k-1} d_{k-1}^t Q d_{k-1}} = \frac{-g_k^t g_k}{g_{k-1}^t g_{k-1}}.$$

In Section 8.6 we present an algorithm summarizing the simplified and modified method and apply the algorithm to Example 8.2.

8.6 CONJUGATE GRADIENT ALGORITHM

Put $x^0 = 0$. Compute $g_0 = -[\nabla f(x^0)]$. If $g_0 = 0$, stop; otherwise, put $d_0 = g_0$,

$$\alpha_0 = \frac{g_0^t g_0}{d_0^t Q d_0}$$

$$x^1 = x^0 + \alpha_0 d_0$$

$$g_1 = -[\nabla f(x^1)]^t.$$

If $g_1 = 0$, stop; otherwise, put

$$\beta_0 = \frac{-g_1^t g_1}{g_0^t g_0}$$

$$d_1 = g_1 - \beta_0 d_0$$

$$\alpha_1 = \frac{g_1^t g_1}{d_1^t Q d_1}$$

$$x^2 = x^1 + \alpha_1 d_1.$$

Iterate; after k steps, compute $g_k = -[\nabla f(x^k)]^t$. If $g_k = 0$, stop; otherwise, put

$$\beta_{k-1} = \frac{-g_k^t g_k}{g_{k-1}^t g_{k-1}}$$

$$d_k = g_k - \beta_{k-1} d_{k-1}$$

$$\alpha_k = \frac{g_k^t g_k}{d_k^t Q d_k}$$

$$x^{k+1} = x^k + \alpha_k d_k.$$

Continue until $g_l = 0$.

Going back to Example 8.2 again to illustrate the algorithm, we have

$$d_0 = g_0 = -[\nabla f(0)]^t = (4,0)$$

$$g_0^t g_0 = 16$$

$$d_0^t Q d_0 = [4,0](16,16) = 64$$

$$\alpha_0 = \tfrac{1}{4}$$

$$x^1 = \alpha_0 d_0 = (1,0)$$

$$g_1 = -Qx_1 + b = (-4,4) + (4,0) = (0,4)$$

$$g_1^t g_1 = 16$$

$$d_1 = g_1 + \frac{g_1^t g_1}{g_0^t g_0} d_0 = (0,4) + (4,0) = (4,4)$$

$$d_1^t Q d_1 = 64$$

$$\alpha_1 = \tfrac{1}{4}$$

$$x^2 = (1,0) + \tfrac{1}{4}(4,4) = (2,1).$$

EXERCISES

8.1 Verify the stated properties of V and D, defined in Section 8.1.

8.2 (a) Verify that Q is positive semidefinite if and only if $\lambda_1 \geq 0$.
(b) Verify that Q is positive definite if and only if $\lambda_1 > 0$.

8.3 Find Q, eigenvalues λ_1 and λ_2, and unit eigenvectors for the following quadratic functions defined on R^2.
(a) $f(x) = x_1^2$.
(b) $f(x) = x_1^2 + x_2$.
(c) $f(x) = x_1^2 + x_2^2$.
(d) $f(x) = x_1^2 - x_2^2$.
(e) $f(x) = x_1 x_2$.

8.4 (a) Do a steepest descent step for Example 8.1 with $x^0 = (1,1)$.
(b) Do two more steps of steepest descent for Example 8.2.

8.5 Find the eigenvalues and corresponding unit eigenvectors for

$$Q = \begin{bmatrix} 4 & -4 \\ -4 & 8 \end{bmatrix}$$

(cf. Example 8.2).

8.6 Show that $\{d_0 = (0,1), d_2 = (2,1)\}$ is a Q-orthogonal basis for Example 8.2; compute x^* as a linear combination of this pair of conjugate directions.

8.7 Do two steps of steepest descent and the conjugate gradient algorithm for the following functions.
(a) $f(x_1,x_2) = x_1^2 + 4x_1 x_2 + 5x_2^2 - 2x_2$.
(b) $f(x_1,x_2) = 2x_1^2 - 6x_1 x_2 + 5x_2^2 + 4x_1 - 4x_2$.
(c) $f(x_1,x_2) = 9x_1^2 - 6x_1 x_2 + 2x_2^2 - 2x_2$.
(d) $f(x_1,x_2) = 8x_1^2 - 4x_1 x_2 + x_2^2 - 2x_2$.

8.8 Given a Q-conjugate basis $\{d_i\}_{i=0}^{n-1}$ for R^n. Show that the minimum value of

$$f(x) = \tfrac{1}{2}x'Qx - b'x \text{ on } L\{d_0, \ldots, d_{k-1}\} \text{ is } \frac{-1}{2} \sum_{i=0}^{k-1} \frac{(d_i'b)^2}{d_i'Qd_i}.$$

8.9 Show that $L(d_0, Qd_0, \ldots, Q^k d_0) = L\{g_0, g_1, \ldots, g_k\} = L\{d_0, \ldots, d_k\}$.

8.10 Suppose that $\{v_1, \ldots, v_n\}$ is an orthogonal basis of unit vectors for R^n. Show that $\sum_{i=1}^{n} v_i v_i'$ is the $n \times n$ identity matrix.

8.11 Suppose that Q is a symmetric, positive definite $n \times n$ matrix and that $\{d_0, \ldots, d_{n-1}\}$ is a Q orthogonal set in R^n. Show that

$$Q^{-1} = \sum_{i=0}^{n-1} \frac{d_i d_i'}{d_i'Qd_i}.$$

8.12 Let Q be a symmetric, positive definite, $n \times n$ matrix with eigenvalues $0 < \lambda_1 \le \lambda_2 \le \cdots \le \lambda_n$ and corresponding Q-orthogonal unit eigenvectors $v_1, \ldots, v_n$. Let $V = [v_1, \ldots, v_n]$ and let u be the $n \times n$ diagonal matrix with $u_{ii} = \sqrt{\lambda_i}$. Put $W = V'UV$. Show that W is a symmetric, positive definite, $n \times n$ matrix with $W^2 = Q$.

9

Network Algorithms

In Section 5.7 we introduced the basic notation and terminology for networks, showed you connections between LP formulations and network formulations of some important problems, and presented illustrative examples. In this chapter we focus on some of the basic ideas used in network algorithms; [T1] is a good reference. Topological ordering of a digraph G is defined in Section 9.1. Network models of transportation problems, and digraph models that can be solved using the method of backward dynamic programming discussed in Section 5.8, can be topologically ordered. Project scheduling, introduced in Section 9.2, can also be modeled by topologically ordered digraphs. The longest path algorithm presented in Section 9.3 to solve project scheduling models introduces the process of *scanning* (or examining) a node relative to its neighboring nodes. Scanning is an important concept in algorithm design. The longest path algorithm scans nodes once in topological order. The shortest path algorithm and the maximum simple path flow algorithm presented in Sections 9.4 and 9.6, respectively, are also based on scanning nodes once, however, in these algorithms the order in which nodes are scanned is determined as the algorithms proceed. Section 9.5 contains a primitive minimum spanning tree algorithm. In Section 9.7 we introduce the residual digraph, which is central to the maximum flow and minimum cost–maximum flow algorithms presented in Sections 9.8 and 9.10.

246

9.1 NOTATION

A digraph G is said to be *topologically ordered* or (equivalently) *lexicographically ordered* if its nodes are labeled so that whenever $[i,j]$ is a route in G, $i < j$.
The digraph model of Example 5.8 that we discussed in Section 5.7 is topologically ordered. Example 5.12 in Section 5.8 provides another convenient illustration; there routes correspond to flights and the ten cities are topologically ordered. Backward dynamic programming is simply a scanning operation that can be applied to the nodes in reverse order. Example 5.11 showed that being able to solve a problem by scanning nodes only once can have significant computational advantages. Thus it is beneficial to know when a digraph can be topologically ordered. Sections 9.2 and 9.3 present another example. Digraph representations for all of these examples contain cycles, but they do not contain order cycles: they are NOcycle digraphs. An order cycle cannot be topologically ordered; consequently, a digraph containing an order cycle cannot be topologically ordered. On the other hand, Proposition 9.1 shows that a digraph can be topologically ordered if it contains no order cycles:

Proposition 9.1.   A NOcycle digraph can be topologically ordered.

Proof: A NOcycle digraph contains at least one node which is not an initial node of a route and at least one node which is not a terminal node of a route: To find a node which is not an initial node, choose a node, say $n(1)$; if $n(1)$ is not an initial node, you are done; otherwise, choose a route $[n(1),n(2)]$; if $n(2)$ is not an initial node you are done; otherwise, choose a route $[n(2),n(3)]$; this process terminates at a node which is not an initial node because there are no order cycles and a finite number of nodes. Let G be a NOcycle digraph with k nodes. Apply the following *Step* to G: Choose a node, say $n(1)$, which is not an initial node of a route in G. Use k as the label for node $n(1)$. Remove $n(1)$ and all the routes which contain $n(1)$ to obtain a new NOcycle digraph $G(1)$ with $k - 1$ nodes. Repeat *Step* for $G(1)$ to obtain a node $n(2)$ labeled $k - 1$ and a NOcycle digraph $G(2)$ with $k - 2$ nodes; $k - 1$ repetitions of *Step* determines a topological ordering on G. With a little thought, you can find more efficient ways to topologically order G.

9.2 PROJECT PLANNING

In this section we describe the problem; in Section 9.3 we describe an algorithm to solve it and apply the algorithm to an example.

A model G for a project planning problem is a NOcycle, topologically ordered digraph. Example 9.3 includes numerical data and a topological ordering for the nodes. Exercise 9.1 asks you to sketch the directed graph for Example 9.1; you should do that as you read this discussion. The nodes in G are n jobs that comprise a project, plus a starting node 1 and a terminal node $k = n + 2$. There are three types of routes in G; these types are described below: (1) Each job has a list, perhaps empty, of predecessor jobs that must be completed before the job can be started. If job i is a predecessor to job j, then there is a route $[i,j]$ in G with $len[i,j] =$ time needed to do job j. (2) If job j has no predecessors, there is a route $[1,j]$ in G with $len[1,j] =$ time needed to do job j. (3) If job j has no successors, there is a route $[j,k]$ in G with $len[j,k] = 0$. Thus a feasible project generates a NOcycle, connected digraph G, which is labeled so that it is topologically ordered. The problem of finding the minimum time necessary to complete the project is solved by finding the length of a longest path from node 1 to node k in G. The longest path algorithm given in the next section solves the problem; indeed, for each job i, it computes the minimum time $l(i)$ after the start of the project at which job i could be completed.

9.3 LONGEST PATH ALGORITHM

To begin, let a NOcycle, connected digraph G be topologically ordered. Two functions, l and p, will be defined on the nodes. The values of l and p will be modified as the algorithm is applied. The final values, $l(i)$ and $p(i)$, of l and p at node i will denote the longest path length from node 1 to node i and the predecessor node to node i in a longest path, respectively. Intermediate values represent estimates based on information assimilated up to that point in the algorithm. After initializing the values of l and p, the nodes will be scanned in topological order: $1,2,\ldots,k-1$. Because the scanning is done in topological order, longest paths to nodes $1,\ldots,i+1$ are known after node i has been scanned. After node $k-1$ is scanned, longest paths to all nodes are known.

Initialization of l and p

Put $l(i) = 0$, $i \geq 1$: at this point we know nothing about the longest path to any node other than node 1, but all longest paths have length at least equal to zero.

Put $p(i) = $ *(empty), $i \geq 1$: initially no node has a predecessor.

The algorithm consists of applying the following scanning step to nodes $1,2,\ldots,k-1$ in sequence. Denote the node to be scanned by u; begin by initializing u to 1.

Initialization of Node to Be Scanned

Put $u = 1$.

Scanning Step

For each route $[u,j]$,
  if $l(j) < l(u) + \text{len}[u,j]$, then
  put $l(j) = l(u) + \text{len}[u,j]$ and *put* $p(j) = u$.

Algorithm

Apply the scanning step.
If $u = k - 1$, then stop;
otherwise, put $u = u + 1$ and apply the scanning step.

Example 9.1. The example begins with numerical data, followed by several scanning steps and a tabulation of values of the functions l and p after each of the scanning steps.

Job list	Top. ord.	Days required	Predecessors	Routes	Length
			Type 1 routes		
A	2	3			
B	3	4	2	[2,3]	4
C	4	2	3	[3,4]	2
D	5	3	3	[3,5]	3
E	6	5	3	[3,6]	5
F	7	3	4,6	[4,7]	3
				[6,7]	3
G	8	4	5	[5,8]	4
H	9	6	8	[8,9]	6
I	10	7	7	[7,10]	7
J	11	5	9,10	[9,11]	5
				[10,11]	5
K	12	2	4,6	[4,12]	2
				[6,12]	2
			Type 2 routes		
				[1,2]	3
			Type 3 routes		
				[11,13]	0
				[12,13]	0

Presumably, you have sketched the NOcycle digraph for this example. The partial digraphs "seen" at a step are drawn for the first four steps shown below.

Step 1

Step 2

Step 3

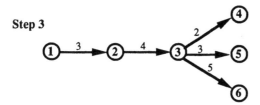

Step 4

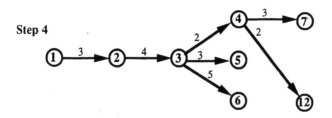

You can do the rest of the steps. The values of the functions l and p after each step appear in the following tableau.

Application of Algorithm to Example 9.1

Node	Initial value	1	2	3	4	5	6	7	8	9	10	11	12
				Node scanned (only changed values of l,p are recorded)									
1	0,*												
2	0,*	3,1											
3	0,*		7,2										
4	0,*			9,3									
5	0,*			10,3									
6	0,*			12,3									
7	0,*				12,4		15,6						
8	0,*					14,5							
9	0,*								20,8				
10	0,*							22,7					
11	0,*									25,9	27,10		
12	0,*				11,4		14,6						
13	0,*											27,11	

The longest path, [1,2,3,6,7,10,11,13], has length 27 days; this path is obtained by tracing backward from node 13: node 13 → node 11 → node 10 . . . , by looking at $p(13) = 11$, $p(11) = 10$, ...

9.4 SHORTEST PATH ALGORITHM

This algorithm applies to both graphs and digraphs. It is similar to the longest path algorithm, but it differs in that the whole scanning order is not known when the algorithm begins: the node to be scanned at step $n + 1$ is determined during step n. Node 1 is scanned first: to find a shortest link–route beginning at node 1; this link–route is a shortest path to the other end, say u, of the link–route. After step 1, shortest paths to nodes 1 and u are known; at step 2, node u is scanned to find another node to which a shortest path is known; step 3 involves scanning that node to obtain one more node to which a shortest path is known. Continuing step by step, shortest paths are determined one node at a step. The algorithm is described below and applied to Example 9.2.

Begin by defining two functions, d and p, on the nodes of G. The values of d and p will be modified as the algorithm is applied. The final values, $d(i)$ and $p(i)$, of d and p at node i will denote the minimal path distance from node 1 to node i and the predecessor of i along the minimal path selected by the algorithm, respectively. After the values of d and p are initialized, node 1 will be scanned. Scanning node 1 will determine another node u to which a minimal path is known; u is added to the list S of nodes to which a minimal path is known. Initially, $S = \{1\}$ because we know that the length of a shortest path to node 1 is zero. Intermediate values of $d(i)$ and $p(i)$ are estimates based on shortest paths to nodes in S followed by a route–link from S to node i. When node u is scanned, shortest paths to node u are included in the process of estimation; updates occur at a node i not in S if a shortest path to u followed by the link (u,i) is shorter than the current estimate, $f(i)$, for the length of a shortest path to node i. When S contains all the nodes, the process is done. Since u denotes the node to be scanned and scanning begins at node 1, u is initialized to 1.

Initialization

Put $u = 1$.
Put $S = \{1\}$.
Put $d(1) = 0$ and $d(i) = M$, $i > 1$, where M is a big (enough) number.
Put $p(i) = *$(empty), $i \ge 1$: initially, no node has a predecessor.

Scanning Step (for graphs; for digraphs, replace link by route)

Put the links (u,v) with v not in S in a list and go through the list once, applying the following "if" statement to each link in the list:
    if $d(u) + \text{len}(u,v) < d(v)$, then
        put $d(v) = d(u) + \text{len}(u,v)$ and
        put $p(v) = u$.
Put $u = \arg\min\{d(v); v$ is not in $S\}$.
Put $S = S + \{u\}$; u is added to the end of the list.

Algorithm

Apply the scanning step. (Node 1 is scanned first.)
If S contains all the nodes, *then* stop;
otherwise, apply the scanning step.

Example 9.2.

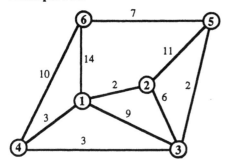

A graphical representation of which paths are determined at the end of each of the first three steps follows.

Step 1

There is no way to get from node 1 to node 2 by a path of lenth less than 2 because the first link in any path from node 1 to node 2 has length at least equal to 2. After step 1, $S = \{1,2\}$, $d(2) = 2$ and $p(2) = 1$.

Step 2

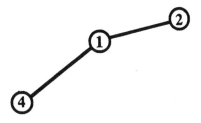

There is no way to get to node 4 by a path of length less than 3 because any path to a node that is not node 4 and not in S has length at least 8. After step 2, $S = \{1,2,4\}$, $d(4) = 3$, and $p(4) = 1$.

Step 3

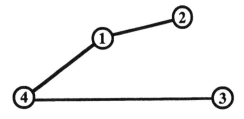

Explain why the path composed of link (1,4) followed by link (4,3) is the unique path of minimal length to node 3.

The following table lists the application of the algorithm to Example 9.2. The table lists the current values $d(i)$, $p(i)$ of d and p for unscanned nodes: for nodes i not in S.

Application of Algorithm to Example 9.2

Node	Initial values of d,p	Node scanned: current values of d,p shown				
		1	2	4	3	5
1	0,*					
2	M,*	2,1				
3	M,*	9,1	8,2	6,4		
4	M,*	3,1	3,1			
5	M,*	M,*	13,2	13,2	8,3	
6	M,*	14,1	14,1	13,4	13,4	13,4

After scanning node

1, $S = \{1,2\}$
2, $S = \{1,2,4\}$
4, $S = \{1,2,4,3\}$
3, $S = \{1,2,4,3,5\}$
5, $S = \{1,2,4,3,5,6\}$

Solutions to the shortest path problem can be written neatly in tabular form. A tabular solution for Example 9.2 follows: in the table, [1,4,3] denotes the route [1,4] followed by the route [4,3].

Shortest Path Solution

Node	Distance to node	Shortest path to node
2	2	[1,2]
4	3	[1,4]
3	6	[1,4,3]
5	8	[1,4,3,5]
6	13	[1,4,6]

Think of Example 9.2 as a network of gravel roads. Now you know the (road) distance between node 1 and any other node. The *minimum spanning tree problem* requires you to pave the minimum possible total distance in a way that will permit travel between any two nodes on pavement. The algorithm that will be used for this situation is easy to describe: Suppose you are required to pave one link per week in a beautiful National Park with great fishing and hiking, both of which you enjoy. You simply plan your paving one week at a time so as to pave as little as possible that week. An algorithm follows.

9.5 MINIMUM SPANNING TREE ALGORITHM

The strategy is to begin somewhere and pave short links first. You can begin at any node. Start at node 1 and pave as little as you can to reach another node. Then pave as little as you can to reach a new node from the pavement. Repeat this procedure until all nodes are accessible by pavement. The algorithm is presented and then applied to Example 9.2, but it is not proved that the algorithm works. Chapter 6 in [T1] is devoted to this problem; three references to proofs of this algorithm are given on page 79 of [T1]. In the algorithm, S denotes the set of nodes currently accessible from node 1 by pavement and T denotes the set of nodes not currently accessible by pavement.

Initialization

Put $S = \{1\}$.
Put $T = \{2,3, \ldots ,k\}$.

Iteration

Put $d = \min\{\text{len}(u,v); u$ is in S, v is in $T\}$.
Put $y = \arg\min\{v; u$ is in S, v is in T, $\text{len}(u,v) = d\}$.
Put $x = \arg\min\{u; u$ is in S, $\text{len}(u,y) = d\}$.
Pave (x,y).
Put $S = S + \{y\}$: add y to S.
Put $T = T - \{y\}$: delete y from T.

Algorithm

If $S = \{1,2, \ldots ,k\}$, then stop
    (all nodes are accessible by pavement);
otherwise, do an iteration.

Tabular Solution for Example 9.2

Iteration	d	y	x	Path paved
1	2	2	1	(1,2)
2	3	4	1	(1,4)
3	3	3	4	(4,3)
4	2	5	3	(3,5)
5	7	6	5	(5,6)

The *maximum* (*simple*) *path flow algorithm* and the *residual digraph* are useful concepts which are presented in the next two sections and used subsequently.

9.6 MAXIMUM (SIMPLE) PATH FLOW ALGORITHM

This algorithm applies to graphs and digraphs. It will be described and then applied to Example 9.3. Suppose that you have a connected graph−digraph G with specified maximum flow capacities, denoted by $\text{cap}(i,j)-\text{cap}[i,j]$, on the links−routes and you wish to find a (simple) path from node 1 to node k on which you can put a flow that is maximal among all such flows.

Here again use is made of two *labeling* functions, f and p, whose values $f(i)$ and $p(i)$ denote the (temporary) maximum flow to node i and the (temporary) predecessor of node i along a (temporary) maximum simple path flow to node i, respectively. Think of $f(i)$ and $p(i)$ as labels attached to node i.

Initialization

Put $f(1) = M$, where M is a big number.
Put $f(i) = 0$, $i > 1$.
Put $p(i) = *$(empty), $i \geq 1$.
Put $S = \{1\}$ = the set of nodes to which a maximum path flow is known.
Put $u = 1$: scan node 1 first.

Explanation (for digraphs; for graphs, replace routes by links).

Scanning node 1 amounts to looking at the capacities of routes $[1,j]$, increasing $f(j)$ to $\text{cap}[1,j]$ and choosing the smallest j for which the capacity of the route is maximal among these routes. Add node j to S because no simple path flow to node j can exceed the capacity of its first route, which is $\leq f(j)$; and put $u = j$ because u is the generic name that has been chosen for the node to be scanned at each step. When node u is scanned, the routes $[u,j]$ with j not in S are examined: the maximum flow to node j on a simple path whose last route is $[u,j]$ is $\min\{f(u), \text{cap}[u,j]\}$; the latter estimate is compared with the current estimate $f(j)$ for the maximum path flow to node j. For each route $[u,j]$ with j not in S, $f(j)$ is increased to $\min\{f(u), \text{cap}[u,j]\}$ where appropriate and both S and u are updated. S is updated by including the first node j with $f(j)$ maximal among the j's not in S; because $f(j)$ is maximal, a maximum flow path to node j is determined at this step in the algorithm. The included node is labeled u and scanned at the next step.

Scanning Step (for digraphs)

Put the routes $[u,j]$, j not in S, in a list and go through the list once, applying the following "if" statement to each route in the list:

if $f(j) < \min\{f(u),\text{cap}[u,j]\}$, then
   put $f(j) = \min\{f(u),\text{cap}[u,j]\}$ and
   put $p(j) = u$.
Put $u = \arg\max\{f(j); j$ not in $S\}$: update u.
Put $S = S + [u]$: update S.

Algorithm

Apply the scanning step.
If $f(k) = f(u)$, then stop
   (we have a maximum simple path flow from node 1 to node k);
otherwise, apply the scanning step.

Example 9.3.

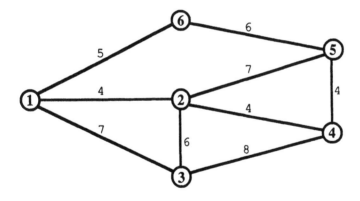

Tabulation of Solution

Node	Initial value of f,p	Node scanned (updates of f,p recorded) 1	3	4	2	5
1	M,*					
2	0,*	4,1	6,3			
3	0,*	7,1				
4	0,*		7,3			
5	0,*			4,4	6,2	
6	0,*	5,1				6,5

Thus the path [1,3,2,5,6] is a path from node 1 to node 6 on which the flow can attain the maximum value of 6: $f(6) = 6$, $p(6) = 5$, $p(5) = 2$, $p(2) = 3$, $p(3) = 1$.

9.7 RESIDUAL DIGRAPH (of a flow on a digraph)

After defining a flow on a digraph G and defining the residual digraph with respect to a flow, Example 5.9 is used to illustrate these concepts.

A *flow* F is a nonnegative function defined on the routes in G. Given a flow $F = \{F[i,j]; [i,j]$ a route in $G\}$; the *residual digraph* $R = R\{G,F\}$ of G and F has the same nodes as G. The routes in R are defined as follows: $[i,j]$ is a route in R if

(1) $[i,j]$ is a route in G and $F[i,j] < \text{cap}[i,j]$,
     then $R \text{ cap}[i,j] = \text{cap}[i,j] - F[i,j]$,

or

(2) $[j,i]$ is a route in G with $1 < i,j < k$ and $F[j,i] > 0$,
     then $R \text{ cap}[i,j] = F[j,i]$.

Thus R is a digraph and R has integral capacities whenever G has integral capacities and F is an integral flow. If F is the zero flow, then $R = G$.

Example 9.4 Referring to Example 5.10, each link in the original graph can be considered as two routes, one in each direction, between the nodes of that link. The residual digraph for the penultimate flow in Example 5.10 is displayed below, followed by a tabulation of application of the maximum path flow algorithm to the residual digraph to produce an augmenting path flow.

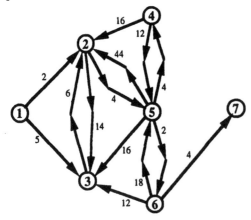

Tabulation of Maximum Path Flow Algorithm

Node	Initial value of f,p	Node scanned (updates of f,p recorded)					
		1	3	2	5	4	6
1	$M,*$						
2	$0,*$	2,1	5,3				
3	$0,*$	5,1					
4	$0,*$				4,5		
5	$0,*$			4,2			
6	$0,*$				2,5		
7	$0,*$						2,6

Thus we get the augmenting path flow

This augmentation is slightly different from the ad hoc augmentation which was found by observation in Example 5.10.

9.8 MAXIMUM FLOW ALGORITHM (for integral flows)

Given a directed graph G with (nonnegative) integral flow capacities; start with the zero flow and use the maximum path flow algorithm on the residual digraph to augment the flow or find that there is no augmenting path flow, in which case the present flow is a maximum flow.

Initialization

Put $F = 0$.

Augment Flow

Compute the residual digraph $R = R(G,F)$.
Apply the maximum path flow algorithm to R to get an augmenting path flow A.

Algorithm

Apply augment flow.
If $A = 0$, then stop (F is a maximum flow);
otherwise, put $F = F + A$ (augment the flow by A), and apply augment flow.

Example 9.5.   The maximum flow algorithm is applied to the following digraph.

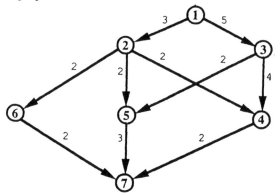

Because the initial flow is zero, the residual digraph R is the digraph G itself. Thus the first step in the algorithm consists of applying the maximum path flow algorithm to the digraph G. The following table displays the updates of f,p as the nodes are scanned by the maximum flow algorithm.

Node	Node scanned			
	1	3	4	2
2	3,1			
3	5,1			
4		4,3		
5		2,3		
6				2,2
7			2,4	

The maximum path flow algorithm generates an augmenting path flow of 2 units of the path [1,3,4,7]. Adding this path flow to the initial flow of zero produces an updated flow F. The residual digraph R for F follows.

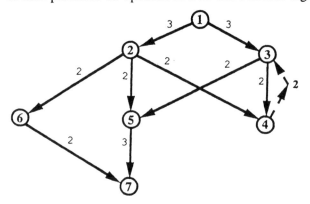

The next table lists updates as we apply the maximum path flow algorithm to the residual digraph R.

		Node scanned			
Node	1	2	3	4	5
2	3,1				
3	3,1				
4		2,2			
5		2,2			
6		2,2			
7					2,5

According to this table, the maximum path flow algorithm generates an augmenting path flow of 2 units on the path [1,2,5,7], which is added to the flow to produce the updated flow F displayed below.

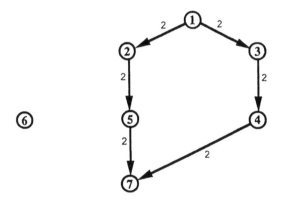

The residual digraph for this flow is given below, followed by a table of updates as nodes are scanned on the maximum path flow algorithm.

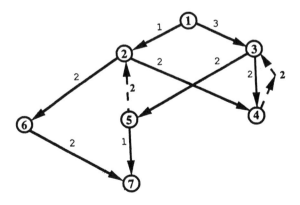

| | Node scanned | | | | | |
Node	1	3	4	5	2	6
2	1,1			2,5		
3	3,1					
4		2,3				
5		2,3				
6					2,2	
7				1,5		2,6

Thus the maximum path flow algorithm generates a maximal augmenting path flow of two units on the path [1,3,5,2,6,7], which is added to the current flow to produce the flow.

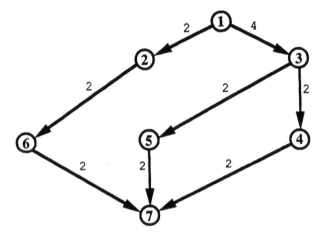

The residual digraph R for this flow appears below.

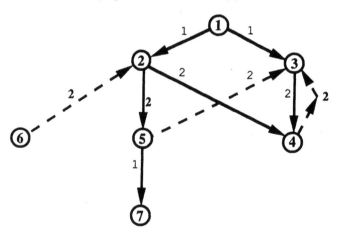

Applying the maximum path flow algorithm to the residual digraph R produces the following table of updates.

Node	Node scanned				
	1	2	3	4	5
2	1,1				
3	1,1				
4		1,2			
5		1,2			
6					
7					1,5

This table shows that an augmenting path flow of 1 unit is generated on the path [1,2,5,7]. Adding this augmenting path flow produces the flow

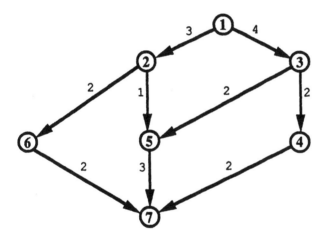

with residual R equal to

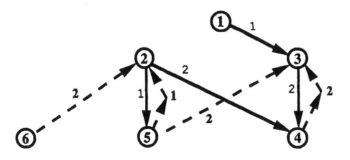

This residual shows that the current flow is a maximum flow; applying the maximum path flow algorithm to the residual will verify this fact. Notice that the corresponding minimum cut given by the maximum flow–minimum cut theorem is composed of the sets {1,3,4} and {2,5,6,7}; there are several minimum cuts for this problem.

9.9 MINIMUM COST–MAXIMUM FLOWS: TRANSPORTATION AND ASSIGNMENT PROBLEMS

In this section the transportation problem is formulated as a minimum cost–maximum flow problem; the *assignment problem* is the special case where every supply and every demand is equal to 1. For examples, a simple assignment problem is solved and an assignment problem with additional scheduling constraints is modeled. These problems can be solved with the algorithm presented in Section 9.10.

Consider the transportation problem with m supply nodes, labeled $2, \ldots, m + 1$, n demand nodes, labeled $m + 1 + 1, \ldots, m + n + 1$, an initial node, labeled 1, and a terminal node, labeled $m + n + 2 = k$. These nodes will be the nodes of our modeling digraph G; G will have three types of routes:

1. Routes $[1,u]$ from the origin to supply node u with capacity $cap[1,u] = Su$, the supply at supply node u, $1 < u < m + 2$, and cost $c[1,u]$ equal to zero.
2. Routes $[v,t]$ from demand node v to the terminal node $t = m + n + 2$ with capacity $cap[v,t] = Dv$, the demand at node v, $m + 1 < v < m + n + 2$, and cost $c[v,t] = 0$.
3. Routes $[u,v]$ with (sufficient) capacity $cap[u,v] = M$ (putting $M = \min\{Su,Dv\}$ will suffice), from supply node u, $1 < u < m + 2$, to demand node v, $m + 1 < v < m + n + 2$, and cost $c[u,v]$ equal to the unit shipping cost from u to v.

Thus G is a NOcycle, connected digraph. The assignment problem is the special case where all the flow capacities are equal to 1.

Solving the transportation problem requires finding *a maximum flow of minimum cost from node 1 to mode k: a minimum cost–maximum flow problem*.

A simple assignment problem is discussed below. Additional starting and terminal nodes are not included in the discussion. Three people, labeled 1, 2, and 3, are assigned to three positions, labeled 4, 5, and 6. The following costs are associated with the assignments:

Route	Cost
[1,4]	1
[1,5]	2
[1,6]	3
[2,4]	3
[2,5]	2
[2,6]	2
[3,4]	2
[3,5]	2
[3,6]	1

An initial assignment: [1,5], [2,6], [3,4], is arbitrarily made to begin. This initial assignment has an associated cost of six. A network diagram for this assignment follows:

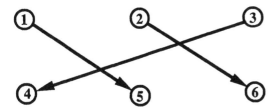

The corresponding residual digraph is

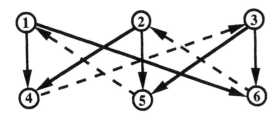

The residual contains the following cycle of cost −1:

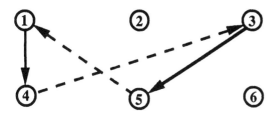

Adding a unit flow on this cycle to the initial assignment results in the following assignment:

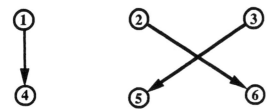

with a cost of 5 and residual

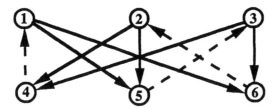

In this residual, the cycle

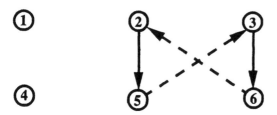

has cost −1. Adding a unit flow on this cycle to the preceding assignment produces the following minimum cost flow:

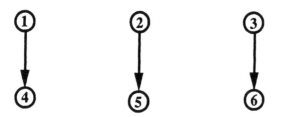

Each of the cycles introduced above resulted in an interchange of two assignments. The following cycle in the residual for the initial assignment has cost −2.

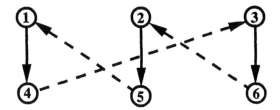

Augmenting the initial assignment by a unit flow on this cycle produces the same optimal assignment in one step. An example of an assignment problem with side conditions follows.

Example 9.6. A company has 12 employees to assign to five skilled jobs. The employees have been rated for each job on a scale of 0 to 10: a rating of zero means that the employee is not qualified to perform the job and a rating of 10 means that the employee is a master of the job. The cost $c\{e,j\}$ of assigning employee e to job j is given by the formula $c\{e,j\} = 10 - r\{e,j\}$, where $r\{e,j\}$ is the rating of employee e for job j. Because of a union agreement, each employee must get a fair share of time on the skilled jobs; it has been negotiated that 3 days during each 10-day period is fair. A maximum flow of minimum cost—a minimum cost–maximum flow—from node S (starting node) to node T (terminal node) on the network G described below solves the problem of determining how to schedule the workers to daily assignments for a 10-day period so as to maximize the total skill assignment to the jobs over the 10-day scheduling period. Another problem would be to assign workers so that the minimum of the total skill assigned on each of the 10 days is maximized.

Model. Nodes

The modeling network G will have several types of nodes;

1. A starting node S
2. Twelve employee nodes, E_1 and E_{12}
3. One hundred twenty assignment nodes A_{IJ}, $1 \le I \le 12$, $1 \le J \le 10$, corresponding to assigning employee I to a skilled job in period J
4. Six hundred assignment nodes B_{IJK}, $I \le 12$, $J \le 10$, $K \le 5$, corresponding to assigning employee I to job K in period J
5. A dump node D, where workers go when they are not assigned to a skilled job
6. A terminal node T

The network is displayed in Fig. 1.

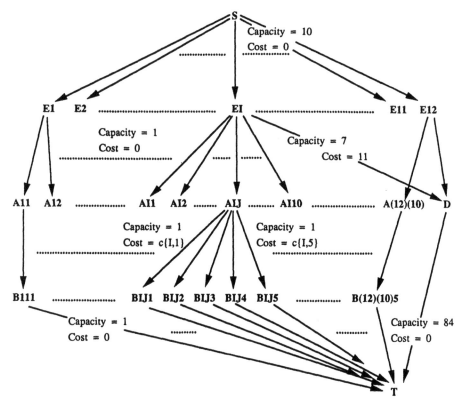

Figure 1   Modeling network G.

Routes   The network has several types of routes:

1. Routes $[S,E_I]$ with capacity $= 10$ and cost $= 0$.
2. Routes $[E_I,A_{IJ}]$ with capacity $= 1$ and cost $= 0$. If the flow on this route is equal to 1, employee I is assigned to some skilled job in period J.
3. Routes $[E_I,D]$ with capacity $= 7$ and cost $= 11$. The flow on this type of route denotes the number of days that employee I is not assigned to a skilled job; putting the cost larger than 10 causes the model to fill all the skilled positions, and putting the capacity equal to 7 forces each employee to be assigned to a skilled job on at least 3 days.
4. Routes $[A_{IJ},B_{IJK}]$ with capacity $= 1$ and cost $= c\{I,K\}$. The flow on these routes is equal to 1 when employee I is assigned to skilled job K on day J.

5. Routes $[B_{IJK},T]$ with capacity = 1 and cost = 0. The flow on these routes is equal to 1 when employee I is assigned to skilled job K on day J.

6. Route $[D,T]$ with cost = 0 and capacity = 84 to permit sufficient flow from node D to node T.

See if you can find a simpler model for this problem.

9.10 MINIMUM COST–MAXIMUM FLOW ALGORITHM
(for integral capacities and flow costs)

Consider a network G. For the case where flow capacities and unit flow costs as are nonnegative integers, an algorithm will be described to find a maximum flow of minimum cost from node 1 to node k. The following general fact about minimum cost flows will be used.

Fact ([T1], page 109)

If F is a minimal cost flow among flows of its size in G, and A is a minimal cost flow among flows of its size in the residual digraph $R = R(G,F)$, then $F + A$ is a minimal cost flow among flows of its size in G.

Unit flow costs on the routes in the residual digraphs are defined below. Recall that if $[i,j]$ is a route in G and the flow on $[i,j]$ is less than the capacity of $[i,j]$, then $[i,j]$ is a route in R.

Flow Costs on Residual Graphs

If a route in R is also a route in G, the unit cost for the route in R is the same as the unit cost in G. If the flow $F[i,j]$ on the route $[i,j]$ in G is positive, the unit cost $Rc[j,i]$ on the route $[j,i]$ in R is equal to $-c[i,j]$, the negative of the unit cost $c[i,j]$ on the route $[i,j]$ in G.

The *capacity*, *cap(P)*, *of a path* P is the minimum of the capacities of the routes in the path P.

Initialization

Put $F = 0$: start with the zero flow.

Augmentation

1. Form the residual graph $R = R(G,F)$.
2. Find a minimal cost unit path P from node 1 to node k in R: this amounts to finding a shortest cost path from node 1 to node k in R.

3. Compute the capacity cap(P) of the minimal cost path P and define an augmenting flow A that is equal to cap(P) on the routes in P and zero on all other routes in G.

Algorithm

Apply augmentation.
*If $A = 0$, then stop: F is the minimum cost–maximum flow;
otherwise*, augment F by A: *put $F = F + A$*, and apply augmentation.

EXERCISES

9.1 Sketch the digraph for Example 9.1.

9.2 (a) Using topological ordering, write a shortest path algorithm for NOcycle connected digraphs.
(b) Apply your algorithm to Example 5.12.

9.3 Apply the shortest path algorithm of Section 9.4 to Example 5.12.

9.4 Consider the following graph:

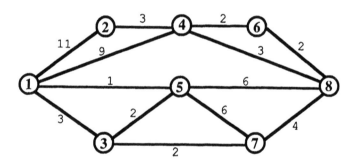

(a) Apply the shortest path algorithm.
(b) Apply the minimum spanning tree algorithm.
(c) Use the maximum flow algorithm to find the maximum flow from node 1 to node 8; then find the corresponding minimum cut.
(d) Suppose that the links (i,j) are replaced by routes $[i,j]$, where $i < j$. Replacing the links by routes produces a NOcycle, connected digraph. Redo part (c) for this digraph.

9.5 Repeat Exercise 9.4(c) for the digraph you sketched in Exercise 9.1.

9.6 Apply the maximum flow algorithm to find the maximum flow from node 1 to node 5 in Example 9.2; then find the corresponding minimum cut.

9.7 Consider Example 9.2.
(a) Suppose that we wish to locate a fire station at a node so that

its maximum distance to any node is minimal. How would you solve this problem?

(b) Suppose that a minimum spanning tree has been paved and that we wish to locate a fire station at a node so that its maximum paved distance to any node is minimal. How would you solve this problem?

9.8 Referring to Example 9.6, suppose that only four workers are needed per day to perform other essential functions; then we have three workers free each day to assign as we wish. Suppose that one day spent observing a skilled job being performed will raise a workers rating on that job to the level of the person that he is watching, if it is not that high already. Suppose that we agree to assign three workers for each day of a 10-day period as observers at the skill jobs. We wish to schedule the observers so as to maximize the increase in the total skill level of the employees; formulate a digraph model to accomplish this objective.

9.9 Given the digraph

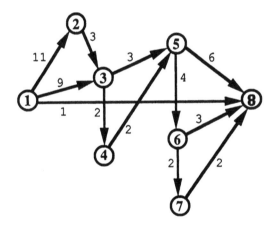

(a) *Beginning at node 1*, apply the longest path algorithm to this digraph: write a tabular solution as in the text and explain what your table entries represent.

(b) Consider the routes to be links which can be paved in either direction; *beginning at node 3*, apply the minimum spanning tree algorithm. List the order and direction in which paving occurs.

(c) Apply the maximum flow algorithm to find the maximum flow from node 1 to node 8 in the given digraph. *Draw the residual digraphs as they occur* at each step in the algorithm. (You can find the maximum simple path flows by inspection if you wish.)

(d) Find the minimum cut corresponding to the maximum flow generated by the maximum flow algorithm.

9.10 Given the digraph (notation = [capacity, cost per unit flow])

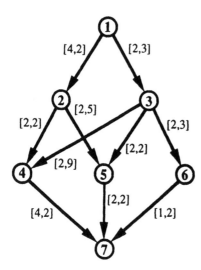

Apply the minimum cost–maximum flow algorithm to find the minimum cost–maximum flow. (I will give you the initial minimum cost
simple path flow below.) Begin by drawing the residual digraph
corresponding to the initial minimum cost path flow:

①—[2,2]→②—[2,2]→④—[2,2]→⑦

(Your next step is to write the table showing the order in which
nodes are scanned and the values of the two functions.)

Appendix 1
Forms of LPs

Each of the three examples in Chapter 2 was discussed from two perspectives, a primal problem and a related dual problem, leading to a primal-dual pair of LPs. Two types of primal-dual pairs emerged; one pair was labeled a standard pair and the other pair was labeled a symmetric pair. I asserted that any linear program L could be written in the form of a standard primal, denoted by P, by doing some algebraic juggling, which will be denoted by $\dot{A}$; thus $P = \dot{A}L$. Let D denote the dual in a standard primal-dual pair (P,D). Let $\dot{A}^{-1}$ denote undoing, or reversing, the algebraic juggling; $\dot{A}^{-1}D$ is defined to be a dual of the linear program L, and $(L,\dot{A}^{-1}D)$ is called a primal-dual pair. Example 2.1 showed how to put the constraints for a symmetric primal or dual in standard primal form (for the dual problem, up to transposing) by introducing surplus or slack variables.

The linear program L could also be written as a symmetric primal P_1 by doing some algebraic juggling, denoted by $\dot{A}_1$: $P_1 = \dot{A}_1 L$; P_1 has a symmetric dual D_1, and reversing this algebra leads to a dual $\dot{A}_1^{-1} D_1$ for L: $(P_1,D_1) = (\dot{A}_1 L,D_1)$ changes to $(L,\dot{A}_1^{-1} D_1)$ when $\dot{A}_1^{-1}$ is applied to it. For a useful concept of duality, $\dot{A}^{-1}D$ and $\dot{A}_1^{-1}D_1$ should be equivalent; they are equivalent. Another useful property of primal-dual pairs is that the dual of the dual is the primal. The following four examples illustrate these concepts.

Example 1. Consider the symmetric primal–dual pair. I will verify that the dual of the dual is the primal as follows. Write the dual

$$\max\ yb$$
$$\text{s.t.}\ \ yA \le c, \qquad y \ge 0$$

in the primal form

$$\min\ -b^t y^t$$
$$\text{s.t.}\ \ -A^t y^t \ge -c^t, \qquad y^t \ge 0.$$

Put $c_1 = -b^t$, $x_1 = y^t$, $A_1 = -A^t$, and $b_1 = -c^t$; then we have a symmetric primal

$$\min\ c_1 x_1$$
$$\text{s.t.}\ \ A_1 x_1 \ge b_1, \qquad x_1 \ge 0.$$

with dual

$$\max\ y_1 b_1$$
$$\text{s.t.}\ \ y_1 A_1 \le c_1, \qquad y_1 \ge 0.$$

Rewrite c_1, A_1, and b_1 in terms of b, A, and c to get

$$\max\ y_1(-c^t)$$
$$\text{s.t.}\ \ y_1(-A^t) \le -b^t, \qquad y_1 \ge 0,$$

which is reorganized to obtain

$$\min\ c y_1^t$$
$$\text{s.t.}\ \ A y_1^t \ge b, \qquad y_1^t \ge 0.$$

This is the primal if we identify x with y_1^t.

Example 2. Consider the standard primal–dual pair. I will show that the dual of the dual is the primal. Put $y = y_1 - y_2$, where y_1 and y_2 are $\ge$ zero, in the dual

$$\max\ yb$$
$$\text{s.t.}\ \ yA \le c,$$

to get

$$\max\ (y_1 - y_2)b = \max[y_1, y_2]\left[\begin{array}{c} b \\ -b \end{array}\right],$$

$$\text{s.t.}\ \ (y_1 - y_2)A = [y_1, y_2]\left[\begin{array}{c} A \\ -A \end{array}\right] \le c.$$

Change the min to a max, add slack variables z, and transpose. Then we have a standard primal

$$\min \ [-b^t, b^t, 0] \begin{bmatrix} y_1^t \\ y_2^t \\ z^t \end{bmatrix}$$

$$\text{s.t.} \ [A^t | -A^t | I] \begin{bmatrix} y_1^t \\ y_2^t \\ z^t \end{bmatrix} = c^t.$$

Put $x_1 = [y_1, y_2, z]^t$, $c_1 = [-b^t, b^t, 0]$, $A_1 = [A^T | -A^T | I]$, and $b = c^t$ and take the dual of

$$\min \ c_1 x_1$$
$$\text{s.t.} \ A_1 x_1 = b_1, \qquad x_1 \geq 0,$$

which is

$$\max \ y_1 b_1$$
$$\text{s.t.} \ y_1 A_1 \leq c_1.$$

Rewrite A_1, b_1, and c_1 in terms of A, b, and c to get

$$\max \ y_1 c^t$$
$$\text{s.t.} \ y_1 [A^T | -A^T | I] \leq [-b^t | b^t | 0],$$

which can be rewritten in the form

$$\min \ c(-y_1)^t$$
$$\text{s.t.} \ A(y_1^t) \leq -b, \qquad A(-y_1)^t \leq b, \qquad y_1 \leq 0$$

or, putting $x = (-y_1)^t$,

$$\min \ cx$$
$$\text{s.t.} \ Ax \geq b, \qquad Ax \leq b, \qquad x \geq 0,$$

which is the standard primal

$$\min \ cx$$
$$\text{s.t.} \ Ax = b, \qquad x \geq 0.$$

Example 3. I will take a symmetric primal, put it in standard primal form, take the (standard) dual, and show that it becomes the (symmetric) dual of the original symmetric primal. Start with a symmetric primal

$$\min cx$$

$$\text{s.t. } Ax \geq b, \qquad x \geq 0.$$

Add surplus variables z and get a standard primal,

$$\min [c,0] \begin{bmatrix} x \\ \overline{z} \end{bmatrix}$$

$$\text{s.t. } [A|-I] \begin{bmatrix} x \\ \overline{z} \end{bmatrix} = b, \qquad \begin{bmatrix} x \\ \overline{z} \end{bmatrix} \geq 0,$$

with (standard) dual

$$\max yb$$

$$\text{s.t. } y[A|-I] \leq [c,0]$$

or, equivalently,

$$\max yb$$

$$\text{s.t. } yA \leq c, \qquad y \geq 0,$$

which is the (symmetric) dual of the symmetric primal.

Example 4. I will repeat the analog of the preceding example for a standard primal

$$\min cx$$

$$\text{s.t. } Ax = b, \qquad x \geq 0.$$

Since $Ax = b \Leftrightarrow Ax \geq b$ and $-Ax \geq -b$, the primal can be rewritten in the symmetric primal form

$$\min cx$$

$$\text{s.t. } \begin{bmatrix} A \\ -A \end{bmatrix} x \geq \begin{bmatrix} b \\ -b \end{bmatrix}, \qquad x \geq 0.$$

The (symmetric) dual of the preceding primal is

$$\max w \begin{bmatrix} b \\ -b \end{bmatrix} = \max [w^1|w^2] \begin{bmatrix} b \\ -b \end{bmatrix} = \max [w^1 - w^2]b$$

$$\text{s.t. } w \begin{bmatrix} A \\ -A \end{bmatrix} = [w^1|w^2] \begin{bmatrix} A \\ -A \end{bmatrix} = [w^1 - w^2]A \leq c, \qquad w \geq 0.$$

Upon putting $y = w^1 - w^2$, we have the standard dual

$$\max \ yb$$
$$\text{s.t.} \ \ yA \le c.$$

The requirement $y \ge 0$ is gone because, while $w^1 \ge 0$ and $w^2 \ge 0$, $y = w^1 - w^2$ is not constrained to be $\ge$ zero.

Appendix 2
Solutions Supplement for Chapter 2

2.2 An LP model for Exercise 2.2 is formulated at the end of Chapter 2; a solution for that model appears below. Notice the phrases "reduced costs," "slack or surplus," and "dual prices." Slack and surplus variables have already been discussed, and reduced costs and dual prices will be discussed subsequently.

OBJECTIVE FUNCTION VALUE

1)   380.000000

VARIABLE	VALUE	REDUCED COST
L	2.000000	.000000
D	.000000	3.000000
M	15.000000	.000000

ROW	SLACK OR SURPLUS	DUAL PRICES
2)	1.170000	.000000
3)	.000000	−40.000000
4)	.000000	−20.000000

2.3 A solution follows for the time focus formulation of Exercise 2.3 presented at the end of Chapter 2.

MIN   30 T1A + 30 T2A + 30 T3A + 30 T4A + 50 T1B + 50 T2B
       + 50 T3B + 50 T4B + 80 T1C + 80 T2C + 80 T3C + 80 T4C

SUBJECT TO

2) $\quad T1A + T2A + T3A + T4A <= 50$
3) $\quad T1B + T2B + T3B + T4B <= 50$
4) $\quad T1C + T2C + T3C + T4C <= 50$
5) $\quad 300\,T1A + 600\,T1B + 800\,T1C = 10000$
6) $\quad 250\,T2A + 400\,T2B + 700\,T2C = 8000$
7) $\quad 200\,T3A + 350\ T3B + 600\,T3C = 6000$
8) $\quad 100\,T4A + 200\,T4B + 300\,T4C = 6000$

OBJECTIVE FUNCTION VALUE

1) $\qquad$ 4047.61900

VARIABLE	VALUE	REDUCED COST
T1A	.000000	5.000000
T2A	.000000	1.428572
T3A	.000000	3.333332
T4A	.000000	5.000000
T1B	16.666670	.000000
T2B	.000000	4.285713
T3B	.000000	3.333332
T4B	30.000000	.000000
T1C	.000000	13.333330
T2C	11.428570	.000000
T3C	10.000000	.000000
T4C	.000000	5.000000

2.4 The rows in the constraint matrix A corresponding to the production focus formulation of the problem at the end of Chapter 2 correspond to land, water, money, and labor constraints. For instance, row 1 says (1) that growing C metric tons of corn, W metric tons of wheat, and S metric tons of soybeans will use $\frac{1}{10}C + \frac{1}{4}W + \frac{1}{6}S$ units of land and (2) that the number of units of land used cannot exceed the number of units of land available.

The land focus formulation of the problem has the advantage that the entries in the constraint matrix are not fractions, which tend to introduce roundoff errors into printouts. Solutions for the land focus formulations of the primal and dual problems follow.

(a) MAX $\quad 100\,C + 24\,W + 48\,S$
$\quad$ SUBJECT TO
$\qquad$ 2) $\quad C + W + S <= 1000$
$\qquad$ 3) $\quad 3\,C + W + 2\,S <= 2000$
$\qquad$ 4) $\quad 15\,C + 10\,W + 12\,S <= 20000$
$\qquad$ 5) $\quad 8\,C + 6\,W + 10\,S <= 8000$

OBJECTIVE FUNCTION VALUE

1)   66666.6600

VARIABLE	VALUE	REDUCED COST
C	666.666700	.000000
W	.000000	9.333332
S	.000000	18.666660

(b) MIN   1000 L + 2000 W + 20000 M + 8000 LA
SUBJECT TO
     2)   L + 3 W + 15 M + 8 LA >= 100
     3)   L + W + 10 M + 6 LA >= 24
     4)   L + 2 W + 12 M + 10 LA >= 48
OBJECTIVE FUNCTION VALUE

1)   66666.6600

VARIABLE	VALUE	REDUCED COST
L	.000000	333.333300
W	33.333330	.000000
M	.000000	10000.000000
LA	.000000	2666.667000

2.5 (a) Let B = no. bugs produced per week
          S = no. superbugs produced per week
          V = no. vans produced per week.

A solution for Exercise 2.5(a) follows.

MAX   1000 B + 1500 S + 2000 V

SUBJECT TO
     2)   15 B + 18 S + 20 V <= 10000
     3)   15 B + 19 S + 30 V <= 15000
     4)   10 B + 20 S + 25 V <= 10080

OBJECTIVE FUNCTION VALUE

1)   861714.300

VARIABLE	VALUE	REDUCED COST
B	276.571400	.000000
S	.000000	157.142900
V	292.571400	.000000

ROW	SLACK OR SURPLUS	DUAL PRICES
2)	.000000	28.571430
3)	2074.286000	.000000
4)	.000000	57.142860

(b) Let E = no. of dollars to offer per hour for engine works time
    W = no. of dollars to offer per hour for body works time
    A = no. of dollars to offer per minute for assembly time.

MIN 10000 E + 15000 W + 10080 A
SUBJECT TO
    2)   15 E + 15 W + 10 A >= 1000
    3)   18 E + 19 W + 20 A >= 1500
    4)   20 E + 30 W + 25 A >= 2000

The constraints for the dual problem say that the offer to rent must be enough to make Volkswagen's rental income at least as much as the profit it would get by using its resources to manufacture vehicles instead of renting them to BMW.

OBJECTIVE FUNCTION VALUE

1)   861714.300

VARIABLE	VALUE	REDUCED COST
E	28.571430	.000000
W	.000000	2074.285000
A	57.142860	.000000

You should have expected the value of W to be zero in the solution to the dual problem because the solution to part (a), the primal, told us that there was unused body works time available.

2.6 (a) A solution is formulated at the end of Chapter 2; a repeat of that formulation and a solution follow.

MAX 500 FL + 640 PL + 700 FP + 800 PP
SUBJECT TO
    2)   140 FP + 160 PP >= 1000
    3)   1000 FL + 800 PL >= 5000
    4)   FL + FP <= 10
    5)   PL + PP <= 10

OBJECTIVE FUNCTION VALUE

1)   14000.0000

VARIABLE	VALUE	REDUCED COST
FL	.000000	.000000
PL	6.250000	.000000
FP	10.000000	.000000
PP	3.750000	.000000

(b) Let T = no. of truckloads of lumber to be produced
    C = no. of cabin kits to be produced.

MAX   4500 T + 6000 C
SUBJECT TO
    2) T >= 5
    3) C >= 3
    4) T + 2 C <= 32
    5) 4 T + 4 C <= 72

OBJECTIVE FUNCTION VALUE

1)   100500.000

VARIABLE	VALUE	REDUCED COST
T	5.000000	.000000
C	13.000000	.000000

2.7 (a) Let D = no. of dishwashers to process per day
        R = no. of refrigerators to process per day
        S = no. of stoves to process per day.

MAX   D + R + S
SUBJECT TO
    2) .75 D + 1.4 R + 1.1 S <= 480
    3) 1.2 D + R + 1.3 S <= 480

OBJECTIVE FUNCTION VALUE

1)   438.709700

VARIABLE	VALUE	REDUCED COST
D	206.451600	.000000
R	232.258100	.000000
S	.000000	.145161

(b) Add the constraints $D \geq 45$, $R \geq 60$, $S \geq 32$ to the LP model for part (a).

MAX   D + R + S
SUBJECT TO
    2) .75 D + 1.4 R + 1.1 S <= 480
    3) 1.2 D + R + 1.3 S <= 480
    4) D >= 45
    5) R >= 60
    6) S >= 32

OBJECTIVE FUNCTION VALUE

1)   434.064500

VARIABLE	VALUE	REDUCED COST
D	181.677400	.000000
R	220.387100	.000000
S	32.000000	.000000

2.8 (a) (Here also I am referring to the model given at the end of Chapter 2.)

MIN $8A + 6B + 2C$
SUBJECT TO
2) $36A + 20B + 16C >= 160$
3) $36A + 20B + 16C <= 180$
4) $25A + 35B + 15C >= 200$
5) $70A + 40B + 25C >= 300$

OBJECTIVE FUNCTION VALUE

1)   29.7916700

VARIABLE	VALUE	REDUCED COST
A	.104167	.000000
B	1.979167	.000000
C	8.541667	.000000

(b) MIN $8A + 6B + 2C$
SUBJECT TO
2) $36A + 20B + 16C >= 160$
3) $36A + 20B + 16C <= 180$
4) $25A + 35B + 15C >= 200$
5) $70A + 40B + 25C >= 300$
6) $.8A - .2B - .2C >= 0$
7) $.5A + .5B - .5C >= 0$

OBJECTIVE FUNCTION VALUE

1)   38.3132600

VARIABLE	VALUE	REDUCED COST
A	1.686747	.000000
B	2.831325	.000000
C	3.915662	.000000

2.9 Let T = no. of toilet paper rolls to produce each day
W = no. of writing pads to produce each day
P = no. of paper towels to produce each day.

MAX   .18 T + .29 W + .25 P
SUBJECT TO
   2)   .5 T + .22 W + .85 P <= 1500
   3)   .2 T + .4 W + .22 P <= 600
   4)   T >= 1000
   5)   W >= 200
   6)   P >= 400

OBJECTIVE FUNCTION VALUE

1)   566.831300

VARIABLE	VALUE	REDUCED COST
T	1000.000000	.000000
W	411.522600	.000000
P	1069.959000	.000000

We have only an approximate solution here because we cannot leave part of a paper towel waiting to be finished the next day.

2.10 (a) Let A = kilograms of almonds to use in the deluxe mix.
         C = kilograms of cashews to use in the deluxe mix.
         P = kilograms of peanuts to use in the deluxe mix.
         W = kilograms of walnuts to use in the deluxe mix.

Exercise 2.10(a) can be modeled either by minimizing the cost of filling the order or by maximizing the profit realized from filling the order. A formulation and solution for the minimization model follow.

MIN   6 A + 5.3 C + 1.4 P + 4.2 W
SUBJECT TO
   2)   A + C + P + W = 400
   3)   C <= 200
   4)   W <= 300
   5)   A >= 40
   6)   C >= 40
   7)   P >= 40
   8)   W >= 40
   9)   A + C >= 200
   10) P <= 80

OBJECTIVE FUNCTION VALUE

1)   1704.00000

VARIABLE	VALUE	REDUCED COST
A	40.000000	.000000
C	160.000000	.000000
P	80.000000	.000000
W	120.000000	.000000

(b) Let AC = kilograms of almonds to use in the companion mix
CC = kilograms of cashews to use in the companion mix
PC = kilograms of peanuts to use in the companion mix
WC = kilograms of walnuts to use in the companion mix
R = kilograms of raisins to use in the companion mix.

An LP model and solution for Exercise 2.10(b) follow; because the quantity of the companion mix is a variable, I will maximize the profit derived from filling both orders.

MAX $4\,A + 4.7\,C + 8.6\,P + 5.8\,W - .1\,AC + .6\,CC$
     $+ 4.5\,PC + 1.7\,WC + 4.1\,R$
SUBJECT TO
   2) $A + AC <= 400$
   3) $C + CC <= 200$
   4) $P + PC <= 600$
   5) $W + WC <= 300$
   6) $A + C + P + W = 400$
   7) $A >= 40$
   8) $C >= 40$
   9) $P >= 40$
  10) $W >= 40$
  11) $A + C >= 200$
  12) $P <= 80$
  13) $AC - 9\,CC + PC + WC + R <= 0$
  14) $3\,AC + 3\,CC + 3\,PC - 7\,WC + 3\,R <= 0$
  15) $4\,AC + 4\,CC + 4\,PC + 4\,WC - 6\,R >= 0$
  16) $2\,AC + 2\,CC + 2\,PC + 2\,WC - 8\,R <= 0$
  17) $AC + CC + PC + WC + R <= 500$

OBJECTIVE FUNCTION VALUE

   1)   3884.00000

VARIABLE	VALUE	REDUCED COST
A	50.000000	.000000
C	150.000000	.000000
P	80.000000	.000000

W	120.000000	.000000
AC	.000000	4.600000
CC	50.000000	.000000
PC	200.000000	.000000
WC	150.000000	.000000
R	100.000000	.000000

If we had filled the first order before we considered the second order, we would have had only enough cashews for 4000 bags of companion mix. We can simplify the percentage constraints and get the solution to tell us how many kilograms of companion mix to make by introducing another decision variable: Let KC = kilograms of companion mix to make.

MAX $4A + 4.7C + 8.6P + 5.8W - .1AC + .6CC$
$+ 4.5PC + 1.7WC + 4.1R$
SUBJECT TO
2) $A + AC <= 400$
3) $C + CC <= 200$
4) $P + PC <= 600$
5) $W + WC <= 300$
6) $A + C + P + W = 400$
7) $A >= 40$
8) $C >= 40$
9) $P >= 40$
10) $W >= 40$
11) $A + C >= 200$
12) $P <= 80$
13) $AC + CC + PC + WC + R - KC = 0$
14) $- CC + .1KC <= 0$
15) $- WC + .3KC <= 0$
16) $- R + .2KC <= 0$
17) $- R + .4KC >= 0$
18) $KC <= 500$

OBJECTIVE FUNCTION VALUE

1)   3884.00000

VARIABLE	VALUE	REDUCED COST
A	50.000000	.000000
C	150.000000	.000000
P	80.000000	.000000
W	120.000000	.000000

AC	.000000	4.600000
CC	50.000000	.000000
PC	200.000000	.000000
WC	150.000000	.000000
R	100.000000	.000000
KC	500.000000	.000000

(c) To model Exercise 2.10(c), simply change the coefficient of A to -30 and the coefficient of AC to -34.1 in Exercise 2.10(b); the model and a solution follow.

MAX $-30\,A + 4.7\,C + 8.6\,P + 5.8\,W - 34.1\,AC + .6\,CC + 4.5\,PC + 1.7\,WC + 4.1\,R$

SUBJECT TO

2)	$A + AC <= 400$
3)	$C + CC <= 200$
4)	$P + PC <= 600$
5)	$W + WC <= 300$
6)	$A + C + P + W = 400$
7)	$A >= 40$
8)	$C >= 40$
9)	$P >= 40$
10)	$W >= 40$
11)	$A + C >= 200$
12)	$P <= 80$
13)	$AC + CC + PC + WC + R - KC = 0$
14)	$- CC + .1\,KC <= 0$
15)	$- WC + .3\,KC <= 0$
16)	$- R + .2\,KC <= 0$
17)	$- R + .4\,KC >= 0$
18)	$KC <= 500$

OBJECTIVE FUNCTION VALUE

1)   2212.00000

VARIABLE	VALUE	REDUCED COST
A	40.000000	.000000
C	160.000000	.000000
P	80.000000	.000000
W	120.000000	.000000
AC	.000000	38.600000
CC	40.000000	.000000
PC	160.000000	.000000

WC	120.000000	.000000
R	80.000000	.000000
KC	400.000000	.000000

We have enough nuts for 4000 bags of deluxe mix and 5000 bags of companion mix, and we would profit from making both 4000 bags of deluxe mix and 5000 bags of companion mix. But in this case we only make 4000 bags of companion mix because making more than 4000 bags of companion mix would force us to use more almonds in the deluxe mix, and the increased cost of deluxe mix due do the change in composition would be greater than the increase in profit from making more than 4000 bags of companion mix.

2.11 Let T_{IJ} = thousands of barrels to use from tank I in order J, $1 \le I \le 3$, $1 \le J \le 4$.

MIN 15 T11 + 15 T12 + 15 T13 + 15 T14 + 17 T21 + 17 T22
   + 17 T23 + 17 T24 + 20 T31 + 20 T32 + 20 T33 + 20 T34
SUBJECT TO
   2) T11 + T12 + T13 + T14 <= 4
   3) T21 + T22 + T23 + T24 <= 3
   4) T31 + T32 + T33 + T34 <= 5
   5) T11 + T21 + T31 = 2
   6) T12 + T22 + T32 = 1.5
   7) T13 + T23 + T33 = 2.5
   8) 6.8 T11 + 7.4 T21 + 8.1 T31 >= 14
   9) 6.8 T12 + 7.4 T22 + 8.1 T32 <= 11.7
   10) 6.8 T13 + 7.4 T23 + 8.1 T33 >= 18
   11) 6.8 T13 + 7.4 T23 + 8.1 T33 <= 19
   12) 6.8 T14 + 7.4 T24 + 8.1 T34 = 22.2
   13) T14 + T24 + T34 = 3

OBJECTIVE FUNCTION VALUE

1)   151.000000

VARIABLE	VALUE	REDUCED COST
T11	1.692308	.000000
T12	.576922	.000000
T13	1.730770	.000000
T14	.000000	.000000
T21	.000000	.000000
T22	.000000	.000000
T23	.000000	.000000

T24	3.000000	.000000
T31	.307692	.000000
T32	.923078	.000000
T33	.769230	.000000
T34	.000000	.000000

Since there are several decision variables with value equal to zero and reduced cost equal to zero, we can expect to have lots of different optimal solutions to this problem. Another optimal solution appears below; can you find more?

VARIABLE	VALUE	REDUCED COST
T11	1.692308	.000000
T12	.000000	.000000
T13	.833334	.000000
T14	1.474358	.000000
T21	.000000	.000000
T22	1.071427	.000000
T23	1.666666	.000000
T24	.261907	.000000
T31	.307692	.000000
T32	.428573	.000000
T33	.000000	.000000
T34	1.263735	.000000

2.12 Let A_I = no. of hours to make size I pipe by process A, $I \le 4$
B_I = no. of hours to make size I pipe by process B, $I \le 4$
T = no. of hours required to process the order.

(a) MIN   T
SUBJECT TO
2) $25 A1 + 42 B1 >= 7000$
3) $20 A2 + 37 B2 >= 6500$
4) $15 A3 + 34 B3 >= 7500$
5) $10 A4 + 31 B4 >= 8500$
6) $A1 + A2 + A3 + A4 - T <= 0$
7) $B1 + B2 + B3 + B4 - T <= 0$

OBJECTIVE FUNCTION VALUE

1)   533.454800

VARIABLE	VALUE	REDUCED COST
A1	280.000000	.000000
A2	253.454800	.000000

A3	.000000	.064499
A4	.000000	.141483
B1	.000000	.059649
B2	38.673100	.000000
B3	220.588200	.000000
B4	274.193500	.000000
T	533.454800	.000000

Thus it takes 14 weeks to process the order.

(b) Since it takes 14 weeks to process the order and 40 hours are available each week, add a constraint, $T \leq 560 = 14 \times 40$, to the model for Exercise 2.12(a):

MIN   16 A1 + 14 A2 + 10 A3 + 8 A4 + 25 B1 + 26 B2
      + 27 B3 + 28 B4
SUBJECT TO
   2)   25 A1 + 42 B1 >= 7000
   3)   20 A2 + 37 B2 >= 6500
   4)   15 A3 + 34 B3 >= 7500
   5)   10 A4 + 31 B4 >= 8500
   6)   A1 + A2 + A3 + A4 − T <= 0
   7)   B1 + B2 + B3 + B4 − T <= 0
   8)   T <= 560

OBJECTIVE FUNCTION VALUE

1)   21859.0500

VARIABLE	VALUE	REDUCED COST
A1	.000000	.150475
A2	302.577000	.000000
A3	257.423000	.000000
A4	.000000	3.096775
B1	166.666700	.000000
B2	12.120530	.000000
B3	107.019300	.000000
B4	274.193500	.000000
T	560.000000	.000000

(c) Let C_I = no. of overtime hours to make type I by process A, $I \leq 4$

D_I = no. of overtime hours to make type I by process B, $I \leq 4$.

MIN   16 A1 + 14 A2 + 10 A3 + 8 A4 + 25 B1 + 26 B2

$$+ 27\,B3 + 28\,B4 + 20\,C1 + 18\,C2 + 14\,C3 + 12\,C4$$
$$+ 29\,D1 + 30\,D2 + 31\,D3 + 32\,D4$$

SUBJECT TO

2)	$25\,A1 + 42\,B1 + 25\,C1 + 42\,D1 >= 7000$
3)	$20\,A2 + 37\,B2 + 20\,C2 + 37\,D2 >= 6500$
4)	$15\,A3 + 34\,B3 + 15\,C3 + 34\,D3 >= 7500$
5)	$10\,A4 + 31\,B4 + 10\,C4 + 31\,D4 >= 8500$
6)	$A1 + A2 + A3 + A4 - T <= 0$
7)	$B1 + B2 + B3 + B4 - T <= 0$
8)	$T <= 480$
9)	$C1 + C2 + C3 + C4 <= 96$
10)	$D1 + D2 + D3 + D4 <= 96$

OBJECTIVE FUNCTION VALUE

1)   22171.5700

VARIABLE	VALUE	REDUCED COST
A1	.000000	.773809
A2	.000000	.000000
A3	480.000000	.000000
A4	.000000	.612904
B1	166.666700	.000000
B2	39.139760	.000000
B3	.000000	2.566666
B4	274.193500	.000000
T	480.000000	.000000
C1	.000000	.773809
C2	74.991440	.000000
C3	20.000000	.000000
C4	.000000	.612904
D1	.000000	.000000
D2	96.000000	.000000
D3	.000000	2.566666
D4	.000000	.000000

2.13 Let F_I = no. of units of food I to put in the diet, $I \leq 5$.

MIN $.15\,F1 + .23\,F2 + .79\,F3 + .47\,F4 + .52\,F2$

SUBJECT TO

2)	$F2 + 5\,F3 + 2\,F4 >= 3$
3)	$7\,F1 + 2\,F4 + 3\,F5 >= 10$
4)	$F1 + 3\,F2 + 4\,F3 + F4 <= 2$
5)	$F1 + F2 + 3\,F4 + 2\,F5 >= 3$
6)	$4\,F2 + F3 + F5 >= 2$

OBJECTIVE FUNCTION VALUE

1)   1.61836100

VARIABLE	VALUE	REDUCED COST
F1	.295082	.000000
F2	.081967	.000000
F3	.000000	.362131
F4	1.459016	.000000
F5	1.672131	.000000

2.14 Let us begin by making some simplifying assumptions. The dryer will be filled with one kind of fruit and operated for 1, 2, 3, or 4 hours, after which the dried fruit will be removed from the dryer and the dryer will be refilled. We will ignore time spent filling the dryer and removing dried fruit. In accord with these assumptions, we have the following decision variables. Let

G_I = no. of cubic meters of grapes to dry for I hours
A_I = no. of cubic meters of apricots to dry for I hours
P_I = no. of cubic meters of plums to dry for I hours.

MIN   G1 + 2 G2 + 3 G3 + 4 G4 + A1 + 2 A2 + 3 A3 + 4 A4
      + P1 + 2 P2 + 3 P3 + 4 P4
SUBJECT TO
    2)   G1 + G2 + G3 + G4 = 20
    3)   A1 + A2 + A3 + A4 = 10
    4)   P1 + P2 + P3 + P4 = 5
    5)   .3 G1 + .23 G2 + .16 G3 + .13 G4 + .46 A1
        + .44 A2 + .34 A3 + .31 A4 + .53 P1 + .51 P2
        + .47 P3 + .42 P4 <= 10

OBJECTIVE FUNCTION VALUE

1)   82.4999900

VARIABLE	VALUE	REDUCED COST
G1	.000000	.333333
G2	.000000	.166667
G3	20.000000	.000000
G4	.000000	.500000
A1	6.250004	.000000
A2	.000000	.666667
A3	3.749996	.000000
A4	.000000	.500000
P1	5.000000	.000000

P2	.000000	.666667
P3	.000000	1.000001
P4	.000000	1.166667

Notice that A_1 and A_3 are not integer valued. However, if we change A_1 to 6 and A_3 to 4, we get an integer-valued feasible solution with objective function value = 83. The objective function value of any integer-valued solution will be an integer. But 83 is the smallest integer, which is ≥ 82.5. Consequently, changing A_1 to 6 and A_3 to 4 provides us with an optimal integer-valued solution.

2.15 Let C_{IJ} = no. of containers of type I to pack with mix J, $I = 1,2$, $J = 1,2$.

MIN   15 C11 + 15 C12 + 25 C21 + 25 C22
SUBJECT TO
  2)   2 C11 + 2 C12 + 3 C21 >= 78
  3)   4 C11 + 2 C12 + 8 C21 + 8 C22 >= 214
  4)   4 C12 + 12 C22 >= 198

OBJECTIVE FUNCTION VALUE

1)   852.500000

VARIABLE	VALUE	REDUCED COST
C11	.000000	.555555
C12	21.000000	.000000
C21	12.000000	.000000
C22	9.500000	.000000

This is not an integer-valued solution, but this solution puts six dresser lamps in the half carton of type 2 packed with mix 2 and no other listed packing option will allow six dresser lamps in one carton. Consequently, I will put $C_{22} = 10$, increase the cost to 865, and claim that I now have the optimal solution. Claiming does not necessarily mean that I am right; is there a better solution?

2.16 (a) (Look at the end of Chapter 2 for definitions of E and R.)

MAX   15 E + 9.5 R
SUBJECT TO
  2)   2 E + R <= 215
  3)   3 E + 2 R <= 525
  4)   4 E + 2 R <= 440
  5)   2 E + 3 R <= 560

OBJECTIVE FUNCTION VALUE

1)   1957.50000

VARIABLE	VALUE	REDUCED COST
E	21.250000	.000000
R	172.500000	.000000

(b) MAX   15 E1 + 10 E2 + 9.5R1 + 6.5 R2
SUBJECT TO
2)   2 E + R <= 215
3)   3 E + 2 R <= 525
4)   4 E + 2 R <= 440
5)   2 E + 3 R <= 560
6)   E − E1 − E2 = 0
7)   R − R1 − R2 = 0
8)   E1 <= 60
9)   R1 <= 50

OBJECTIVE FUNCTION VALUE

1)   1667.50000

VARIABLE	VALUE	REDUCED COST
E	60.000000	.000000
R	95.000000	.000000
E1	60.000000	.000000
E2	.000000	3.000000
R1	50.000000	.000000
R2	45.000000	.000000

2.18 (b) Let $a - 1 = P_1 - N_1$ and $b = P_2 - N_2$, where $P_1, P_2, N_1, N_2 \geq 0$.

MIN   P1 + N1 + 3 P2 + 3 N2
SUBJECT TO
2)   P1 − N1 + 2 P2 − 2 N2 >= 1
3)   P1 − N1 + P2 − N2 = 1

Constraint 2 says that $(a - 1) + 2b \geq 1$: $a + 2b \geq 2$.

OBJECTIVE FUNCTION VALUE

1)   1.00000000

VARIABLE	VALUE	REDUCED COST
P1	1.000000	.000000
N1	.000000	2.000000
P2	.000000	1.000000
N2	.000000	5.000000

You can write the second constraint in symmetric primal form.

(c) In addition to the variables introduced in part (b), let $a = P_3 - N_3$, where P_3 and $N_3 \geq 0$.

MIN   P1 + N1 + 3 P2 + 3 N2
SUBJECT TO
2)   P1 − N1 + 2 P2 − 2 N2 >= 1
3)   P2 + N2 + P3 + N3 <= 2
4)   P1 − N1 − P3 + N3 = −1

Constraint 3 says that $|a| + |b| \leq 2$, and constraint 4 says that $(a - 1) - a = -1$.

OBJECTIVE FUNCTION VALUE

1)   1.00000000

VARIABLE	VALUE	REDUCED COST
P1	1.000000	.000000
N1	.000000	2.000000
P2	.000000	1.000000
N2	.000000	5.000000
P3	2.000000	.000000
N3	.000000	.000000

You can rewrite the second and third constraints in symmetric primal form.

2.19 Put $(D/45 - R/60) = P_1 - N_1$
$(D/45 - S/32) = P_2 - N_2$
$(R/60 - S/32) = P_3 - N_3$.

MAX   D + R + S − 20 P1 − 20 N1 − 20 P2 − 20 N2 − 20 P3 − 20 N3
SUBJECT TO
2)   .75 D + 1.4 R + 1.1 S <= 480
3)   1.2 D + R + 1.3 S <= 480
4)   4 D − 3 R − 180 P1 + 180 N1 = 0
5)   32 D − 45 S − 1440 P2 + 1440 N2 = 0
6)   8 R − 15 S − 480 P3 + 480 N3 = 0

OBJECTIVE FUNCTION VALUE

1)   422.622100

VARIABLE	VALUE	REDUCED COST
D	138.817500	.000000
R	185.090000	.000000
S	98.714650	.000000
P1	.000000	20.000000

N1	.000000	20.000000
P2	.000000	22.544990
N2	.000000	17.455010
P3	.000000	12.827760
N3	.000000	27.172240

2.20 I will give three solutions to Exercise 2.20; the first solution is similar to the solution I gave for Exercise 2.16(b) and the second solution follows the pattern of Exercise 2.19.

Solution 1. Put $D_1 = \min\{D,50\}$, and put $D_2 = D - D_1$

MAX   R + S + D1 + .5 D2
SUBJECT TO
   2) 1.4 R + 1.1 S + .75 D <= 480
   3) R + 1.3 S + 1.2 D <= 480
   4) D − D1 − D2 = 0
   5) D1 <= 50

OBJECTIVE FUNCTION VALUE

1)   409.375000

VARIABLE	VALUE	REDUCED COST
R	157.291700	.000000
S	202.083400	.000000
D	50.000000	.000000
D1	50.000000	.000000
D2	.000000	.312500

Solution 2. Put $D - 50 = P - N$.

MAX   R + S + D − .5 P
SUBJECT TO
   2) 1.4 R + 1.1 S + .75 D <= 480
   3) R + 1.3 S + 1.2 D <= 480
   4) D − P + N = 50

OBJECTIVE FUNCTION VALUE

1)   409.375000

VARIABLE	VALUE	REDUCED COST
R	157.291700	.000000
S	202.083400	.000000
D	50.000000	.000000
P	.000000	.312500
N	.000000	.187500

Solution 3. Introduce a new decision variable, T.

MAX   T
SUBJECT TO
2) $1.4\,R + 1.1\,S + .75\,D <= 480$
3) $R + 1.3\,S + 1.2\,D <= 480$
4) $T - D - R - S <= 0$ $(T <= D + R + S)$
5) $T - .5D - R - S <= 25$
       $(T <= D + R + S - .5\{D - 50\}$ if $D >= 50)$

Check that if $D \le 50$, constraint 5 is satisfied automatically whenever constraint 4 is satisfied.

2.21 (a) There are two types of decisions in this problem: (1) whether or not to send a truck from a warehouse to a lumberyard; and (2) if we decide to send a truck, how many bags of mix should be put on the truck? We return to this problem in Chapter 5.

 (b) A model is formulated at the end of Chapter 2; a solution follows.

MIN   P1 + N1 + P2 + N2 + P3 + N3 + P4 + N4 + P5 + N5 + P6
       + N6 + 2 P7 + 2 N7 + 2 P8 + 2 N8
SUBJECT TO
2) $- P1 + N1 + P3 - N3 = 98$
3) $P3 - N3 - P5 + N5 = 99$
4) $P5 - N5 - P7 + N7 = 34$
5) $P2 - N2 - P4 + N4 = 15$
6) $P4 - N4 - P6 + N6 = 67$
7) $- P6 + N6 + P8 - N8 = 48$

OBJECTIVE FUNCTION VALUE

1)   269.000000

VARIABLE	VALUE	REDUCED COST
P1	1.000000	.000000
N1	.000000	2.000000
P2	34.000000	.000000
N2	.000000	2.000000
P3	99.000000	.000000
N3	.000000	2.000000
P4	19.000000	.000000
N4	.000000	2.000000
P5	.000000	1.000000
N5	.000000	1.000000
P6	.000000	2.000000

N6	48.000000	.000000
P7	.000000	4.000000
N7	34.000000	.000000
P8	.000000	3.000000
N8	.000000	1.000000

2.22 Let M_I = thousands of feet leased for month I, $1 \leq I \leq 12$
Q_I = thousands of feet leased for quarter I, $1 \leq I \leq 4$
H_I = thousands of feet leased for half-year I, $I = 1,2$
Y = thousands of feet leased for the year.

MIN $10\,Y + 7\,H1 + 7\,H2 + 5\,Q1 + 5\,Q2 + 5\,Q3 + 5\,Q4$
$+ 3\,M1 + 3\,M2 + 3\,M3 + 3\,M4 + 3\,M5 + 3\,M6 + 3\,M7$
$+ 3\,M8 + 3\,M9 + 3\,M10 + 3\,M11 + 3\,M12$
SUBJECT TO
2) $Y + Q1 + M1 >= 10$
3) $Y + H1 + Q1 + M2 >= 15$
4) $Y + H1 + Q1 + M3 >= 23$
5) $Y + H1 + Q2 + M4 >= 32$
6) $Y + H1 + Q2 + M5 >= 43$
7) $Y + H1 + Q2 + M6 >= 52$
8) $Y + H1 + Q3 + M7 >= 50$
9) $Y + H2 + Q3 + M8 >= 56$
10) $Y + H2 + Q3 + M9 >= 40$
11) $Y + H2 + Q4 + M10 >= 25$
12) $Y + H2 + Q4 + M11 >= 15$
13) $Y + H2 + Q4 + M12 >= 10$

OBJECTIVE FUNCTION VALUE

1)   510.000000

Y	25.000000	.000000
H1	10.000000	.000000
H2	.000000	4.000000
Q1	.000000	5.000000
Q2	8.000000	.000000
Q3	15.000000	.000000
Q4	.000000	5.000000
M1	.000000	3.000000
M2	.000000	3.000000
M3	.000000	3.000000
M4	.000000	3.000000
M5	.000000	1.000000

M6	9.000000	.000000
M7	.000000	1.000000
M8	16.000000	.000000
M9	.000000	3.000000
M10	.000000	3.000000
M11	.000000	3.000000
M12	.000000	3.000000

ROW	SLACK OR SURPLUS	DUAL PRICES
2)	15.000000	.000000
3)	20.000000	.000000
4)	12.000000	.000000
5)	11.000000	.000000
6)	.000000	−2.000000
7)	.000000	−3.000000
8)	.000000	−2.000000
9)	.000000	−3.000000
10)	.000000	.000000
11)	.000000	.000000
12)	10.000000	.000000
13)	15.000000	.000000

Your solution to Exercise 2.22 may be different from this solution, but it will have the same objective function value if it is correct. Multiple solutions to this problem will be discussed in more detail later.

2.23 (a) Let R_1 = tens of thousands of gallons of 150-ppm filtered river water used per day

R_2 = tens of thousands of gallons of 75-ppm filtered river water used per day

W_1 = tens of thousands of gallons of 50-ppm unfiltered well water used per day

W_2 = tens of thousands of gallons of 10-ppm filtered well water used per day.

MIN   R1 + 3 R2 + 4 W1 + 5.5 W2
SUBJECT TO
  2)   W1 + W2 <= 4
  3)   R1 + R2 + W1 + W2 = 10
  4)   15 R1 + 7.5 R2 + 5 W1 + W2 <= 100

OBJECTIVE FUNCTION VALUE

  1)   23.3333300

VARIABLE	VALUE	REDUCED COST
R1	3.333333	.000000
R2	6.666667	.000000
W1	.000000	.333333
W2	.000000	.766667

(b) MIN R1 + 3 R2 + 4 W1 + 5.5 W2
SUBJECT TO
    2)   W1 + W2 <= 4
    3)   R1 + R2 + W1 + W2 = 10
    4)   15 R1 + 7.5 R2 + 5 W1 + W2 <= 100
    5)   R2 = 0

OBJECTIVE FUNCTION VALUE

   1)   25.7500000

VARIABLE	VALUE	REDUCED COST
R1	6.000000	.000000
R2	.000000	.000000
W1	1.500000	.000000
W2	2.500000	.000000

2.24 (a) There are lots of ways to formulate a model for this problem; here is one. Let

B = no. of bodies to make per 8-hour day
A = no. of axles to make per 8-hour day
W = no. of wheels to make per 8-hour day.

Put $A = 2B$ and put $W = 4B$; then the mix of parts is optimal and the number of kits produced is equal to the number of bodies produced. Consequently, we can simply solve the LP

MAX B
S.T. (.8 + .2 + 2)B <= 480   (saw time)
     (1.2 + .5 + 3.2)B <= 960   (sander time)
     (.6 + .6 + 1.6)B <= 480   (sprayer time)
or

MAX B
S.T.   3B <= 480
      4.9B <= 960
      2.8B <= 480

and find that $B = 480/3 = 160$ is the maximum number of kits that can be made in an 8-hour day.

(b) Let E_1 = no. of minutes employee 1 spends on the saw per day
E_2 = no. of minutes employee 2 spends on the sprayer per day.

MAX B
S.T.   3B <= E1   (saw time)
        4.9B <= 480 − E1 + 480 − E2   (sander time)
        2.8B <= E2   (sprayer time)
or

MAX B
SUBJECT TO
    2) 3 B − E1 <= 0
    3) 4.9 B + E1 + E2 <= 960
    4) 2.8 B − E2 <= 0

OBJECTIVE FUNCTION VALUE

1)   89.7196300

VARIABLE	VALUE	REDUCED COST
B	89.719630	.000000
E1	269.158900	.000000
E2	251.215000	.000000

2.25 Let H = no. of horses to buy
C = no. of cows to buy
G = no. of goats to buy
F = no. of feeder calves to buy.
The solutions to the LP models formulated below do not have all integer values and we cannot buy parts of an animal, so we are getting only approximate solutions to the problems.

(a) MAX   320 H + 140 C + 20 G + 210 F
SUBJECT TO
    2) 1.5 H + .8 C + .25 G + F <= 35
    3) H >= 3
    4) C >= 1
    5) G >= 4

OBJECTIVE FUNCTION VALUE

1)   7302.66700

VARIABLE	VALUE	REDUCED COST
H	22.133330	.000000
C	1.000000	.000000

G	4.000000	.000000
F	.000000	3.333328

(b) Let FF = no. of feeder calves to buy and put in feedlot.

MAX   320 H + 140 C + 20 G + 210 F + 100 FF
SUBJECT TO
    2)   1.5 H + .8 C + .25 G + F + .1 FF <= 35
    3)   H >= 3
    4)   C >= 1
    5)   G >= 4
    6)   80 H + 60 C + 15 G + 40 F + 150 FF <= 12000
    7)   FF <= 100

OBJECTIVE FUNCTION VALUE

1)   12884.1800

VARIABLE	VALUE	REDUCED COST
H	3.000000	.000000
C	1.000000	.000000
G	4.000000	.000000
F	21.513720	.000000
FF	71.863010	.000000

2.26 Let T = no. of tricycles to manufacture during the month
     W = no. of wagons to manufacture during the month.
(a) To minimize the discount, we wish to minimize

$$2(1000 - T) + 2(500 - W) = 3000 - 2T - 2W,$$

or, equivalently, we wish to

MAX   2 T + 2 W
SUBJECT TO
    2)   17 T + 14 W <= 20000
    3)   8 T + 6 W <= 10000
    4)   T <= 1000
    5)   W <= 500
         OBJECTIVE FUNCTION VALUE
    1)   2529.41200

VARIABLE	VALUE	REDUCED COST
T	764.705900	.000000
W	500.000000	.000000

(b) MAX   11 T + 5 W
    SUBJECT TO
      2)   17 T + 14 W <= 20000
      3)   8 T + 6 W <= 10000
      4)   T <= 1000
      5)   W <= 500

OBJECTIVE FUNCTION VALUE

1)   12071.4300

VARIABLE	VALUE	REDUCED COST
T	1000.000000	.000000
W	214.285700	.000000

2.27 Let FL = no. of truckloads of fir used to make lumber
      FP = no. of truckloads of fir used to make plywood
      PL = no. of truckloads of pine used to make lumber
      PP = no. of truckloads of pine used to make plywood.

The objective we choose depends on the immediate situation of the Douglas mill. For example, if immediate income is an overriding priority, we can let the cares of today suffice and maximize $500FL + 700FP + 640PL + 800PP$. On the other hand, we can focus on minimizing future loss of income due to the discount and maximize $200FL + 140FP + 160PL + 160PP$. Solutions for these two cases follow.

MAXIMIZING INCOME

MAX   500 FL + 640 PL + 700 FP + 800 PP
SUBJECT TO
  2)   140 FP + 160 PP <= 1000
  3)   FL + FP <= 3
  4)   PL + PP <= 2
  5)   1000 FL + 800 PL <= 5000

OBJECTIVE FUNCTION VALUE

1)   3700.00000

VARIABLE	VALUE	REDUCED COST
FL	.000000	200.000000
PL	.000000	160.000000
FP	3.000000	.000000
PP	2.000000	.000000

MINIMIZING DISCOUNT

MAX   200 FL + 160 PL + 140 FP + 160 PP
SUBJECT TO
  2)   140 FP + 160 PP <= 1000
  3)   FL + FP <= 3
  4)   PL + PP <= 2
  5)   1000 FL + 800 PL <= 5000
OBJECTIVE FUNCTION VALUE

1)   920.000000

VARIABLE	VALUE	REDUCED COST
FL	3.000000	.000000
PL	.000000	.000000
FP	.000000	60.000000
PP	2.000000	.000000

2.28 This problem may look like an LP problem, and one can solve it using LP; but it is really simply a "cost/benefit ratio" problem: For each type of tire, compare the ratio of the cost of each type of press to the number of that type of tire which can be made per hour. For example, consider small tractor tires: since $40000/1.5 < 80000/2.5$, the cost/benefit ratio is smaller for type A presses with regard to small tractor tires—it is more efficient to make small passenger tires on type A presses; consequently, we will buy type A presses on which to make small pasenger tires. Continue this process for the other types of tires. Then tabulate your results and order the presses; but you cannot order fractions of a press. If you set the problem up as an LP, your software will do all the computations and print out the answers for you. Let

A = no. of \$40,000 presses to buy
B = no. of \$80,000 presses to buy
A_I and B_I = no. of type A and B presses to be allocated to production of the six types of tires, $I \leq 6$.

There are 7488 hours in a work year.

MIN   A + 2 B
SUBJECT TO
  2)   1.5 A1 + 2.5 B1 = 1.335 = 10000/7488
  3)   .5 A2 + 1.5 B2 = 1.335
  4)   5 A3 + 8 B3 = 40.06
  5)   3 A4 + 5 B4 = 26.71

6) $2A5 + 4B5 = 13.35$
7) $A6 + 3B6 = 32.05$
8) $A - A1 - A2 - A3 - A4 - A5 - A6 = 0$
9) $B - B1 - B2 - B3 - B4 - B5 - B6 = 0$

OBJECTIVE FUNCTION VALUE

1)   47.6270000

VARIABLE	VALUE	REDUCED COST
A	24.480330	.000000
B	11.573330	.000000
A1	.890000	.000000
B1	.000000	.333333
A2	.000000	.333333
B2	.890000	.000000
A3	8.012000	.000000
B3	.000000	.400000
A4	8.903333	.000000
B4	.000000	.333333
A5	6.675000	.000000
B5	.000000	.000000
A6	.000000	.333333
B6	10.683330	.000000

Notice that $B_5 = 0$ and the reduced cost of $B_5 = 0$; that reflects the fact that the cost/benefit ratios for producing large passenger tires on type A and type B presses are equal. Hence we can transfer production of some large passenger tires to type B presses without changing the total cost of the presses needed to do the job. Consequently, there are five optimal solutions to this problem:

$$\begin{array}{l} A \\ = \\ B \end{array} \quad \begin{array}{ccccc} 24 & 22 & 20 & 18 & 16 \\ 12 & 13 & 14 & 15 & 16 \end{array}$$

2.29 (You will find the decision variables for the solution given below defined in the model formulated at the end of Chapter 2.)

MAX $L1 + L2 + L3 + L4$
SUBJECT TO
2) $L1 + 4L2 + 3L3 <= 50$
3) $2L3 + 2L4 <= 100$
4) $3L1 + L2 + 2L4 <= 30$

OBJECTIVE FUNCTION VALUE

1)   31.6666700

VARIABLE	VALUE	REDUCED COST
L1	.000000	.833333
L2	.000000	.833333
L3	16.666670	.000000
L4	15.000000	.000000

2.30 Let P_I = no. of tons of ingredient I to use in 2000-pound pallet mix,
I = 1,2,3
Q_I = no. of tons of ingredient I to use in 3000-pound pallet mix,
I = 1,2,3
P = no. of 2000-pound pallets to produce
Q = no. of 3000-pound pallets to produce.

MAX   7.5P + 15Q − 2P3 − 7.5P4 − .5P5 − 6Q1 − 9Q2 − 2Q3
SUBJECT TO
P   = P3 + P4 + P5   (# tons 2000 lb. mix)
1.5Q = Q1 + Q2 + Q3   (# tons 3000 lb. mix)

3P3 + 10P4   >= 5P   (5% A reqt.)
30P3 + 10P4   >= 18P   (18% B reqt.)
P4 + 100P5 >= 2P   (2% C reqt.)
P4   >= .1P   (10% P4 reqt.)

12Q1 + 20Q2 +   3Q3 >= (33/2)Q   (11% A reqt.)
10Q1 +   8Q2 + 30Q3 >= (45/2)Q   (15% B reqt.)

P + .6Q <= 40   (mixing time)
.4P + 1.2Q <= 40   (packaging time)

Alternatively, one can state and solve three small LPs and obtain a solution to this problem. The three problems deal with the following three considerations: (1) Determine the composition of 2000-pound pallet mix, (2) determine the composition of 3000-pound pallet mix, and (3) determine the optimum mix of pallets to produce. Because these three considerations are not linked in Exercise 2.30, the solutions to the three small problems compose a solution to Exercise 2.30. A solution to our original formulation appears below.

MAX   7.5 P + 15 Q − 2 P3 − 7.5 P4 − .5 P5 − 2 Q3
− 6 Q1 − 9 Q2
SUBJECT TO
2)   − P + P3 + P4 + P5 = 0
3)   − 1.5 Q + Q3 + Q1 + Q2 = 0
4)   − 5 P + 3 P3 + 10 P4 >= 0
5)   − 18 P + 30 P3 + 10 P4 >= 0

$$6) \quad -2\,P + P4 + 100\,P5 >= 0$$
$$7) \quad -.1\,P + P4 >= 0$$
$$8) \quad -16.5\,Q + 3\,Q3 + 12\,Q1 + 20\,Q2 >= 0$$
$$9) \quad -22.5\,Q + 30\,Q3 + 10\,Q1 + 8\,Q2 >= 0$$
$$10) \quad P + .6\,Q <= 40$$
$$11) \quad .4\,P + 1.2\,Q <= 40$$

OBJECTIVE FUNCTION VALUE

1)   274.319100

VARIABLE	VALUE	REDUCED COST
P	25.000000	.000000
Q	25.000000	.000000
P3	17.247510	.000000
P4	7.325747	.000000
P5	.426743	.000000
Q3	19.852940	.000000
Q1	.000000	.294118
Q2	17.647060	.000000

2.31 Recall that it was decided to maximize the savings due to producing parts rather than buying them when this problem was discussed at the end of Chapter 2. Let

B = no. of bases produced
S = no. of shafts produced
T = no. of tops produced.

MAX $.7\,B + .2\,S + T$
SUBJECT TO
$$2) \quad 10\,B + .2\,S + 5\,T <= 9600$$
$$3) \quad 7\,B + .25\,S + 7.5\,T <= 9600$$
$$4) \quad B <= 1000$$
$$5) \quad S <= 4000$$
$$6) \quad T <= 1000$$

OBJECTIVE FUNCTION VALUE

1)   1910.00000

VARIABLE	VALUE	REDUCED COST
B	157.142900	.000000
S	4000.000000	.000000
T	1000.000000	.000000

2.32 Let C = no. of cars to buy

V = no. of vans to buy
B = no. of buses to buy.

The ratio constraints on vans and buses, $\frac{1}{4} \leq V/B \leq \frac{1}{2}$, can be rewritten in terms of two linear inequalities, $B \leq 4V$ and $2V \leq B$. The following LP models the problem.

```
MAX   C + 2 V + 3 B
SUBJECT TO
      2)    12 C + 16 V + 40 B <= 1800
      3)    C >= 20
      4)    B >= 20
      5)    B <= 30
      6)    -4 V + B <= 0
      7)    2 V - B <= 0
```

OBJECTIVE FUNCTION VALUE

1)   150.000000

VARIABLE	VALUE	REDUCED COST
C	70.000000	.000000
V	10.000000	.000000
B	20.000000	.000000

Thus we have an optimal solution. But I asked you to find all six solutions, and there does not seem to be any obvious indication by the printed solution that there might be multiple solutions: There are no visible variables (the slack and surplus variable values are not visible) with value equal to zero and with reduced cost equal to zero. So what shall we do? Well, my LP package has a sensitivity analysis feature that I can use; it will print out the following information:

RANGES IN WHICH THE BASIS IS UNCHANGED:

OBJ COEFFICIENT RANGES

VARIABLE	CURRENT COEF	ALLOWABLE INCREASE	ALLOWABLE DECREASE
C	1.000000	.500000	.000000
V	2.000000	.000000	.666667
B	3.000000	.000000	INFINITY

Noting that there is a zero allowable decrease in C, we can decrease the coefficient of C slightly in the objective function and see what happens. If you do not have such a feature or do not want to use it, you can think as follows: Buses are at their lowest acceptable level, so let us increase the utility of buses slightly in the model and see what happens. The results of both of these changes are shown below.

DECREASING THE COEFFICIENT OF C

MAX .99 C + 2 V + 3 B
SUBJECT TO
2)   12 C + 16 V + 40 B <= 1800
3)   C >= 20
4)   B >= 20
5)   B <= 30
6)   − 4 V + B <= 0
7)   2 V − B <= 0

OBJECTIVE FUNCTION VALUE

1)   149.700000

VARIABLE	VALUE	REDUCED COST
C	30.000000	.000000
V	15.000000	.000000
B	30.000000	.000000

INCREASING THE COEFFICIENT OF B

MAX   C + 2 V + 3.001 B
SUBJECT TO
2)   12 C + 16 V + 40 B <= 1800
3)   C >= 20
4)   B >= 20
5)   B <= 30
6)   − 4 V + B <= 0
7)   2 V − B <= 0

OBJECTIVE FUNCTION VALUE

1)   150.030000

VARIABLE	VALUE	REDUCED COST
C	30.000000	.000000
V	15.000000	.000000
B	30.000000	.000000

How about that—both changes provided us with a second solution, (30,15,30)! Now you only have to find four more! But you know that any point on the interval between (70,10,20) and (30,15,30) with all integer coordinates will be an optimal solution to the problem. You can find the other four solutions easily; the objective function value of V must be an integer between 10 and 15, thus there are only four new values to try for V: $V = 11, 12, 13,$ and 14. Moreover, you can observe that $B = 2V$ on the interval. All four new values for V give you a new optimal solution.

Now I will analyze this problem by looking at the benefit/cost ratios:

Car	1/12	(mediocre)
Van	2/16	(best)
Bus	3/40	(worst)

Because vans are a better value than buses. I will consider solutions with $2V = B$: I will consider solutions involving cars and multiples of a "one van plus two bus" package; each such package costs $96,000 and has a utility of 8. The number of these packages becomes a decision variable for this formulation of the problem. Let $P = $ no. of "one van plus two bus" packages to buy.

The benefit/cost ratios of cars and packages are both $\frac{1}{12}$, so it suffices to consider integer solutions of the following LP:

MAX $C + 8P$
S.T. $C + 8P <= 150$
        $C \quad >= \quad 20$
        $P >= \quad 10$
        $P <= \quad 15.$

Clearly, the maximum occurs when $C + 8P = 150$; thus there are six optimal solutions:

$$P = \begin{array}{cccccc} 10 & 11 & 12 & 13 & 14 & 15 \\ 70, & 62, & 54, & 46, & 38, & 30. \end{array}$$
$$C$$

2.34 Below A corresponds to truck 1 and B corresponds to truck 2. Let

$A_I = $ no. of truckloads of type I to transport to the dock on truck 1, $I \le 4$

$B_I = $ no. of truckloads of type I to transport to the dock on truck 2, $I \le 4$.

(a) MIN   T
     SUBJECT TO
         2) $25 A1 + 42 B1 >= 7000$
         3) $20 A2 + 37 B2 >= 6500$
         4) $15 A3 + 34 B3 >= 7500$

　　5)　10 A4 + 31 B4 >= 8000
　　6)　A1 + A2 + A3 + A4 − T <= 0
　　7)　B1 + B2 + B3 + B4 − T <= 0

OBJECTIVE FUNCTION VALUE

1)　　　　　　　522.985100

VARIABLE	VALUE	REDUCED COST
A1	280.000000	.000000
A2	242.985100	.000000
A3	.000000	.064499
A4	.000000	.141483
B1	.000000	.059649
B2	44.332410	.000000
B3	220.588200	.000000
B4	258.064500	.000000
T	522.985100	.000000

Thus it takes 14 weeks to process the order. Consequently, include the constraint $T \leq 560$ to find the minimal cost of transporting the order to the dock in 14 weeks below.

(b)　MIN　18 A1 + 16 A2 + 12 A3 + 10 A4 + 26 B1 + 26 B2
　　　　　　　+ 18 B3 + 16 B4
　SUBJECT TO
　　2)　25 A1 + 42 B1 >= 7000
　　3)　20 A2 + 37 B2 >= 6500
　　4)　15 A3 + 34 B3 >= 7500
　　5)　10 A4 + 31 B4 >= 8000
　　6)　A1 + A2 + A3 + A4 − T <= 0
　　7)　B1 + B2 + B3 + B4 − T <= 0
　　8)　T <= 560

OBJECTIVE FUNCTION VALUE

1)　　　　　　　17994.7100

VARIABLE	VALUE	REDUCED COST
A1	143.336600	.000000
A2	325.000000	.000000
A3	.000000	2.188234
A4	.000000	3.470967
B1	81.347260	.000000
B2	.000000	.640001

B3	220.588200	.000000
B4	258.064500	.000000
T	560.000000	.000000

ROW	SLACK OR SURPLUS	DUAL PRICES
2)	.000000	−.720000
3)	.000000	−.800000
4)	.000000	−.654118
5)	.000000	−.652903
6)	91.663400	.000000
7)	.000000	4.240002
8)	.000000	4.240002

We cannot send parts of a truck to the dock; so we do not have a solution, but we do have some information that we can use to get an approximate solution as follows. The slack in constraint 6 tells me that we can increase the number of loads that we send to the dock on truck A. Also, we can round one of the B_Is up if we round the other two down. Rounding B_3 down will require sending truck A twice; but we can decrease B_1 to 81 and increase A_1 to 144, and we can decrease B_4 to 258 and increase A_4 to 1. Consequently, I will choose the approximate solution

VARIABLE	VALUE
A1	144
A2	325
A3	0
A4	1
B1	81
B2	0
B3	221
B4	258
T	560

with cost equal to 18014. Can you find a better solution?

(c) Let C_I = no. of overtime truckloads of type I to transport to the dock on truck A

D_I = no. of overtime truckloads of type I to transport to the dock on truck B.

MIN   18 A1 + 16 A2 + 12 A3 + 10 A4 + 26 B1 + 26 B2
      + 18 B3 + 16 B4 + 22 C1 + 20 C2 + 16 C3 + 14 C4 + 30 D1
      + 30 D2 + 22 D3 + 20 D4

SUBJECT TO
  2) $25\,A1 + 42\,B1 + 25\,C1 + 42\,D1 >= 7000$
  3) $20\,A2 + 37\,B2 + 20\,C2 + 37\,D2 >= 6500$
  4) $15\,A3 + 34\,B3 + 15\,C3 + 34\,D3 >= 7500$
  5) $10\,A4 + 31\,B4 + 10\,C4 + 31\,D4 >= 8000$
  6) $A1 + A2 + A3 + A4 - T <= 0$
  7) $B1 + B2 + B3 + B4 - T <= 0$
  8) $T <= 480$
  9) $C1 + C2 + C3 + C4 <= 96$
 10) $D1 + D2 + D3 + D4 <= 96$

OBJECTIVE FUNCTION VALUE

1)   18310.8700

VARIABLE	VALUE	REDUCED COST
A1	116.456700	.000000
A2	325.000000	.000000
A3	.000000	2.188234
A4	.000000	3.470967
B1	1.347229	.000000
B2	.000000	.640001
B3	220.588200	.000000
B4	258.064500	.000000
T	480.000000	.000000
C1	.000000	4.000000
C2	.000000	4.000000
C3	.000000	6.188234
C4	.000000	7.470967
D1	96.000000	.000000
D2	.000000	.640001
D3	.000000	.000000
D4	.000000	.000000

ROW	SLACK OR SURPLUS	DUAL PRICES
2)	.000000	−.720000
3)	.000000	−.800000
4)	.000000	−.654118
5)	.000000	−.652903
6)	38.543330	.000000
7)	.000000	4.240002
8)	.000000	4.240002
9)	96.000000	.000000
10)	.000000	.240002

Here again we get information that you can use to construct an approximate solution.

Remark: The model for Exercise 2.34 is intentionally very similar to the model for Exercise 2.12. In Exercise 2.12 you could replace the inequalities in constraints 6 and 7 by equalities in parts (a) and (c) (i.e., replace " ≤ " by "=") and get the correct answers, but here you get a different answer if you do that in part (c). In Exercise 2.12 you can also argue that the information that was given did not make it clear that by the number of weeks, I meant to allow only integral numbers of weeks. In many problems it is implicit that we are dealing with discrete-valued variables; part of your job is to decide which values to permit the decision variables to take. In Exercise 2.34 the restriction to an integer number of weeks is clear; in Exercise 2.12 the same restriction was implied by the discrete time term "week." In Exercise 2.12 I accepted production of fractions of a meter of pipe; here I will not permit sending fractions of a truck to the dock.

Appendix 3
Solutions Supplement for Chapter 5

5.7 (a) Demand during October, November, and December totals 450,000 units; but only 360,000 units can be produced during that period of time. Consequently, 90,000 units must be in storage at the beginning of October in order to meet excess demand during the fourth quarter of the year. However, 90,000 units require at least 180,000 cubic feet of storage space, whereas only 150,000 cubic feet are available.

(b) Careful examination of the problem lets us assert the solution, but I will formulate an LP to do the job below. From reading the problem, it is clear that we can produce enough extra in months 7 to 9 to meet the excess demand in months 10 to 12, so let

A_I = thousands of item 1 to produce in month I, $7 \leq I \leq 12$
B_I = thousands of item 2 to produce in month I
C_I = thousands of item 1 in inventory at end of month I
D_I = thousands of item 2 in inventory at end of month I
I_J = thousands of cubic feet of inventory space used during month $J + 1$
S_J = thousands of cubic feet of storage space rented during month $J + 1$.

MIN   .1 I7 + .1 I8 + .1 I9 + .1 I10 + .1 I11 + .15 S7
       + .15 S8 + .15 S9 + .15 S10 + .15 S11
SUBJECT TO
       2)   A7 − C7 = 30
       3)   A8 + C7 − C8 = 30
       4)   A9 + C8 − C9 = 30
       5)   A10 + C9 − C10 = 100
       6)   A11 + C10 − C11 = 100
       7)   A12 + C11 = 100
       8)   B7 − D7 = 15
       9)   B8 + D7 − D8 = 15
      10)   B9 + D8 − D9 = 15
      11)   B10 + D9 − D10 = 50
      12)   B11 + D10 − D11 = 50
      13)   B12 + D11 = 50
      14)   A7 + B7 <= 120
      15)   A8 + B8 <= 120
      16)   A9 + B9 <= 120
      17)   A10 + B10 <= 120
      18)   A11 + B11 <= 120
      19)   A12 + B12 <= 120
      20)   2 C7 + 4 D7 − I7 − S7 = 0
      21)   2 C8 + 4 D8 − I8 − S8 = 0
      22)   2 C9 + 4 D9 − I9 − S9 = 0
      23)   2 C10 + 4 D10 − I10 − S10 = 0
      24)   2 C11 + 4 D11 − I11 − S11 = 0
      25)   I7 <= 150
      26)   I8 <= 150
      27)   I9 <= 150
      28)   I10 <= 150
      29)   I11 <= 150

OBJECTIVE FUNCTION VALUE

      1)   40.5000000

VARIABLE	VALUE	REDUCED COST
A7	30.000000	.000000
A8	45.000000	.000000
A9	105.000000	.000000
A10	70.000000	.000000
A11	70.000000	.000000
A12	70.000000	.000000

B7	15.000000	.000000
B8	15.000000	.000000
B9	15.000000	.000000
B10	50.000000	.000000
B11	50.000000	.000000
B12	50.000000	.000000
C7	.000000	.000000
C8	15.000000	.000000
C9	90.000000	.000000
C10	60.000000	.000000
C11	30.000000	.000000
D7	.000000	.000000
D8	.000000	.200000
D9	.000000	.300000
D10	.000000	.200000
D11	.000000	.200000
I7	.000000	.100000
I8	30.000000	.000000
I9	150.000000	.000000
I10	120.000000	.000000
I11	60.000000	.000000
S7	.000000	.150000
S8	.000000	.050000
S9	30.000000	.000000
S10	.000000	.050000
S11	.000000	.050000

5.8 (a) The " = 4% fat" requirement implies that we must use equal percentages of milk from farms 1 and 3; there is 6% protein in this mix. Let M denote the mix composed of equal parts from farms 1 and 3. To have a final blend with at least 5% protein, we must use at least as much mix M as we use milk from farm 3. However any such blend will have at least 6.25% lactose, which is too much.

(c) Let A = fraction of milk from farm 1 in the mix
B = fraction of milk from farm 2 in the mix
C = fraction of milk from farm 3 in the mix
W = fraction of water in the mix.

An additional variable M will be used to denote the price of milk in the mix. This variable will permit us to change models easily below. First, let us minimize W.

MIN   W
SUBJECT TO
2) $3A + 4B + 5C = 4$
3) $5A + 4B + 7C >= 5$
4) $8A + 5B + 7C <= 6$
5) $-M + 6.8A + 7.5B + 8.3C = 0$
6) $A + B + C + W = 1$

OBJECTIVE FUNCTION VALUE

1)   .357142800E-01

VARIABLE	VALUE	REDUCED COST
M	7.364286	.000000
A	.178571	.000000
B	.464286	.000000
C	.321429	.000000
W	.035714	.000000

(d) MIN   M
SUBJECT TO
2) $3A + 4B + 5C = 4$
3) $5A + 4B + 7C >= 5$
4) $8A + 5B + 7C <= 6$
5) $-M + 6.8A + 7.5B + 8.3C = 0$
6) $A + B + C + W = 1$

OBJECTIVE FUNCTION VALUE

1)   6.64000000

VARIABLE	VALUE	REDUCED COST
M	6.640000	.000000
A	.000000	1.820000
B	.000000	.860000
C	.800000	.000000
W	.200000	.000000

(e) You can read the answer from the solution to part (c).

5.9 (a) Let R_I = no. of barrels of component I to put in regular,
$1 \le I \le 3$
P_I = no. of barrels of component I to put in premium
R = no. of barrels of regular to produce
P = no. of barrels of premium to produce.

MAX   9.8 R + 12 P
SUBJECT TO
    2)   R − R1 − R2 − R3 = 0
    3)   P − P1 − P2 − P3 = 0
    4)   − 7 R + 8 R1 + 20 R2 + 4 R3 <= 0
    5)   − 6 P + 8 P1 + 20 P2 + 4 P3 <= 0
    6)   − 80 R + 83 R1 + 109 R2 + 74 R3 >= 0
    7)   − 100 P + 83 P1 + 109 P2 + 74 P3 >= 0
    8)   R1 + P1 <= 2800
    9)   R2 + P2 <= 1400
    10)   R3 + P3 <= 4000

OBJECTIVE FUNCTION VALUE

  1)   73575.3800

VARIABLE	VALUE	REDUCED COST
R	7507.692000	.000000
P	.000000	.000000
R1	2800.000000	.000000
R2	707.692400	.000000
R3	4000.000000	.000000
P1	.000000	.000000
P2	.000000	.000000
P3	.000000	.022597

(b) To get the octane number up to 100, we must use more than 50% component 2, which would cause the vapor pressure to be greater than 12.

(c) Let P_I = fraction of component I used in a blend of gasoline O denote the octane number of the blend.

MAX   O
SUBJECT TO
    2)   8 P1 + 20 P2 + 4 P3 <= 6
    3)   − O + 83 P1 + 109 P2 + 74 P3 >= 0
    4)   P1 + P2 + P3 = 1

OBJECTIVE FUNCTION VALUE

  1)   78.5000000

VARIABLE	VALUE	REDUCED COST
O	78.500000	.000000
P1	.500000	.000000
P2	.000000	1.000000
P3	.500000	.000000

(d) Let V denote the vapor pressure of the blend.

MIN   V
SUBJECT TO
   2) 8 P1 + 20 P2 + 4 P3 − V <= 0
   3) 83 P1 + 109 P2 + 74 P3 >= 100
   4) P1 + P2 + P3 = 1

OBJECTIVE FUNCTION VALUE

1)   15.8461500

VARIABLE	VALUE	REDUCED COST
P1	.346154	.000000
P2	.653846	.000000
P3	.000000	.153845
V	15.846150	.000000

5.11 Let L = thousands of feet of lumber to manufacture
     P = thousands of feet of plywood to manufacture.
(a) A model for Exercise 5.11(a) follows.

MAX   L + .75 P
SUBJECT TO
   2) L + .8 P <= 40
   3) .3 L + .12 P <= 10

This model is so simple that we could easily solve it graphically, but I will use an LP package:

OBJECTIVE FUNCTION VALUE

1)   39.1666700

VARIABLE	VALUE	REDUCED COST
L	26.666660	.000000
P	16.666670	.000000

(c) The following model for Exercise 5.11(c) uses the same decision variables. The objective function maximizes the "bonus" that we get for delivering the products early.

MAX   .15 L + .15 P
SUBJECT TO
   2) L + .8 P <= 40
   3) .3 L + .12 P <= 10
   4) L >= 15
   5) P >= 20

6) P <= 40
7) L <= 30

OBJECTIVE FUNCTION VALUE

1)   6.93750000

VARIABLE	VALUE	REDUCED COST
L	15.000000	.000000
P	31.250000	.000000

5.12 Let R_J = hundreds of refrigerators to manufacture in quarter J, $J \leq 4$
S_J = hundreds of stoves to manufacture in quarter J, $J \leq 4$
D_J = hundreds of dishwashers to manufacture in quarter J, $J \leq 4$.

I will use constraints like the following to deal with backlogging and inventory costs:

$$R_1 - 15 = P_{R1} - N_{R1}$$

or

$$N_{R1} - P_{R1} + R_1 = 15 \quad \text{(see constraint 7 below)};$$

here P_{R1} represents the number of hundreds of refrigerators which will be in inventory at the end of period 1, similarly N_{R1} represents the number of hundreds of refrigerators to be backlogged.

Constraints of the type introduced above are used for each period and for each product in the following model.

MIN   20 NR1 + 20 NR2 + 20 NR3 + 20 NS1 + 20 NS2 + 20 NS3
      + 10 ND1 + 10 ND2 + 10 ND3 + 5 PR1 + 5 PR2 + 5 PR3
      + 5 PS1 + 5 PS2 + 5 PS3 + 5 PD1 + 5 PD2 + 5 PD3
SUBJECT TO
      2)   2 R1 + 4 S1 + 3 D1 <= 150
      3)   2 R2 + 4 S2 + 3 D2 <= 150
      4)   2 R3 + 4 S3 + 3 D3 <= 150
      5)   2 R4 + 4 S4 + 3 D4 <= 150
      6)   R2 = 0
      7)   NR1 − PR1 + R1 = 15
      8)   NR2 − PR2 + R1 + R2 = 25
      9)   NR3 − PR3 + R1 + R2 + R3 = 45
     10)   R1 + R2 + R3 + R4 = 58.5
     11)   NS1 − PS1 + S1 = 15
     12)   NS2 + S1 + S2 − PS2 = 30
     13)   NS3 + S1 + S2 + S3 − PS3 = 42

14) S1 + S2 + S3 + S4 = 58.5
15) ND1 + D1 − PD1 = 10
16) ND2 + D1 + D2 − PD2 = 30
17) ND3 + D1 + D2 + D3 − PD3 = 45
18) D1 + D2 + D3 + D4 = 71.5

OBJECTIVE FUNCTION VALUE

1)   85.0000000

VARIABLE	VALUE	REDUCED COST
NR1	.000000	25.000000
NR2	.000000	17.500000
NR3	.000000	22.500000
NS1	.000000	20.000000
NS2	.000000	25.000000
NS3	.000000	25.000000
ND1	.000000	10.000000
ND2	.000000	13.750000
ND3	.000000	13.750000
PR1	10.000000	.000000
PR2	.000000	7.500000
PR3	.000000	2.500000
PS1	.000000	5.000000
R1	25.000000	.000000
S1	15.000000	.000000
D1	10.000000	.000000
R2	.000000	.000000
S2	16.375000	.000000
D2	20.000000	.000000
R3	20.000000	.000000
S3	16.250000	.000000
D3	15.000000	.000000
R4	13.500000	.000000
S4	10.875000	.000000
D4	26.500000	.000000
PS2	1.375000	.000000
PS3	5.625000	.000000
PD1	.000000	5.000000
PD2	.000000	1.250000
PD3	.000000	1.250000

There is another way of writing the constraints which is useful when there is a large number of periods to consider. For example, we can replace constraints 8 and 9 by the equivalent recursive versions:

$$8)\text{-}7)\text{:} \quad NR2 - NR1 - PR2 + PR1 + R2 = 10$$

and

$$9)\text{-}8)\text{:} \quad NR3 - NR2 - PR3 + PR2 + R3 = 20,$$

respectively.

5.13 Let C_I = no. of workers to make cheddar during week I, $1 \le I \le 8$

S_I = no. of workers to make Swiss during week I

O_I = no. of workers to work overtime during week I

T_I = no. of workers to begin training new workers during week I

N_I = no. of new workers to begin training during week I

A_I = no. of thousands of pounds of cheddar to backlog at end of week I

B_I = no. of thousands of pounds of Swiss to backlog at end of week I.

MIN   540 O1 + 540 O2 + 540 O3 + 540 O4 + 540 O5 + 540 O6
      + 540 O7 + 540 O8 + 500 A1 + 500 A2 + 500 A3 + 500 A4
      + 500 A5 + 500 A6 + 500 A7 + 600 B1 + 600 B2 + 600 B3
      + 600 B4 + 600 B5 + 600 B6 + 600 B7 + 2880 N1 + 2520 N2
      + 2160 N3 + 1800 N4 + 1440 N5 + 1080 N6 + 720 N7

SUBJECT TO

2)	A1 + .4 C1 = 10	
3)	A2 + .4 C1 + .4 C2 = 20	
4)	A3 + .4 C1 + .4 C2 + .4 C3 = 32	
5)	A4 + .4 C1 + .4 C2 + .4 C3 + .4 C4 = 44	
6)	.4 C1 + .4 C2 + .4 C3 + .4 C4 + .4 C5 + A5 = 60	
7)	.4 C1 + .4 C2 + .4 C3 + .4 C4 + .4 C5 + .4 C6 + A6 = 76	
8)	.4 C1 + .4 C2 + .4 C3 + .4 C4 + .4 C5 + .4 C6 + .4 C7 + A7 = 96	
9)	.4 C1 + .4 C2 + .4 C3 + .4 C4 + .4 C5 + .4 C6 + .4 C7 + .4 C8 = 116	
10)	.24 S1 + B1 = 6	
11)	.24 S1 + .24 S2 + B2 = 13.2	
12)	.24 S1 + .24 S2 + .24 S3 + B3 = 21.6	
13)	.24 S1 + .24 S2 + .24 S3 + .24 S4 + B4 = 32.4	
14)	.24 S1 + .24 S2 + .24 S3 + .24 S4 + .24 S5 + B5 = 43.2	
15)	.24 S1 + .24 S2 + .24 S3 + .24 S4 + .24 S5 + .24 S6 + B6 = 55.2	

16)   .24 S1 + .24 S2 + .24 S3 + .24 S4 + .24 S5
       + .24 S6 + B7 + .24 S7 = 67.2
17)   .24 S1 + .24 S2 + .24 S3 + .24 S4 + .24 S5
       + .24 S6 + .24 S7 + .24 S8 = 79.2
18)   − O1 + C1 + S1 + T1 <= 50
19)   − O2 + C2 + S2 + T1 + T2 <= 50
20)   − O3 + C3 + S3 + T2 + T3 − N1 <= 50
21)   − O4 + C4 + S4 + T3 + T4 − N1 − N2 <= 50
22)   − O5 + C5 + S5 + T4 + T5 − N1 − N2 − N3 <= 50
23)   − O6 + C6 + S6 + T5 + T6 − N1 − N2 − N3
       − N4 <= 50
24)   − O7 + C7 + S7 + T6 + T7 − N1 − N2 − N3
       − N4 − N5 <= 50
25)   − O8 + C8 + S8 + T7 − N1 − N2 − N3 − N4 − N5
       − N6 <= 50
26)   N1 + N2 + N3 + N4 + N5 + N6 + N7 = 50
27)   − 3 T1 + N1 <= 0
28)   − 3 T2 + N2 <= 0
29)   − 3 T3 + N3 <= 0
30)   − 3 T4 + N4 <= 0
31)   − 3 T5 + N5 <= 0
32)   − 3 T6 + N6 <= 0
33)   − 3 T7 + N7 <= 0

OBJECTIVE FUNCTION VALUE

1)   131118.800

VARIABLE	VALUE	REDUCED COST
C1	25.000000	.000000
C2	25.000000	.000000
C3	30.000000	.000000
C4	30.000000	.000000
C5	40.000000	.000000
C6	40.000000	.000000
C7	50.000000	.000000
C8	50.000000	.000000
S1	25.000000	.000000
S2	30.000000	.000000
S3	35.000000	.000000
S4	45.000000	.000000
S5	45.000000	.000000
S6	50.000000	.000000

S7	50.000000	.000000
S8	50.000000	.000000
O1	6.979167	.000000
O2	14.791670	.000000
O3	.000000	135.000000
O4	.000000	135.000000
O5	.000000	168.750000
O6	.000000	168.750000
O7	.000000	177.187500
O8	.000029	.000000
A1	.000000	500.000000
A2	.000000	162.500000
A3	.000000	500.000000
A4	.000000	415.625000
A5	.000000	500.000000
A6	.000000	478.906300
A7	.000000	942.968800
B1	.000000	600.000000
B2	.000000	37.500000
B3	.000000	600.000000
B4	.000000	459.375000
B5	.000000	600.000000
B6	.000000	564.843800
B7	.000000	1338.281000
T1	6.979167	.000000
T2	2.812501	.000000
T3	3.124996	.000000
T4	1.250005	.000000
T5	2.499993	.000000
T6	.000008	.000000
T7	.000000	708.750000
N1	20.937500	.000000
N2	8.437503	.000000
N3	9.374987	.000000
N4	3.750015	.000000
N5	7.499979	.000000
N6	.000015	.000000
N7	.000000	.000000

This solution is not satisfactory because we cannot train parts of workers. After trying a few integer constraints, I settled for the following additional constraints:

        34)   T1 = 7
        35)   T2 = 3
        36)   T3 = 3
        37)   T4 = 1
        38)   T5 = 3.

With these additional constraints, we get the following solution:

OBJECTIVE FUNCTION VALUE

1)   131688.000

VARIABLE	VALUE	REDUCED COST
C1	25.000000	.000000
C2	25.000000	.000000
C3	30.000000	.000000
C4	30.000000	.000000
C5	40.000000	.000000
C6	40.000000	.000000
C7	50.000000	.000000
C8	50.000000	.000000
S1	25.000000	.000000
S2	29.000000	.000000
S3	35.000000	.000000
S4	46.000000	.000000
S5	45.000000	.000000
S6	50.000000	.000000
S7	50.000000	.000000
S8	50.000000	.000000
O1	7.000000	.000000
O2	14.000000	.000000
O3	.000000	144.000000
O4	.000000	288.000000
O5	.000000	.000000
O6	1.000000	.000000
O7	.000000	144.000000
O8	.000000	216.000000
A1	.000000	500.000000
A2	.000000	140.000000

A3	.000000	140.000000
A4	.000000	1220.000000
A5	.000000	500.000000
A6	.000000	140.000000
A7	.000000	320.000000
B1	.000000	600.000000
B2	.240000	.000000
B3	.240000	.000000
B4	.000000	1800.000000
B5	.000000	600.000000
B6	.000000	.000000
B7	.000000	.000000
T1	7.000000	.000000
T2	3.000000	.000000
T3	3.000000	.000000
T4	1.000000	.000000
T5	3.000000	.000000
T6	.000000	.000000
T7	.000000	720.000000
N1	21.000000	.000000
N2	9.000000	.000000
N3	9.000000	.000000
N4	3.000000	.000000
N5	8.000000	.000000
N6	.000000	348.000000
N7	.000000	.000000

5.14 (a) Let A = kilograms of apples to use in apple jelly
C = kilograms of cherries to use in cherry jelly
A_J = kilograms of apples to use in jam
C_J = kilograms of cherries to use in jam
A_M = kilograms of apple mash to use in jam
C_M = kilograms of cherry mash to use in jam.

I included the following variables so that the printout would tell us how much of each product to produce. Let

A_A = kilograms of apple jelly to produce
C_C = kilograms of cherry jelly to produce
J = kilograms of jam to produce.

Notice that

6/10 × 10/7 kg of apples + 4/10 kg of sugar produces 1 kg of apple jelly
4/10 × 10/5 kg of cherries + 6/10 kg of sugar produces 1 kg of cherry jelly.

One kilogram of apply jelly costs (6/7) × (.6) + (.4) × (.43) + .4 dollars. One kilogram of cherry jelly costs (4/5) × (1.1) + (.6) × (.43) + .4 dollars. Total production costs for jam are .5(.32)J + .6A_J + 1.1C_J + .4J dollars. The following model maximizes profit.

MAX   1.5327 A + 1.5775 C + 2.5266 AJ + 1.1333 CJ
SUBJECT TO
2) .15 A + .15 AJ − AM > = 0   (apple mash)
3) .1 C + .1 CJ − CM > = 0   (cherry mash)
4) .7 AJ − .5 CJ + AM − CM = 0   (equal weights in jam)
5) .7 AJ + .5 CJ − 1.5 AM − 1.5 CM = 0 (30% juice and 20% mash)
6) .7 AJ + .5 CJ − .3 J = 0   (30% juice)
7) A + AJ <= 4000
8) C + CJ <= 1200
9) 7 A − 6 AA = 0
19) 5 C − 4 CC = 0

OBJECTIVE FUNCTION VALUE

1)   8210.12600

VARIABLE	VALUE	REDUCED COST
A	3785.714000	.000000
C	1140.000000	.000000
AJ	214.285700	.000000
CJ	60.000000	.000000
AM	.000000	.755536
CM	120.000000	.000000
J	600.000000	.000000
AA	4416.667000	.000000
CC	1425.000000	.000000

(b) I will solve the model with the additional constraint $C_M = 0$ below and then compare the objective function values of the two solutions to determine whether or not it is profitable to make any jam.

MAX   1.5327 A + 1.5775 C + 2.5266 AJ + 1.1333 CJ

SUBJECT TO

2) $.15 A + .15 AJ - AM >= 0$
3) $.1 C + .1 CJ - CM >= 0$
4) $.7 AJ - .5 CJ + AM - CM = 0$
5) $.7 AJ + .5 CJ - 1.5 AM - 1.5 CM = 0$
6) $.7 AJ + .5 CJ - .3 J = 0$
7) $A + AJ <= 4000$
8) $C + CJ <= 1200$
9) $7 A - 6 AA = 0$
10) $5 C - 4 CC = 0$
11) $CM = 0$

OBJECTIVE FUNCTION VALUE

1)   8023.80000

VARIABLE	VALUE	REDUCED COST
A	4000.000000	.000000
C	1200.000000	.000000
AJ	.000000	.000000
CJ	.000000	.000000
AM	.000000	.000000
CM	.000000	.000000
J	.000000	.151107
AA	4666.667000	.000000
CC	1500.000000	.000000

Thus we conclude that we should pay the $50 to make cherry mash.

5.23 Let N = no. of products that can be assembled per 16-hour day. Label the three machines of type A, label the five machines of type B, and define 21 zero–one integer variables as follows:

$A_{IJ} = 1$ if we make part J on the Ith machine of type A
$B_{IJ} = 1$ if we make part J on the Ith machine of type B

MAX   N
SUBJECT TO

2) $192 A11 + 192 A21 + 192 A31 + 96 B11 + 96 B21$
     $+ 96 B31 + 96 B41 + 96 B51 - N >= 0$
3) $240 A12 + 240 A22 + 240 A32 + 192 B12 + 192 B22$
     $+ 192 B32 + 192 B42 + 192 B52 - 2N >= 0$
4) $400 B13 + 400 B23 + 400 B33 + 400 B43 + 400 B53$
     $- N >= 0$
5) $A11 + A12 = 1$

  6)   A21 + A22 = 1
  7)   A31 + A32 = 1
  8)   B11 + B12 + B13 = 1
  9)   B21 + B22 + B23 = 1
10)   B31 + B32 + B33 = 1
11)   B41 + B42 + B43 = 1
12)   B51 + B52 + B53 = 1
INTEGER-VARIABLES = 21

Unfortunately, this model may run and run and run! During the spring 1989 semester, I assigned this problem to a class of 28 students. Some people got the solution in a reasonable amount of computer time; some did not. Some models never arrived at an optimal solution, and others did so only after several hours. One person let his model run for over an hour on a PC, then moved it to a mainframe computer, where he got a solution. So let us try a different approach. Begin by letting

A_I = no. of type A machines on which to make part I, $I \leq 2$
B_I = no. of type B machines on which to make part I, $I \leq 3$.

Next model the problem without the integer variable requirements as follows:

MAX   N
SUBJECT TO
  2) $-N + 192\,A1 + 96\,B1 >= 0$
  3) $-2\,N + 240\,A2 + 192\,B2 >= 0$
  4) $-N + 400\,B3 >= 0$
  5) $A1 + A2 = 3$
  6) $B1 + B2 + B3 = 5$

Now extend the constraints by using the binary expansion $B_3 = I_1 + 2I_2 + 4I_3$, where I_1, I_2, and I_3 are 0–1 integer variables, to force B_3 to be integer valued.

MAX   N
SUBJECT TO
  2) $-N + 192\,A1 + 96\,B1 >= 0$
  3) $-2\,N + 240\,A2 + 192\,B2 >= 0$
  4) $-N + 400\,B3 >= 0$
  5) $A1 + A2 = 3$
  6) $B1 + B2 + B3 = 5$
  7) $-I1 - 2\,I2 - 4\,I3 + B3 = 0$
INTEGER-VARIABLES = 3 : I1,I2,I3

OBJECTIVE FUNCTION VALUE

1)   400.000000

VARIABLE	VALUE	REDUCED COST
I1	1.000000	− 400.000000
I2	.000000	− 1200.000000
I3	.000000	− 1600.000000
N	400.000000	.000000
A1	.777778	.00000
B1	2.611111	.000000
A2	2.222222	.000000
B2	1.388889	.000000
B3	1.000000	.000000

We now know that we cannot make more than 400 assemblies. Consequently, an integer solution that makes 400 is surely optimal. We also know that it suffices to use one type B machine to make part 3. Thus I put $B_3 = 1$ below and used the binary representations of A_1 and B_1 as integer variables.

MAX   N
SUBJECT TO
     2)   − N + 192 A1 + 96 B1 >= 0
     3)   − 2 N + 240 A2 + 192 B2 >= 0
     4)   − N + 400 B3 >= 0
     5)   A1 + A2 = 3
     6)   B1 + B2 + B3 = 5
     7)   − I1 − 2 I2 − 4 I3 + B1 = 0
     8)   − I4 − 2 I5 + A1 = 0
     9)   B3 = 1
INTEGER-VARIABLES = 5 : I1,I2,I3,I4,I5

OBJECTIVE FUNCTION VALUE

1)   400.000000

VARIABLE	VALUE	REDUCED COST
I1	1.000000	.000000
I2	.000000	.000000
I3	.000000	.000000
I4	.000000	.000000
I5	1.000000	.000000
N	400.000000	.000000
A1	2.000000	.000000

B1	1.000000	.000000
A2	1.000000	.000000
B2	3.000000	.000000
B3	1.000000	.000000

You may have noticed that we can model the problem in one step by incorporating eight 0–1 integer variables as follows:

MAX   N
SUBJECT TO
2) $- N + 192\,A1 + 96\,B1 >= 0$
3) $- 2\,N + 240\,A2 + 192\,B2 >= 0$
4) $- N + 400\,B3 >= 0$
5) $A1 + A2 = 3$
6) $B1 + B2 + B3 = 5$
7) $- I1 - 2\,I2 + A1 = 0$
8) $- I3 - 2\,I4 - 4\,I5 + B1 = 0$
9) $- I6 - 2\,I7 - 4\,I8 + B2 = 0$
INTEGER-VARIABLES = 8 : I1,I2,I3,I4,I5,I6,I7,I8.

OBJECTIVE FUNCTION VALUE

1)   400.000000

VARIABLE	VALUE	REDUCED COST
I1	.000000	.000000
I2	1.000000	.000000
I3	1.000000	400.000000
I4	.000000	800.000000
I5	.000000	1600.000000
I6	1.000000	400.000000
I7	1.000000	800.000000
I8	.000000	1600.000000
N	400.000000	.000000
A1	2.000000	.000000
B1	1.000000	.000000
A2	1.000000	.000000
B2	3.000000	.000000
B3	1.000000	.000000

5.24 Let $P =$ no. of rolling pins that can be manufactured in a 144-hour workweek

$C_I =$ no. of 8-hour shifts scheduled to produce cylinders on machine I

E_I = no. of 8-hour shifts scheduled to produce ends on machine
I.

Define four integer variables, A_1 to A_4, to regulate the setup of the first four machines as follows: If $A_I = 1$, cylinders can be produced on machine I; otherwise, ends can be produced on machine I.

MAX P
SUBJECT TO

2)	$-P + 30\,C1 + 40\,C2 + 50\,C3 + 60\,C4 + 165\,C5$ $+ 235\,C6 >= 0$
3)	$-2\,P + 50\,E1 + 70\,E2 + 100\,E3 + 130\,E4$ $+ 300\ E5 + 480\,E6 >= 0$
4)	$-18\,A1 + C1 <= 0$
5)	$-18\,A2 + C2 <= 0$
6)	$-18\,A3 + C3 <= 0$
7)	$-18\,A4 + C4 <= 0$
8)	$18\,A1 + E1 <= 18$
9)	$18\,A2 + E2 <= 18$
10)	$18\,A3 + E3 <= 18$
11)	$18\,A4 + E4 <= 18$
12)	$C5 + E5 <= 18$
13)	$C6 + E6 <= 18$

INTEGER-VARIABLES = 4 : A1,A2,A3,A4

OBJECTIVE FUNCTION VALUE

1)   5308.10500

VARIABLE	VALUE	REDUCED COST
A1	1.000000	−50.210510
A2	1.000000	−52.105220
A3	1.000000	−9.473663
A4	.000000	33.157900
P	5308.105000	.000000
C1	18.000000	.000000
C2	18.000000	.000000
C3	18.000000	.000000
C4	.000000	.000000
C5	18.000000	.000000
C6	.757894	.000000
E1	.000000	.000000
E2	.000000	.000000
E3	.000000	.000000

E4	18.000000	.000000
E5	.000000	9.157890
E6	17.242110	.000000

LP OPTIMUM IS IP OPTIMUM

5.26 First, I will split each demand point into two demand points; the first represents the minimal supply needed and the second represents the optional amount, in addition to the minimal supply, that the demand point would like to receive.

The minimum allocations total $140 million and the total amount available from the agencies totals $180 million, so $40 million is available for optional allocation. Consequently, even though each agency would like to receive an infinite amount of money, the most each can reasonably hope for is $40 million. Hence, I put an upper limit of $40 (million) on each optional demand. I also added a fifth (dummy) agency with a supply of $120 million to balance supply and demand.

I put the problem in minimization format by subtracting each of the benefits from 10—this gives all nonnegative numbers; you can use the negatives of the entries in the table if you prefer. I used 90 for the big M in the tableau model after I put it in minimization form:

Tableau

Agency	1	2	3	4	5	6	7	8	Supply
1	8	8	5	5	3	3	3	3	40
2	4	4	7	7	7	7	4	4	30
3	3	3	1	1	4	4	6	6	30
4	5	5	3	3	6	6	0	0	80
5	90	0	90	0	90	0	90	0	120
Demand	20	40	30	40	40	40	50	40	300

I used a TRANS package to obtain the following solution.
Solution: Total benefit = 270

Agency	1	2	3	4	5	6	7	8	Total allocation
1	0	0	0	0	40	0	0	0	40
2	20	10	0	0	0	0	0	0	30
3	0	0	30	0	0	0	0	0	30
4	0	0	0	0	0	0	50	30	80
5	0	30	0	40	0	40	0	10	120

5.28 Let X_I = no. of units of water to transport along route I, $I \leq 16$.

MIN $12\,X2 + 14\,X7 + 9\,X13 + 17\,X14 + 5\,X15 + 4\,X16$
SUBJECT TO
$$
\begin{array}{rl}
2) & X2 + X1 - X3 - X6 - X8 - X12 = 13 \\
3) & X3 + X4 + X5 - X9 = 11 \\
4) & X8 + X9 + X10 + X11 = 14 \\
5) & X13 + X14 + X12 - X11 = 17 \\
6) & -X7 - X13 + X15 - X1 >= -4 \\
7) & -X7 - X13 + X15 - X1 <= 8 \\
8) & -X2 - X14 - X15 + X16 = 6 \\
9) & X7 + X6 - X4 - X10 = 9 \\
10) & X1 <= 20 \\
11) & X2 <= 20 \\
12) & X3 <= 10 \\
13) & X4 <= 10 \\
14) & X5 <= 10 \\
15) & X6 <= 10 \\
16) & X7 <= 10 \\
17) & X8 <= 10 \\
18) & X9 <= 10 \\
19) & X10 <= 10 \\
20) & X11 <= 10 \\
21) & X12 <= 10 \\
22) & X13 <= 10
\end{array}
$$

OBJECTIVE FUNCTION VALUE

1)   752.000000

VARIABLE	VALUE	REDUCED COST
X2	20.000000	.000000
X7	.000000	2.000000
X13	10.000000	.000000
X14	4.000000	.000000
X15	26.000000	.000000
X16	56.000000	.000000
X1	20.000000	.000000
X3	1.000000	.000000
X6	9.000000	.000000
X8	10.000000	.000000
X12	7.000000	.000000
X4	.000000	.000000
X5	10.000000	.000000

X9	.000000	.000000
X10	.000000	.000000
X11	4.000000	.000000
ROW	SLACK OR SURPLUS	DUAL PRICES
2)	.000000	−21.000000
3)	.000000	−21.000000
4)	.000000	−21.000000
5)	.000000	−21.000000
6)	.000000	−9.000000
7)	4.000000	.000000
8)	.000000	−4.000000
9)	.000000	−21.000000
10)	.000000	12.000000
11)	.000000	5.000000
12)	9.000000	.000000
13)	10.000000	.000000
14)	.000000	21.000000
15)	1.000000	.000000
16)	10.000000	.000000
17)	.000000	.000000
18)	10.000000	.000000
19)	10.000000	.000000
20)	6.000000	.000000
21)	3.000000	.000000
22)	.000000	3.000000

5.29 Begin by ordering the nodes as follows: $S_1,S_2,S_3,T_1,T_2,T_3,T_4,D_1,D_2$. Then list the routes lexicographically according to this ordering of the nodes; thus X_1 denotes the flow along route (1,4), which goes from S_1 to T_1, and so on. A solution to part (a) appears on pages 199–200.

(b) The following LP models the maximum flow problem.

MAX F
SUBJECT TO
2) X1 + X2 <= 20
3) X3 + X4 + X5 <= 10
4) X6 + X7 + X8 <= 30
5) X1 + X3 + X6 − X9 = 0
6) X4 + X7 − X10 − X11 − X12 = 0
7) X2 + X9 + X10 − X13 = 0
8) X5 + X11 − X14 − X15 = 0

9) $X13 + X14 - I1 = 0$
10) $X8 + X12 + X15 - I2 = 0$
11) $I1 + I2 - F = 0$
12) $X1 <= 8$
13) $X2 <= 17$
14) $X3 <= 7$
15) $X4 <= 6$
16) $X5 <= 12$
17) $X6 <= 8$
18) $X7 <= 6$
19) $X8 <= 22$
20) $X9 <= 10$
21) $X10 <= 8$
22) $X11 <= 5$
23) $X12 <= 15$
24) $X13 <= 4$
25) $X14 <= 9$
26) $X15 <= 8$

OBJECTIVE FUNCTION VALUE

1)   42.000000

VARIABLE	VALUE	REDUCED COST
X1	.000000	.000000
X2	4.000000	.000000
X3	.000000	1.000000
X4	2.000000	.000000
X5	8.000000	.000000
X6	.000000	.000000
X7	6.000000	.000000
X8	22.000000	.000000
X9	.000000	.000000
X10	.000000	1.000000
X11	.000000	.000000
X12	8.000000	.000000
X13	4.000000	.000000
X14	.000000	.000000
X15	8.000000	.000000
I1	4.000000	.000000
I2	38.000000	.000000
F	42.000000	.000000

(c) The following LP models the minimum cost flow problem with a 5% toll at each transshipment point.

MIN   C
SUBJECT TO
  2) $X1 + X2 <= 20$
  3) $X3 + X4 + X5 <= 10$
  4) $X6 + X7 + X8 <= 30$
  5) $.95\,X1 + .95\,X3 + .95\,X6 - X9 = 0$
  6) $.95\,X4 + .95\,X7 - X10 - X11 - X12 = 0$
  7) $.95\,X2 + .95\,X9 + .95\,X10 - X13 = 0$
  8) $.95\,X5 + .95\,X11 - X14 - X15 = 0$
  9) $X13 + X14 - I1 = 0$
 10) $X8 + X12 + X15 - I2 = 0$
 11) $I1 + I2 - F = 0$
 12) $- C + 8\,X1 + 17\,X2 + 7\,X3 + 6\,X4 + 12\,X5$
        $+ 8\,X6 + 6\,X7 + 22\,X8 + 10\,X9 + 8\,X10$
        $+ 5\,X11 + 15\,X12 + 4\,X13 + 9\,X14 + 8\,X15 = 0$
 13) $I1 >= 25$
 14) $I2 >= 15$

OBJECTIVE FUNCTION VALUE

1)   786.421100

VARIABLE	VALUE	REDUCED COST
C	786.421100	.000000
X1	.000000	1.350000
X2	4.105264	.000000
X3	.000000	2.900000
X4	10.000000	.000000
X5	.000000	.550000
X6	.000000	3.900000
X7	30.000000	.000000
X8	.000000	1.813156
X9	.000000	.000000
X10	22.210530	.000000
X11	15.789470	.000000
X12	.000000	1.263157
X13	25.000000	.000000
X14	.000000	1.842105
X15	15.000000	.000000
I1	25.000000	.000000
I2	15.000000	.000000
F	40.000000	.000000

The following LP models the maximum flow problem with a 5% toll at each transshipment point.

MAX F
SUBJECT TO

2) X1 + X2 <= 20
3) X3 + X4 + X5 <= 10
4) X6 + X7 + X8 <= 30
5) .95 X1 + .95 X3 + .95 X6 − X9 = 0
6) .95 X4 + .95 X7 − X10 − X11 − X12 = 0
7) .95 X2 + .95 X9 + .95 X10 − X13 = 0
8) .95 X5 + .95 X11 − X14 − X15 = 0
9) X13 + X14 − I1 = 0
10) X8 + X12 + X15 − I2 = 0
11) I1 + I2 − F = 0
12) X1 <= 8
13) X2 <= 17
14) X3 <= 7
15) X4 <= 6
16) X5 <= 12
17) X6 <= 8
18) X7 <= 6
19) X8 <= 22
20) X9 <= 10
21) X10 <= 8
22) X11 <= 5
23) X12 <= 15
24) X13 <= 4
25) X14 <= 9
26) X15 <= 8

OBJECTIVE FUNCTION VALUE

1)   41.2000000

VARIABLE	VALUE	REDUCED COST
X1	.000000	.000000
X2	4.210526	.000000
X3	.000000	.950000
X4	6.000000	.000000
X5	4.000000	.000000
X6	.000000	.000000
X7	6.000000	.000000
X8	22.000000	.000000
X9	.000000	.000000
X10	.000000	1.000000
X11	.000000	.050000

X12	11.400000	.000000
X13	4.000000	.000000
X14	.000000	.000000
X15	3.800000	.000000
I1	4.000000	.000000
I2	37.200000	.000000
F	41.200000	.000000

5.36 Let A_I = thousands of product A to produce in quarter I
B_I = thousands of product B to produce in quarter I
C_I = thousands of product A to inventory in quarter $I + 1$
D_I = thousands of product B to inventory in quarter $I + 1$
F_I = thousands of product A to inventory in rented space
H_I = thousands of product B to inventory in rented space.

MAX $3 A1 + A2 + 2 A3 + 2 A4 + 6 B1 + 2 B2 + 4 B3 + 4 B4$
$- 1.5 C1 - 1.5 C2 - 1.5 C3 - 2.5 D1 - 2.5 D2$
$- 2.5 D3 - .5 F1 - .5 F2 - .5 F3 - .5 H1 - .5 H2 - .5 H3$
SUBJECT TO
   2) $A1 - C1 = 40$
   3) $A2 + C1 - C2 = 10$
   4) $A3 + C2 - C3 = 30$
   5) $A4 + C3 = 100$
   6) $B1 - D1 = 35$
   7) $B2 + D1 - D2 = 25$
   8) $B3 + D2 - D3 = 45$
   9) $B4 + D3 = 100$
 10) $A1 + B1 <= 120$
 11) $A2 + B2 <= 120$
 12) $A3 + B3 <= 120$
 13) $A4 + B4 <= 120$
 14) $C1 - F1 - E1 = 0$
 15) $C2 - F2 - E2 = 0$
 16) $C3 - F3 - E3 = 0$
 17) $D1 - H1 - G1 = 0$
 18) $D2 - H2 - G2 = 0$
 19) $D3 - H3 - G3 = 0$
 20) $2 E1 + 4 G1 <= 100$
 21) $2 E2 + 4 G2 <= 100$
 22) $2 E3 + 4 G3 <= 100$

OBJECTIVE FUNCTION VALUE
  1)   1050.00000

VARIABLE	VALUE	REDUCED COST
A1	60.000000	.000000
A2	25.000000	.000000
A3	75.000000	.000000
A4	20.000000	.000000
B1	60.000000	.000000
B2	.000000	.750000
B3	45.000000	.000000
B4	100.000000	.000000
C1	20.000000	.000000
C2	35.000000	.000000
C3	80.000000	.000000
D1	25.000000	.000000
D2	.000000	.000000
D3	.000000	.000000
F1	.000000	.250000
F2	.000000	.500000
F3	30.000000	.000000
H1	10.000000	.000000
H2	.000000	1.750000
H3	.000000	1.000000
E1	20.000000	.000000
E2	35.000000	.000000
E3	50.000000	.000000
G1	15.000000	.000000
G2	.000000	1.250000
G3	.000000	1.500000

(b) Let P = thousands of additional units of demand for product A to generate in quarter 3

Q = thousands of additional units of demand for product B to generate in quarter 3.

MAX $3\,A1 + A2 + 2\,A3 + 2\,A4 + 6\,B1 + 2\,B2 + 4\,B3 + 4\,B4$
$- 1.5\,C1 - 1.5\,C2 - 1.5\,C3 - 2.5\,D1 - 2.5\,D2$
$- 2.5\,D3 - .5\,F1 - .5\,F2 - .5\,F3 - .5\,H1 - .5\,H2$
$- .5\,H3 - P - 1.5\,Q$

SUBJECT TO

2) $A1 - C1 = 40$
3) $A2 + C1 - C2 = 10$
4) $A3 + C2 - C3 - P = 30$
5) $A4 + C3 = 100$
6) $B1 - D1 = 35$

7) $B2 + D1 - D2 = 25$
8) $B3 + D2 - D3 - Q = 45$
9) $B4 + D3 = 100$
10) $A1 + B1 <= 120$
11) $A2 + B2 <= 120$
12) $A3 + B3 <= 120$
13) $A4 + B4 <= 120$
14) $C1 - F1 - E1 = 0$
15) $C2 - F2 - E2 = 0$
16) $C3 - F3 - E3 = 0$
17) $D1 - H1 - G1 = 0$
18) $D2 - H2 - G2 = 0$
19) $D3 - H3 - G3 = 0$
20) $2\,E1 + 4\,G1 <= 100$
21) $2\,E2 + 4\,G2 <= 100$
22) $2\,E3 + 4\,G3 <= 100$

OBJECTIVE FUNCTION VALUE

1) ·   1050.00000

VARIABLE	VALUE	REDUCED COST
A1	60.000000	.000000
A2	40.000000	.000000
A3	60.000000	.000000
A4	20.000000	.000000
B1	60.000000	.000000
B2	.000000	.750000
B3	60.000000	.000000
B4	100.000000	.000000
C1	20.000000	.000000
C2	50.000000	.000000
C3	80.000000	.000000
D1	25.000000	.000000
D2	.000000	.000000
D3	.000000	.000000
F1	.000000	.250000
F2	.000000	.500000
F3	30.000000	.000000
H1	10.000000	.000000
H2	.000000	1.750000
H3	.000000	1.000000
E1	20.000000	.000000

E2	50.000000	.000000
E3	50.000000	.000000
G1	15.000000	.000000
G2	.000000	1.250000
G3	.000000	1.500000
P	.000000	1.500000
Q	15.000000	.000000

Notice that there is *no* additional profit from advertising, even though we sell 15,000 more units of product *B*: we are just generating money to pay for advertising. So you see that this situation has multiple optimal solutions. When the estimated unit demand generating cost for product *B* is lower than $1.50, it is advantageous to advertise. Advertising cost for product *B* has another critical value at $1 per unit. To see how the solution changes as this cost becomes less than $1, I modified the objective function by changing the coefficient of *Q* to −.9 as indicated below and reran the problem to find the solution that follows.

MAX 3 A1 + A2 + 2 A3 + 2 A4 + 6 B1 + 2 B2 + 4 B3
+ 4 B4 − 1.5 C1 − 1.5 C2 − 1.5 C3 − 2.5 D1
− 2.5 D2 − 2.5 D3 − .5 F1 − .5 F2 − .5 F3
− .5 H1 − .5 H2 − .5 H3 − P − .9 Q

OBJECTIVE FUNCTION VALUE

1)   1065.00000

VARIABLE	VALUE	REDUCED COST
A1	60.000000	.000000
A2	100.000000	.000000
A3	.000000	.100000
A4	20.000000	.000000
B1	60.000000	.000000
B2	.000000	.750000
B3	120.000000	.000000
B4	100.000000	.000000
C1	20.000000	.000000
C2	110.000000	.000000
C3	80.000000	.000000
D1	25.000000	.000000
D2	.000000	.000000
D3	.000000	.000000
F1	.000000	.250000

F2	60.000000	.000000
F3	30.000000	.000000
H1	10.000000	.000000
H2	.000000	1.150000
H3	.000000	1.100000
E1	20.000000	.000000
E2	50.000000	.000000
E3	50.000000	.000000
G1	15.000000	.000000
G2	.000000	1.650000
G3	.000000	1.600000
P	.000000	2.000000
Q	75.000000	.000000

5.41 If you got a different solution to Exercise 2.22 than I got, your printout might indicate a variable with value equal to zero and reduced cost equal to zero. I discussed how that case indicates the possible presence of multiple solutions in Chapter 4. My solution in Appendix 2 does not give us that clue, but we do have another clue. According to the discussion of dual variables in Section 5.9, note that the surplus variables corresponding to constraints 10 and 11 are equal to zero and they have reduced costs equal to zero: remember, those reduced costs are equal to the corresponding dual prices (up to a + or − sign, depending on how the software package formats problems). Thus there may be other optimal solutions in which these two surplus variables are positive.

There are three basic feasible solutions to Exercise 2.22; I will label them A, B and C. Their values are tabulated below.

SOLUTION	A	B	C
VARIABLE	VALUE	VALUE	VALUE
Y	25.000000	33.000000	43.000000
H1	10.000000	10.000000	.000000
H2	.000000	.000000	.000000
Q1	.000000	.000000	.000000
Q2	8.000000	.000000	.000000
Q3	15.000000	7.000000	7.000000
Q4	.000000	.000000	.000000
M1	.000000	.000000	.000000
M2	.000000	.000000	.000000
M3	.000000	.000000	.000000
M4	.000000	.000000	.000000

M5	.000000	.000000	.000000
M6	9.000000	9.000000	9.000000
M7	.000000	.000000	.000000
M8	16.000000	16.000000	6.000000
M9	.000000	.000000	.000000
M10	.000000	.000000	.000000
M11	.000000	.000000	.000000
M12	.000000	.000000	.000000

Every convex combination X of A, B, and C [i.e., $X = (aA + bB + cC)$, where $a \geq 0$, $b \geq 0$, $c \geq 0$, and $a + b + c = 1$] is a solution to Exercise 2.22. Which of these solutions are integer valued?

Reading List

The following list is intended to provide you with an entree to the literature pertaining to the topics which we have discussed and related topics.

[BJ]   Bazaraa, Mokhtar S., and Jarvis, John J., *Linear Programming and Network Flows*, John Wiley & Sons, New York, 1977.

[C]   Chvatal, Vasek, *Linear Programming*, W. H. Freeman and Co., New York, 1983.

[D]   Dantzig, George B., *Linear Programming and Extensions*, Princeton University Press, Princeton, NJ, 1963.

[DL]   Dreyfus, Stuart E., and Law, Averill M., *The Art and Theory of Dynamic Programming*, Academic Press, New York, 1977.

[F]   Franklin, Joel, *Methods of Mathematical Economics Linear and Nonlinear Programming: Fixed Point Theorems*, Springer-Verlag, New York, 1980.

[H1]   Hadley, G., *Linear Algebra*, Addison-Wesley, Reading, MA, 1961.

[H2]   Hadley, G., *Linear Programming*, Addison-Wesley, Reading, MA, 1962.

[H3]   Harvey, Charles M., *Operations Research: An Introduction to Linear Optimization and Decision Analysis*, Elsevier-North Holland, New York, 1979.

[G]   Gass, Saul I., *Linear Programming*, McGraw-Hill, New York, 1975.

[JB]   Jensen, Paul A., and Barnes, J. Wesley, *Network Flow Programming*, John Wiley & Sons, New York, 1980.

[K]   Kaplan, Edward L., *Mathematical Programming and Games*, John Wiley & Sons, New York, 1982.

[KB]   Kolman, Bernard, and Beck, Robert E., *Elementary Linear Programming with Applications*, Academic Press, New York, 1980.

[L]   Luenberger, David G., *Linear and Nonlinear Programming*, Addison-Wesley, Reading, MA, 1984.

[M]   Murty, Katta G., *Linear and Combinatorial Programming*, John Wiley & Sons, New York, 1976.

[N]   Nemhauser, George L., *Introduction to Dynamic Programming*, John Wiley & Sons, New York, 1966.

[NW]   Nemhauser, George L., and Wosley, Laurence A., *Integer and Combinatorial Optimization*, John Wiley & Sons, New York, 1988.

[S1]   Schrage, Linus, *Linear, Integer, and Quadratic Programming with LINDO*, The Scientific Press, Palo Alto, CA, 1986.

[S2]   Schrijver, Alexander, *Theory of Linear and Integer Programming*, John Wiley & Sons, New York. 1986.

[S3]   Shapiro, Roy D., *Optimization Models for Planning and Allocation: Text and Cases in Mathematical Programming*, John Wiley & Sons, New York, 1984.

[S4]   Strang, Gilbert, *Linear Algebra and Its Applications*, Academic Press, New York, 1980.

[T1]   Tarjan, Robert E., *Data Structures and Network Algorithms*, SIAM, Philadelphia, 1983.

[T2]   Thie, Paul R., *An Introduction to Linear Programming and Game Theory*, John Wiley & Sons, New York, 1988.

Index

For Product Safety Concerns and Information please contact our EU
representative GPSR@taylorandfrancis.com
Taylor & Francis Verlag GmbH, Kaufingerstraße 24, 80331 München, Germany